AUGMENTED URBAN SPACES

Augmented Urban Spaces
Articulating the Physical and Electronic City

Edited by
ALESSANDRO AURIGI
Newcastle University, UK

and

FIORELLA DE CINDIO
University of Milan, Italy

ASHGATE

Published by
Ashgate Publishing Limited
Gower House
Croft Road
Aldershot
Hampshire GU11 3HR
England

Ashgate Publishing Company
Suite 420
101 Cherry Street
Burlington, VT 05401-4405
USA

www.ashgate.com

British Library Cataloguing in Publication Data
Augmented urban spaces : articulating the physical and electronic city. - (Design and the built environment series)
1. Sociology, Urban 2. Information technology - Social aspects 3. Public spaces 4. Wireless communication systems
I. Aurigi, Alessandro II. De Cindio, Fiorella
307.7'6

Library of Congress Cataloging-in-Publication Data
Augmented urban spaces : articulating the physical and electronic city / [edited] by Alessandro Aurigi and Fiorella De Cindio.
p. cm. -- (Design and the built environment)
Includes bibliographical references and index.
ISBN 978-0-7546-7149-7
1. City planning. 2. Cities and towns--Effect of technological innovations on. 3. Public spaces. I. Aurigi, Alessandro. II. De Cindio, Fiorella.

HT166.A849 2008
307.1'216--dc22

2008015808

ISBN 978-0-7546-7149-7

Printed and bound in Great Britain by
MPG Books Ltd, Bodmin, Cornwall.

Contents

List of Figures

List of Tables

List of Contributors

Patrick Allen is a Lecturer in Creative Technology at the University of Bradford. His research encompasses the application of framing, space and visual representation, to a wide variety of multimodal texts, including film, news discourse and urban screens.

Alessandro Aurigi works as a senior lecturer at Newcastle University, where he is the Director of Architecture and member of the Global Urbanism Research Unit (GURU). In Newcastle he teaches architectural design and theories on the information age, high technologies and architecture. Previously, he has worked as a lecturer at the Bartlett School of the Built Environment, University College London and as a research fellow in the Centre for Advanced Spatial Analysis (CASA), UCL. His main research interests are studying the relationships between the emergence of the information society and the ways we imagine, conceive, design, and manage buildings and cities. Alex has produced several publications on the topic, among which the 2005 book *Making the Digital City: The Early Shaping of Urban Internet Space* (Ashgate).

Diane Dechief has a BA English/Psychology from the University of Alberta and a MA Media Studies from Concordia University. She is a Ph.D student in the Faculty of Information studies at the University of Toronto. Her research interests include information-seeking and use by recent immigrants, ICT policy, rural immigration, digital inequalities/communication rights, and civic participation. She is currently researching how and why people's names may change during immigration to and settlement in Canada. She received a Social Sciences and Humanities Research Council Doctoral Fellowship (2007-2010) to support this work.

Fiorella De Cindio is Associate Professor at the Department of Information and Communication of the University of Milan (Italy), where she teaches and carries out research mainly on Distributed Systems Design, Online Communities, Community Networks and e-participation. She is the Director of the Civic Informatics Laboratory (LIC), that she created in 1994, and the Founder and President of the Milan Community Network (RCM) and of the Association for Informatics and Civic Networking of Lombardy (A.I.Re.C.) which groups Community Networks in the Lombardy Region. In 2001 she received the 'Ambrogino d'Oro', the civic top award assigned by the Milan Municipality to citizens who have contributed to the 'city's development.

Tom Denison is a research associate with the Centre for Community Networking Research within Monash University's Faculty of Information Technology. His main research focus is in the areas of social informatics and community informatics. He is currently undertaking research, towards a doctorate, into the take up and use of information and communications technology by community sector organizations. With a background in library automation and electronic publishing, Tom has also consulted widely in Australia and Vietnam.

Ines Di Loreto graduated in Philosophy at the University of Milan (Italy) where she is now a Ph.D candidate in Computing Science. Her research interests include Social Media and their societal impact. In particular she investigates the relationship between ICT and representation of the self, analyzing how representations - and the resulting relationships built through them - are constructed in the web 2.0 age.

Eleonora Di Maria is researcher in Business management ad the University of Padua and Senior Researcher at TEDIS, Venice International University. Her research focuses on the evolution of business models and local systems related to ICT, internazionalization and competitiveness.

Susan Drucker, Juris Doctor, is Professor of Communication in the department of Journalism/ Media Studies/ Public Relations of the School of Communication, Hofstra University. She is an attorney and her work examines freedom of expression, media regulation and the relationship between media technology and human factors, particularly as viewed from a legal perspective. She is the treasurer of the Urban Communication Foundation and the author/editor of 8 books and over 100 articles and book chapters. Among her publications are two editions of *Real Law @ Virtual Space: The Regulation of Cyberspace* (Hampton Press) and *The Urban Communication Reader* (Hampton Press).

Rodrigo José Firmino is Lecturer in Urban Management at Pontifical Catholic University (PUCPR) in Curitiba, Brazil. He worked as Postdoctoral Fellow at University of São Paulo, Brazil, from 2004 and 2007, researching the co-development of urban and technological strategies for cities in developing countries, while sponsored by FAPESP. He trained as an architect and planner at State University of São Paulo, and obtained his Ph.D. in Urban Planning from the School of Architecture, Planning and Landscape at the University of Newcastle upon Tyne, U.K. in 2004. He also holds an MPhil in Architecture and Urbanism from the University of São Paulo, 2000. He can be contacted at rodrigo.firmino@pucpr.br.

Romano Fistola graduated at the Faculty of Architecture of Naples. He has been research fellow at the National Resarch Council of Italy and then permanent researcher at the Land Planning and Management Institute in Naples. At the end of the 1990s he has been Assistant Teacher for the course in “Technological Innovation and Territorial Transformations” at the Faculty of Engineering of the University of Naples “Federico II”, where he now teaches as professor at the Ph.D course “Engineering of the territorial systems and networks”. He also works as professor of “Town and Country Planning” and “Geographical Information Science” at the University of Sannio. He is an elected member of the steering committee of the Regional Science Association of Italy. His research interests are mainly focused on the modification of the urban activities system with special attention to the relationship between territorial transformations and ICT. He is the author of more than 60 academic publications.

Marcus Foth is a Senior Research Fellow at the Institute for Creative Industries and Innovation, Queensland University of Technology (QUT), Brisbane, Australia. He received a BCompSc(Hon) from Furtwangen University, Germany, a BMultimedia from Griffith University, Australia and an MA and Ph.D in digital media and urban sociology from QUT. Dr Foth is the recipient of an Australian Postdoctoral Fellowship supported under the Australian Research Council's Discovery funding scheme. He was a 2007 Visiting Fellow at the Oxford Internet Institute, University of Oxford, UK. Employing participatory design and action research, he is working on cross-disciplinary research and development at the intersection of people, place and technology with a focus on urban informatics, locative media and mobile applications. Dr Foth has published over forty articles in journals, edited books, and conference proceedings in the last four years and is among the top 25 authors with the highest number of research paper downloads from QUT's ePrints archive. He is a member of the Australian Computer Society and the Executive Committee of the Association of Internet Researchers. His publications are listed at http://www.urbaninformatics.net/publications/

Mark Gaved is a Research Fellow in the Institute of Educational Technology at The Open University. His recent Ph.D research has explored how internet technologies have been appropriated at a local level to create hybrid virtual/physical networked communities, and whether these may offer a model for future socio-technical developments. Mark's research interests and publications include community informatics, hacker cultures, and educational technology.

Gary Gumpert, Ph.D. is Emeritus Professor of Communication at Queens College of the City University of New York, and President of the Urban Communication Foundation. His primary research focuses on the nexus of communication technology and social relationships, particularly looking at urban and suburban development, the alteration of public space, and the changing nature of community. Among his 10 authored and edited books include *Talking Tombstones and Other Tales of the Media Age* (Oxford University Press) and *The Urban Communication Reader* (Hampton Press).

Thomas A. Horan is Associate Professor and Institute Director at Claremont Graduate University's School of Information Systems and Technology. Dr. Horan has over twenty years experience in studying societal impacts of information technology. He currently directs two interdisciplinary technology institutes: Claremont Information and Technology Institute and Kay Center for E-Health Research. In this capacity, Dr. Horan leads numerous research studies funded by agencies such as the National Science Foundation and the U.S. Department of Transportation. He has published over two dozen technical articles and authored two books on digital technologies (*Digital Places, Digital Infrastructures*). Prior to joining the faculty at Claremont

Graduate University (CGU), Dr. Horan worked as a Senior Analyst for the U.S. General Accounting Office in Washington, DC. He has his Masters and Ph.D. degrees from CGU.

Esperanza Huerta is Assistant Professor at the College of Business Administration at the University of Texas at El Paso (UTEP). She holds a Ph.D. and a M.S. in the Management of Information Systems from Claremont Graduate University in California. Her research focuses on the behavioural aspects of human-computer interaction. She has conducted research on the impact of the Internet on education and social interaction. She has also explored the role of human-computer interaction in the willingness to share information in business settings. Esperanza has published several articles in academic journals, conferences, and practitioner magazines. She is also the author of several chapters in edited books. She serves as a member of the Editorial Board of *Information and Management*.

Graeme Johanson is Director of the Centre for Community Networking Research (www.ccnr.net), Johanson has run many research and practical projects about the value of ICTs for community improvement locally and internationally. Current projects deal with Chinese migrants and social inclusion in cities; improving networks between community-based organisations, home businesses, and local government; developing a knowledge management standard for community-based organizations; the formation of a National Non-profit ICT Coalition; prototyping an annotation system for capturing and sustaining indigenous oral culture; managing protocols and data for e-Research repositories; launching the International Development Informatics Association (IDIA) in 2007; analysis of the impact of interactive ICTs on policy development by Malaysian public servants; and studying libraries as knowledge commons.

Heesang Lee is Lecturer in the Department of Geography Education at Kyungpook National University, South Korea, where he lectures in urban and cultural geography, after finishing his Ph.D. thesis in the Department of Geography at Durham University, UK in 2006. His main research interests focus on media geographies, technocultural spaces and cyborg cities, particularly being concerned with hybrid spaces such as human-machine spaces, actual-virtual spaces and so on.

Graham Longford is an Assistant Professor in the Department of Politics at Trent University and a Co-Investigator with the Community Wireless Infrastructure Research Project (CWIRP). Prior to teaching at Trent, he held a postdoctoral research fellowship in the Faculty of Information Studies at the University of Toronto, where he worked with the Canadian Research Alliance for Community Innovation and Networking (CRACIN). Dr. Longford's research interests include the social and political aspects of new information and communication technologies, community networking, and telecommunications policy.

Malcolm McCullough is the author of *Digital Ground - Architecture Pervasive Computing and Environmental Knowing* (2004) and *Abstracting Craft* (1996). He is Associate Professor of Architecture at the Taubman College of Architecture and Urban Planning, at the University of Michigan, and previously served on the design faculty at Carnegie Mellon and at Harvard.
He has lectured in many countries on the urbanism of pervasive computing.

Stefano Micelli is Associate Professor of Business Management and Innovation management at the Ca'Foscari University of Venice and Director of TEDIS and Dean of Venice International University. His research includes the impact of ICT on firms and local manufacturing systems of SMEs, innovation and design, knowledge management.

Adrian Millward (adrian.millward@adrianmillwardassociates.co.uk) has a degree in Engineering and an MBA specialising in Business strategy for the Public Sector. He has successfully led major projects including regeneration programmes in Sheffield, Southampton and Leeds, culminating in project managing the creation of a £2b Public-Private Joint Venture Partnership. He is currently advising Local Authorities across the UK on devising Carbon Reduction programmes including best practice in managing information and community involvement.

Paul Mulholland is a Research Fellow in the Knowledge Media Institute, at The Open University. His research interests and publications have covered topics including knowledge management, online communities, museum informatics and educational technology. He has published over 40 papers in the form of journal articles, full conference papers and book chapters. He has been a member of over 20 conference programme committees and is an Associate Editor of the *International Journal of Human-Computer Studies*.

David Murakami Wood is a Lecturer in the School of Architecture Planning & Landscape, and researcher in the Global Urban Research Unit (GURU) at Newcastle University, UK. He is an urban geographer who specialises in the study of the history, technologies, practices and ethics of surveillance. He is a co-founder and Managing Editor of the international journal of surveillance studies, Surveillance & Society and a co-founder and trustee of the Surveillance Studies Network (SSN), the new charitable global association for surveillance studies. He co-ordinated and edited the acclaimed Report on the Surveillance Society, for the Office of the Information Commissioner, published in November 2006. He is currently working on a book on the geography of surveillance.

Celene Navarrete is a Ph.D. candidate in the program of Management of Information Systems in Claremont Graduate University (CGU) in California. She holds a Bachelors degree in Information Science from the University of Aguascalientes in México and a Masters degree in the Management of Information Systems from CGU. Her particular research interests centers in the role that societal and cultural aspects play in the diffusion and adoption of information and communication technologies by nations, communities, and individuals. She is currently conducting research in Community Informatics and Electronic Government.

Nancy Odendaal is currently a Senior Lecturer in Urban and Regional Development Planning at the School of Architecture, Planning and Housing at the University of KwaZulu-Natal in Durban, South Africa. She is also a Ph.D Candidate in the School of Architecture and Planning at Wits University in Johannesburg. Before joining the University full-time in 2001, she worked on the EU-funded Cato Manor Development project in Durban for five years as a planner and manager of the Cato Manor Development Association's Geographic Information System. Prior work experience includes involvement in a number of planning and development projects in Namibia and Swaziland, working in the public and private sectors. Her current research interests include local governance, Information and Communication Technology and cities, metropolitan planning whilst she also teaches planning theory and integrated development planning. Her academic interests are supplemented by consulting input for National and Local Government into a number of initiatives that include strategic planning, land use management and institutional development.

Natalie Pang is a Ph.D candidate in Monash University's Faculty of Information Technology. Her research focuses on the knowledge commons in the contemporary media environment using case studies from cultural institutions such as libraries and museums. Her teaching interests include information management, usability evaluations, peer to peer technologies and approaches, and digital libraries. Other than her involvement in ongoing research with communities from China, Italy, Australia, Singapore and Malaysia, she has also previously served as an Honorary Research Associate of Museum Victoria in Australia and a Visiting Scholar at the School of Communication and Information, Nanyang Technological University in Singapore. Natalie is currently a contributing member of the Foundation for Peer to Peer Alternatives, with interests in participatory technologies in Web 2.0, social networks and discourse on open content licenses applicable to knowledge resources in the digital environment.

Annalisa Pelizza graduated in Media Studies at University of Bologna with a dissertation in semiotics dealing with the strategies of media representation of early 2000s social movements. She is currently Ph. D. candidate at QUA_SI. Quality of Life in the Information Society interdisciplinary project at the University of Milan – Bicocca. Her current research interests focus on digital communities and innovation and use a framework derived from Science and Technology Studies. Pelizza has also been jointly leading academic research and on field techno-social experimentation. She has been active in the Italian Telestreet movement, in national and international contexts. She attended art residencies, workshops and developed projects at institutions like Transmediale Media Art Festival, Prix Ars Electronica, V2 Center for Unstable Media, International Film Festival Rotterdam. As a researcher, she gave lectures at, among others, the University of Bologna; the Advanced School of Journalism of Bologna; the Boltzmann Institute Media.Art.Research, Linz; the Complutense University of Madrid.

Alison Powell is a Ph.D candidate in the Communication Studies department at Concordia University in Montreal, where she holds a Canada Graduate Scholarship. She is a member of the Canadian Research Alliance for Community Innovation and Networking. Her work focuses on the social and political impacts of grassroots technology development. She currently studies community and municipal WiFi networks and is also interested in other forms of alternative, non-commercial technological development, including open-source software development and open hardware hacking. She publishes for and presents to audiences including telecommunications policy specialists, geeks and hackers, and scholars in the social sciences and humanities.

Michael Powell (m.powell@pop3.poptel.org.uk) has a degree in History, an MA in Information Technology Management and also trained as a nurse. Since 1994 he has worked as a consultant on information management for development organisations – both local and international – and been active in initiatives such as the Open Information Project, a community and voluntary network in Sheffield started in 1994, the Labour Telematics Centre, UK Communities on Line, and OTIS. He is currently director of the 'Information and Knowledge Management and International Development: Emergent Issues Programme' (http://www.ikmemergent.net).

Laura Anna Ripamonti is Assistant Professor at the Department of Information and Communication of the University of Milan (Italy), where she teaches "Economics and Enterprise Management" to Computer Science undergraduate students and "Laboratory of Informatics" to students graduating in Biosciences. She graduated in Engineering and Managerial Sciences at Politecnico di Milano and she got a Ph.D in Informatics from the University of Milan. Her research interests focus on the Community Informatics, with a special interest for knowledge representation and social interaction supported by ICT. Due to her multidisciplinary background, she is interested both in the technological and in the organizational aspects of Community Informatics, which she prefers to investigate through an "action research" approach.

Paul Sanders, MArch (Natal), BA(Hons), DipArch (Kingston), is Senior Lecturer and the Architecture Co-ordinator in the School of Design at QUT. He graduated from Kingston Polytechnic in the UK in 1986. During his professional career he has been responsible for a number of major building projects as well as an involvement in urban design and housing issues. Paul began his academic career at the University of Natal in 1998 and following a period as visiting lecturer to QUT in Australia, joined the staff on an ongoing basis in 2003. He has published in international journals in the fields of architecture and urban design, and was awarded a MArch by research in 2003. He is currently enrolled in a Ph.D, in the field of urban morphology; researching the correlation of historical urban analysis and models of sustainable urban form.

Don Schauder (MA Sheffield, MEd, Ph.D Melbourne, FALIA, MACS) is Emeritus Professor of Information Management, and former Associate Dean (Research), in the Faculty of Information Technology at Monash University. He was a pioneer of

Australian electronic publishing as founder of INFORMIT Electronic Publishing, and of community networking as co-founder of VICNET: Victoria's Network. He has been Director of several libraries, the first being the South African Library for the Blind and the most recent RMIT University Library. He is Honorary Chair of two Monash University research teams - the Centre for Community Networking Research (CCNR), and the Information & Telecommunication Needs Research Group (ITNR).

Anthony Townsend is a technology forecaster and strategist with the Institute for the Future, an independent non-profit research group based in Palo Alto, California. Prior to joining the Institute, Anthony enjoyed a decade-long career in academia, where his research focused on the role of telecommunications in urban development and design. Anthony worked and lived in South Korea on and off between 2001 and 2004, as an urban planning consultant and Fulbright scholar. During this period, he directed several major research projects funded by the National Science Foundation and Department of Homeland Security. Anthony holds a Ph.D. in urban and regional planning from Massachusetts Institute of Technology. He spends most of his time and enormous amounts of bandwidth telecommuting form his home in New York City.

Kenneth C. Werbin is a Postdoctoral Fellow with the Infoscape Research Laboratory at Ryerson University in Toronto, Canada, where he continues his investigations into contemporary conjunctions of technology, culture, and power, including ongoing research into the discursive space of community-networking initiatives. Kenneth recently completed his Ph.D at Concordia University in Montreal, Canada, entitled "The list serves: the apparatuses of security and governmentality" which investigates modern and contemporary manifestations of "watch-list" culture. His contribution to the chapter presented in this book was funded by the Canadian Research Alliance for Community Innovation and Networking (www.cracin.ca), a broad network of devoted and determined Canadian and International academic, government, and community partners, whose collective efforts made this research possible.

Kirsty Williamson is the Director of the research group, Information and Telecommunications Needs Research (ITNR), a joint initiative of Monash and Charles Sturt Universities, in Australia. Since the early 1990s, she has undertaken many research projects, with her principal area of research being "human information behaviour". Research contexts have been varied, with libraries having been the focus of several studies. Her research has been funded by a range of different organisations including the principal funding body of Australian Universities, the Australian Research Council (ARC), as well as libraries of various types - public, school and Australian State Libraries.

Katharine Willis is a researcher, artist and architect whose interests lie in exploring the ways through which we interact with our spatial environment. This work investigates navigation, wayfinding and identity and the transformative possibilities of mobile and wireless technologies. A key aim of the work is to propose approaches to understanding how we can create legible environments when urban public space is experienced through new media. She is currently an EU Marie Curie research fellow on the MEDIACITY project, Bauhaus University of Weimar and was previously a DAAD researcher in the Spatial Cognition program at the University of Bremen. Prior to this she worked on interactive site-specific art projects and installations in UK and internationally. Katharine's background and training is in architecture and she is a qualified Architect.

Introduction

Augmented Urban Spaces

Alessandro Aurigi and Fiorella De Cindio

It is becoming increasingly evident that the gradual development of an enriched media environment, ubiquitous computing, mobile and wireless communication technologies, as well as the Internet as a non-extraordinary part of our everyday lives, are changing the ways people use cities and live in them. Architects and urban planners, human geographers, sociologists and other scholars have been looking at this in the past few decades, some of them well pre-dating the internet with their reflections. Computer scientists obviously directly contribute with their work to this re-definition and extension of space, though they are not always fully aware of this circular relation between space, society and technology. And this occurs to the point of them getting often surprised by unpredicted and unplanned uses and transformations of the very technologies they develop and deploy.

What is less clear, however, is how these technologies are actually modifying city living and the fruition of urban spaces. Influential scholars and critics from computing, media and social studies, and built environment disciplines have depicted scenarios that would range from the demise of the city altogether to leave the scene to a displaced "cyber" society based on nearly total fluidity of its spaces and related functions, to the establishment of a parallel, "virtual" city enhancing, rather than annihilating, the physical one. Both utopia and dystopia have been heavily involved in predictions on mega-scale transformations of society, cities and space. And yet, when we observe things around us we can notice how cities are not giving any sign of obsolescence, how people are still people who just happen to use technologies, how space has not been replaced by virtual environments – despite the recently renewed hype on phenomena like Second Life – and basically how deeply "embedded" digital technologies have become, and how "real" their functions feel. There are no entirely digital lifestyles – or if there are we are talking about a fairly irrelevant minority – yet at the same time most lifestyles are digitally-enhanced, in many different as well as subtle ways. In the augmented city, 'virtual' and 'physical' spaces are no longer two separate dimensions, but just parts of a continuum, of a whole. The physical and the digital environment have come to define each other and concepts such as public space and "third place", identity and knowledge, citizenship and public participation are all inevitably affected by the shaping of the reconfigured, augmented urban space.

Popular wisdom suggests that it is easier to look far away into the horizon than observe our eyebrows, which are yet so near. Similarly, it is not trivial at all understanding what happens to our ability to understand and "live" urban space when this, right here, right now, "intersects" with everyday usage of our mobile phone or wireless networks, or what software and code, embedded within so many objects and practices, could mean to us even if we are not programmers or computer geeks.

To try and understand better all of this immanent "augmentation" we believe that we need to look at the "micro", local but real and significant aspects of the articulation of physical and digital, and that in doing this a truly cross-disciplinary approach is needed. However, a really effective cross-fertilisation of disciplines is hard to achieve. Often, different professionals and scholars have looked at "other" perspectives as some fancy add-ons helping to boost their discourse, rather than a challenge to literally contaminate, question, modify and enrich their own disciplinary niche.

This book was born as a humble but passionate attempt to contribute to this "contamination" of perspectives. It is the joint effort of two different editors. On the one hand, an architect/urban planner who since his own first graduation thesis has been looking at ICT as much more than "infrastructure", focusing on high technologies as potentially part of that urban "glue" that keeps people together in social as well as business life, and cannot be separated from space and place. On the other hand, a computer scientist who made of the cultural clout of architecture as well as sociology a fundamental informer of her approach to designing software technologies and applications for the urban dimension. All other chapter contributors, as well as their diversity within the book, are also testimonials of such contamination, together with the often overlapping themes of their papers.

For all these reasons, this is a book of questions, experiences, suggestions and – in one word – exploration. The contributions in Augmented Urban Spaces consider the emergence of the phenomenon from three main perspectives which become the sections the book is organised into.

The first section on "augmented spaces" is informed by a theoretical and conceptual analysis of the nature of the relationship and combination of physical and digital environments, supported however by "here and now" empirical observations and studies. Some of the chapters look at how to interpret and reflect on the ways these combinations could redefine the public sphere of the city as well as people's individual ways to live and relate with urban space. These are all useful elements to try and understand ICT-enriched spaces' potential to support successful interactions, public-ness as well as meaning-making in cities, and the numerous issues that are raised by these "enhancements".

The second perspective – and section – looks at the social constituents of urban space: the communities living in and shaping them. This part of the book explores ways in which communities themselves are becoming "augmented" and indeed "augmenting" themselves from the grassroots, and what this can mean for the enhancement of democratic participation, public discourse, and shared knowledge.

The third section of the book aims at considering issues and challenges which economic regeneration strategies, planning and urban design face as civic spaces are getting increasingly augmented. On the one hand it looks at what institutions, municipalities and designers can or could do, in a proactive and structured way, and presents some different approaches and ideas on ICT-based regeneration. On the other hand considers how fine-grained, everyday changes cannot be entirely controlled and become part of the spontaneous transformations of urban fabric and society, asking what could be done to wisely and positively react to this new condition. It addresses where built environment practitioners stand in this situation, and how they can start successfully addressing this changing arena.

Each of the sections of the book is individually introduced by one of the editors, so we refer the reader to those specific introductions for a more detailed discussion. It is important to remark – in a way obviously – that dividing up a text into sections such as these serves well for giving an overall framework to the reader and "pace" to the volume. But – yet again – it can make us fall into the trap of creating a degree of "boxing" of themes, contributors and discourses. We just hope that readers will agree with us in appreciating how difficult dealing with interdisciplinary books can be, and reading the "story" presented here as a progression in which themes, concepts, ideas constantly overlap and complement each other despite any specific disciplinary or professional "labelling".

Finally, we would like to thank our colleagues – some of whom have taken part in this book – who helped organising or participated in the "Digital Cities: the Augmented Public Space" workshop, held in Milan within the wider "Communities and Technologies 2005" international conference. Although this volume does not represent the proceedings of that event – having gone well beyond it in many different ways, and with only six chapters somehow related to work presented in that occasion – the idea itself of working on such a project started as a consequence of the workshop, and was strongly inspired by it.

PART 1
Augmented Spaces

Alessandro Aurigi

The section that opens the book is – not by chance – very diverse in its contents, examined technologies and raised issues. This reflects the complexity of the topic and the range of different ways in which contemporary cities get "augmented" by ICT. It is a natural consequence of the fact that we are not talking about some precisely defined "product" or "solution", such as the so-called virtual or digital cities conceived in the 1990s and consisting of more or less sophisticated and interactive internet sites. Augmentation is a much more capillary and complex phenomenon, embracing the city as a whole and its components: spaces – at different scales – people, businesses and so on. It is something that goes well beyond the remit of the single initiative, actor, and concept, though of course specific case studies can still be examined, and specific actions envisaged.

Ubiquitous computing and mobile technologies are blurring the distinction between physical space and digital media, with the latter also combining in complex ways, sometimes resulting in very physical, public and visible setups, whilst other times being hardly visible and strictly personal. Urban ICT can therefore materialise into the very conspicuous presence of big screens in prime civic locations, remain partially hidden in the mobile phones within people's pockets or simply exist as a non-physical, but highly local and potentially very meaningful, geo-referenced database of spatial tags.

In this section Allen claims that 'the majority of city centres in the UK now fall into the category of augmented urban spaces rather than this being confined to major international and global cities'. The capillarity and "everyday" character of the intersection of physical and digital makes the debate shift further – or maybe creates room for complementary reflections – from the simple consideration of high technologies as one of the supporting factors for the affirmation of global, mega-cities and financial hubs, or from the efficiency-inspired vision of e-government innovation, to the very local, small scale of ordinary urban space, in ordinary towns, enhanced by very "ordinary" telecommunications networks.

And it is not just the distinction between the different urban scales interested by the information revolution which is put into question. Looking at the deployment of large displays in public spaces, Allen argues that 'both the building upon which an urban screen is placed, and the space in which the building is located can be considered as a form of multimodal text'. This leads to considering how only a blurred distinction seems to exist between space and information, as elements of space increasingly are powerful conveyors of information, whilst information – materialising into them – becomes more and more spatially-related. This is in a way not a new phenomenon. McCullough for instance takes readers on a journey that starts from ancient inscriptions and their epigraphy, and brings us to thinking how urban "markup", the ability to "layer" the city with digital inscription, tagging places and buildings in a participatory, interactive way can be a way to support new cultural productions which exist within 'a new domain [which] emerges between the authority of broadcast and the defiance of graffiti'.

Relationships are extended and multiplied, and urban space gets literally "augmented" by this, as more becomes possible. This extension of relationships and presence is not just quantitatively characterised by "more" possibilities. It can also have an important qualitative side. Heesang Lee's study on the spatiality of the mobile phone in Korea notes that existing in a public or private space can also depend on being connected to it through a switched-on phone, rather than being physically present. Connection and disconnection – and desire to be connected or disconnected – become important aspects of our ability to inhabit spaces, and of the quality of this condition. Augmented space, Heesang Lee argues, is not just 'intelligent' and 'efficient' but 'emotional' too, and this introduces a qualitative, very personal variable in this relationship between us and digitally-enhanced space, which can make connection assume the positive connotation of belonging to a place and to a network associated to it, or the negative one of being controlled and subdued to the network itself, seen in its wide sense of digital infrastructure as well as that of the people who share a connection.

However, as a consequence of these extensions of relationships, it is inevitable to consider that the re-definition of urban space and the extended range of possibilities that ICT may introduce, may also carry an increased risk of uncertainty and disorientation. Some of the authors highlight the tension between place and non-place as an important issue here. Mediated space can often be compared to globally-characterised non-places like airports and shopping malls. The tension could also suggest a "displacement" effect of digital technologies, taking their user "somewhere else", which is indeed a non-place denying or competing with the actual physical place they are at. Urban ICT, seen by many as a force of displacement and global – spatially indifferent – access, needs to get itself grounded again, in situations where space does matter.

This attention for ways of "framing" the augmented spatial experience and making it meaningful is – in different ways and at different scales – expressed by several of the contributors to this section.

Allen sees the body – present and involved in the augmented situation – as what can frame or re-frame this experience. But he also argues about the power of local context and culture, when this becomes one of the guiding constraints to the design and deployment of digital and media-rich urban initiatives. As it will also be highlighted in the other sections of the book – from different viewpoints – the efforts in "grounding" projects as well as making them "owned" and truly participated seem to be the other crucial factor for "framing" augmented space and allowing it to avoid the "non-place" trap.

This is also confirmed by Willis's observations on a Wi-Fi endowed café in London, and on the fact that rather than it becoming a non-place for anonymous, casual and entirely nomadic patrons, a mere physical platform to shelter "surfers", it would be appropriated by regular customers looking exactly for the augmented "bonus" of the place itself, providing a social physical environment with a digital edge. McCullough also notes how libraries – a physical space often seen as under threat from the ability to consult electronic information in a de-spatialised way – have become places of open digital access, providing a sense-rich spatial context for these activities, and support for social networking. As mentioned earlier, even mobile phones, one of the emblems of individualistic, displaced existence and virtually sustained relationships, end up being much more bound to local places and communities than expected, Heesang Lee notes.

But making all of this work positively for our cities requires a degree of proactiveness in shaping and designing grounded, meaningful intersections of space and ICT. Doing this implies questioning and refreshing our theoretical as well as practical approaches to space, as the simple juxtaposition of physical and digital – or the retro-fitting of physical space with digital facilities – is not a guarantee for the obtainment of properly, and usefully, augmented spaces. Willis argues that 'The two domains are operating on different structures, layered one on top of another but in many instances not working as a unified domain. In order to resolve these disparities it will be necessary to rethink some of the ways we act, occupy and also construct our physical world', calling therefore for the need of some "augmented" knowledge.

This prompts us to review a series of basic concepts and frameworks through which we read and interpret urban space. Willis again addresses how concepts of separation, bounded-ness, bodily presence, linkage and temporality are impacted upon by the presence and use of ICT. She also draws people – and people's changing habits – into the picture and argues that 'spatializing these communication technologies and reconnecting them to spatial settings requires new views on the inter-connectedness of location and behaviour'.

Such issues are also central to Foth and Sanders's research work on the augmentation of residential buildings and complexes, and the intersection of physical settings and digitally-enhanced community. They remark the importance of considering 'opportunities to inform residential architecture through advanced understandings of social networks and communicative ecologies will be essential in order to create public space that accommodates the needs of urban residents and their new social formations'. Their chapter ideally links this first book section on space with the following one on community augmentation, and shows how – despite the rather artificial distinction forced by the book's economy – these are really the two inseparable sides of an "augmented place-making" coin and will have to be dealt with in jointly.

But there is a third – obvious maybe, but sometimes overlooked by those who deal with the built environment – factor in the equation. Back in 1999, and addressing an architectural evergreen issue such as the relationships between form and function, Mitchell noted how 'in the design of smart things and places, form may still follow function – but only up to a point. For the rest, function follows code'[1].

---

1 Mitchell, W.J. (1999) *E-topia : urban life, Jim - but not as we know it*, (Cambridge, MA: MIT Press).

David Murakami Wood focuses on this by highlighting the software dimension of augmented space and looking at the emergence of "spatial protocols" as 'highly restrictive and controlling rules embedded within the materiality of urban space, which produce all kinds of new liberatory and repressive possibilities'. Murakami Wood points at the fact that the software providing the "intelligence" in spatial augmentation is certainly not neutral. It cannot just be seen as a platform which can be used in virtually infinite ways to extend relationships, but as something that embeds rules, behaviours and the ability – or lack of – of establishing relations in certain ways. The relative invisibility of the "soft" elements of augmentation, coupled with a degree of lack of transparency on how "code" is written, should prompt designers and shapers of augmented space – especially when this is public – to consider open source philosophies and approaches to 'create spatial protocols that actually open up better forms of civility and interaction and new domains of possibility for a wider range of citizens'.

All in all, this opening section provides a complex picture and a series of questions which are somehow left open – not a surprising thing for an emerging topic – though the different chapters suggest – if not precise solutions – at least lines of enquiry and issues for practice. On the one hand augmented space is seen as something potentially re-defining behaviours in ways we do not fully know, whilst on the other hand ways of re-framing and making sense – and positive usage – of this new range of possibilities are suggested. Grounding the experience spatially as well as culturally and socially seems a common thread – or set of themes – to focus on. Making it as much as possible open, transparent and participated is another one.

Do we need new rules and theories to tackle this? Maybe we need to re-frame our existing spatial concepts to accommodate this increased range of possibilities. Augmentation might end up "remediating" – as Bolter and Grusing would have argued about cyberspace[2] – our established spatial practices. Research and reflection is already under way, as some of the contents in this section – amongst certainly a much wider and richer global production – prove.

Finally, practice, experiments and designs – with all their potential limits – are essential to understanding more. Learning by doing is essential to explore new territories. Foth and Sanders quote in their chapter Hornecker *et al.* who examine opportunity spaces where 'there is no urgent problem to be solved, but much potential to augment and enhance practice in new ways'[3] and highlight why this is a worthy topic for urban designers as well activists in social informatics. It is not about reacting to problems to solve – hence having to wait for a major change to happen or materialise – but proactively putting forward new ideas for improving the spaces we live in.

---

2 Bolter, J.D. and Grusin, R. (1999) *Remediation: Understanding New Media*, (Cambridge, MA: MIT Press).

3 Hornecker, E., Halloran, J., Fitzpatrick, G., Weal, M., Millard, D., Michaelides, D., et al. (2006) 'UbiComp in Opportunity Spaces: Challenges for Participatory Design', paper presented at the Participatory Design Conference (PDC), Trento, Italy.

Chapter 1

Places, Situations and Connections

Katharine S. Willis

Introduction

> I then shouted into M (the mouthpiece) the following sentence:
>
> 'Mr. Watson--come here--I want to see you.'
>
> To my delight he came and declared that he had heard and understood what I said.'.[1]

In Graham Bell's first experimental call enabling communication between two people in separate locations (Bell 1876, 40-41) his opening instruction was to ask the person at the other end of the line to come to where he was. This underlines how the way of understanding the world so very often requires visual presence to authenticate social experience. In so many aspects of our everyday life we tend to "believe it when we see it". Our visual experience of the physical environment we inhabit therefore guides a great deal of how we perceive, remember and act in the world. Our spatial perception is also to a great extent influenced by the visual features and characteristics of physical space. In the 1960's the urban planner Kevin Lynch underlined the extent to which we essentially visually perceive and categorise the world in his seminal work, *The Image of the City* (Lynch 1960). In this study he established that individuals construct mental imagery about the space in which they move, which he proposed was broken down into a series of five key elements; landmarks, edges, districts, paths and nodes. He introduced the term "image-ability" to describe the qualities of a city which make it understandable to any citizen, again underlining the effect of the visual form of the city on perception and memory of physical space.

As far back as the introduction of the telephone, evidence can be found of how communications technology has had a significant impact on the structure and social use of the city. According to the work of de Sola Pool and his co-authors in *The Social Impact of the Telephone (*de Sola Pool 1977), the telephone contributed considerably to urban decentralisation and mass migration to suburbia, and also helped to create the specific architectural forms of the skyscraper and skyline (Gottmann, in de Sola Pool 1977, 310). But the telephone was and still is a fixed wire technology; it has to start somewhere and end somewhere with wires in-between. As we move into the twenty-

---

1 Alexander Graham Bell's notebook entry of March 10, 1876, in which he describes the first successful experiment with the telephone.

first century we are experiencing a growth in a whole range of new communication technologies which enable not just wireless but also mobile communication. Mobile phones, wireless internet, Bluetooth, GPS and all their associated applications enable the oft repeated ideal of communication "any time, any place". We no longer need to be sat at a computer in an office building to send an email, or hold a wired receiver to make a phone call. Consequently a number of authors have highlighted how the predominant visuo-spatial way of understanding the city is being fundamentally affected by such technologies which have very little visual presence. Graham (Graham *et al.* 1986, 50) highlighted the fact that: 'Given this visual pre-occupation , it is easy to diagnose the virtual invisibility of telecommunication in cities as a key reason for the curious neglect of telecommunications issues in cities'. Batty (1990, 128) further noted that 'cities are becoming invisible to us in certain important ways', and in another paper set out a research agenda which would look at a series of methods for enabling a visualisation of these nodes and networks (Batty in Hodge et al. 2000) which has also been addressed by Dodge and Kitchin (Dodge et al. 2000). Townsend (Townsend 2000) and others have also highlighted how the temporal quality of wireless and mobile networks reconfigure the spatial and visual qualities of the city, and so should cause us to question the nature of city infrastructures and how we plan out cities and physical and social sites of activity.

This chapter will investigate how we perceive and act in the city, and how this is affected by the presence of mobile and wireless technologies. In particular it will focus on the notion of public space in the city, and how these spaces can be occupied when citizens experience them through practices that are not necessarily visual or spatial in the terms that we traditionally accept urban experience. It will draw on a case study of the presence and social impact of a wireless network in an urban setting, and then conclude by proposing a series of ways in which we may need to adapt out notion of urban public space to better enable us to enact our experience of the city whilst inhabiting places and moving between flows.

Places

In addition to our visual memories and experiences we also experience space as set of social settings and places which have come to have meaning for us. Despite the visual dominance of many elements that make up our idea of the city, our behaviour in space and memories of past events also categorise and organize our actions. In addition to the landmarks, paths and edges of the city we also have an acute awareness of socially defined structures; neighbourhoods, the place we call home, even our favourite meeting place or the spot in the park where we can catch the afternoon sun. In urban public space there is an emphasis on the social domain of the street as a means of organising or even coordinating space. It is no accident that a city map is a street map.

The fundamental role of space in structuring activities in the digital world is underlined by the widespread use of spatial metaphors in digital worlds: café, library, electronic highway, home and rooms are common ways of describing the properties of digital 'spaces'. Indeed the use of space as an organising metaphor for interaction

(Lakoff and Johnson 1980) is a natural one. The capabilities of technologies have been implicitly designed around an understanding of the relationship between the structure of the environment and understandings of the activities that take places there. In terms of structure, the established physical model considers space as a three-dimensional structure, where qualities such as location can be defined in terms of exact metric co-ordinates. Alongside this domain of spatial settings is a world of social settings, which take their sense from configurations of social actions. The Euclidean qualities of spaces afford essentially physical constraints, based on the fact that humans cannot walk through walls, that objects do not float, and that light creates shadows on surfaces. In parallel, space also provides social affordances, based on socially acceptable norms such as the fact that we have different types of conversation on street corners than in bedrooms, and we would feel comfortable shouting at a football match, but not in a church. Space frames human action and interaction in multiple and varied ways. In fact our understanding of space underpins many aspects of how we in turn inhabit and act, and can be interpreted in terms of the corresponding spatial capabilities of individuals. If we are to better understand how our perception of urban public space is changed through our experience of mobile and wireless networks it is critical to uncover how our actions in the world are affected by our concepts of metric space. Many of these are based on the Euclidean or topological understanding of space, and can be interpreted in terms of the corresponding spatial capabilities of individuals can be summarised as follows:

Physical separation The concept of distance describes the state of two objects which cannot by definition be physically present in the same location, and correspondingly that physical separation or absolute position is a key property of any person or thing. This can be extended towards the relational aspect of spatial proximity, where concepts of closeness and remoteness are grounded.

Bounded-ness The idea that space has extents which are not infinite affects how structures of spatial separation and constraints are understood. Space is typically conceived as having some form of definable extent, which enables it to be sub-divided in a range of ways into units with particular properties. Out of such concepts arise sociological frameworks such as territory, neighbourhood and even personal space.

Presence If we take the case of individuals, then any experience of space is framed by a subjective awareness that they have a physical or bodily presence. In many cases this awareness is formed through perception, and vision tends to dominate. In the context of any activity this awareness is constantly measured, evaluated and updated, in processes such as orientation and navigation.

Linkage Although this is implicit in the previous descriptions, the concept of linkage or relation is inherent in space. For example, an effect in space, such as light and shadow, is felt universally; light doesn't just fall on one side of an object, but illuminates all sides in a proportional manner. Objects or people in space are understood as being either more or less connected, a factor which is realised in a diverse range of ways; such as ideas of scale, social networks or even economics.

Temporality In three-dimensional space, time is seen as the fourth dimension. Space can simultaneously be understood as a fundamentally stable environment, undergoing little change (e.g. buildings), or conversely a changeful state in almost constant flux (e.g. a journey).

Situations

The Euclidean notion of space is just one part of a complex jigsaw of how we understand the city. Places are also understood as sites of social activity, and this is particularly true for urban public space. These are places without fixed patterns of use, and as such offer possibilities for more than one social activity. Social interactions and activities are dependent on settings or situations which are guided by the physical setting (Goffmann 1969, 20). Interestingly many of the relationships which structure our social world are not analogous, but operate on network type structures consisting of strong and weak ties, which are primarily relational in nature. This means they do not exist as entities in themselves, but only exist in context of the link to other entities in the network. If we think about a building in Euclidean terms, it is an enclosed space with a particular function. But if we look at it in terms of its social structure it can be seen as a complex series of loosely and strongly connected links and nodes making up a social network, which even have the possibility for total disconnection from all other nodes. For instance a place such as a church or a classroom is physically designed and socially designated to support a radial topology of communication (see Fig 1.1), with uni-directional links from one person to many. On the other hand, a café

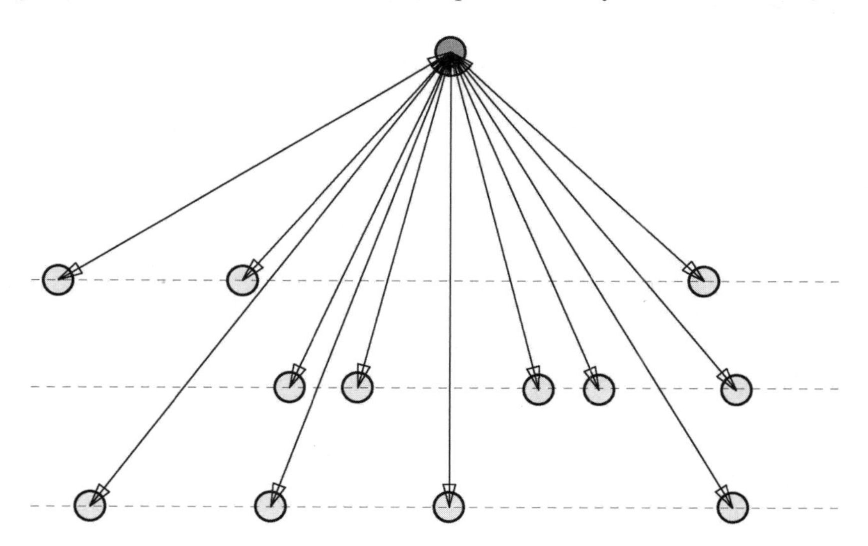

Figure 1.1 Church Network Topology, Credit: Author

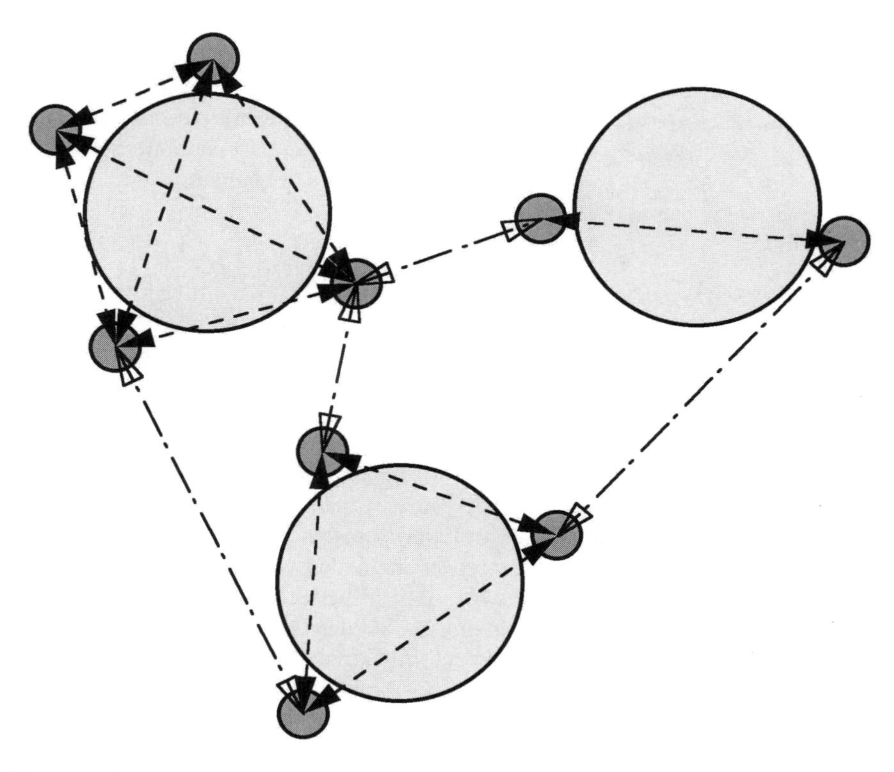

Figure 1.2 Café Network Topology, Credit: Author

topology (see Fig 1.2) consists of clusters of highly interconnected nodes (the tables) which are loosely associated to form a large network (the cafe), with optional one-way (eavesdropping) links (Adams 1998, 91).

The social use of a space is similarly temporarily fluid. A church may appear to visually be a church building with fixed walls, but for the people that visit it, it may be a place to pray privately, a place to congregate with a group, or a place to mourn on a one-off occasion. All these ways of occupying the space are valid, and they do not conflict. This implies that despite the relatively rigid notions of Euclidean space by which we perceive a church building we are far more flexible in accepting the church as a place of many simultaneous and even conflicting social settings.

The way we communicate with others is therefore influenced both by the physical location and also the patterns of behaviour that occur in the location; two conditions which are fundamentally interlinked and dependent on one another. In this manner our actions in space can be considered as situated in that they are guided by a rich and complex background of social relations and behaviours. Communication technologies in urban settings further enable multiple social realities to occur in one place, since they can be understood as overriding the boundaries and definitions of situations supported by physical settings. Since "where" you are no longer defines "who" you are, new media eliminate a traditional dimension of civic legibility

(Mitchell 1995, 101). The underlying dichotomy of public versus private in public space is rendered more fluid by new mobile and wireless technologies, which in turn imply a fundamental transformation in the norms of public action and conduct. The same physical space may be caught within the domain of two different social occasions.

Connections

Mobile and wireless media have specialized infrastructures, and as these technologies emerge in the city, they become overlaid with existing urban infrastructures. The individual's image of the city, which they use to navigate and orientate themselves within urban space, is no longer simply confined to physical elements and configurations. The proliferation of wireless internet or Wi-Fi[2] nodes in urban environments is creating a dense communications infrastructure, which re-draws existing spatial thresholds and territories. The nodes, which are essentially black box transmitters operating on the frequency of 802.11b standard create a region of access of between forty-five metres indoors and ninety metres outdoors. Due to the extent of this region, or territory, wireless access to the node can be available well beyond the physical borders and thresholds that traditionally delineate the boundary between private and public space. The information flow ignores the material thresholds of walls and doors and extends beyond traditional materially bounded notions of space.

The nature of the frameworks in which nodes are understood by urban citizens can be further investigated through the practical ways in which they are identified. In order to instantiate a node in a location it must be named or labeled with a tag. Despite the fact that both the hardware and signals from such nodes have a very distinct spatial extent and position node names rarely any bear relation to the static physical location. Instead the names of nodes predominantly reference either the modem hardware which enables access to the node or the name of the network, both parts of the technology infrastructure Indeed the very vagueness of the word "node" indicates the loss of a language for naming environmental value (Sennett in Carter *et al.* 1993, 319). The identity of the node is not perceived as delineating spatial territory or having temporal qualities, whether physical or digital. Wi-Fi nodes are understood more along the metaphor of a switch, which simply establish and break linkages, and as such they equate with access to information flow.

The infrastructures of mobile communication technologies offer significant issues for the way in which urban space is perceived and constructed. Mackenzie (Mackenzie 2005, 3) highlights how wireless internet explicitly re-frames the cultural relevance of infrastructure, inverting the traditional paradigm of technical structure as a site from which relations to the exterior environment run in and out. Thus, physical infrastructures which have traditionally been viewed as the dominant form in the city as a spatial construction, are being re-mapped through the ubiquity

2 Wi-Fi is a wireless technology intended to improve the interoperability of wireless local area network products. A Wi-Fi enabled device can connect to the Internet when within range of a wireless node.

of communications networks. This highlights the importance of understanding how physical space can better integrate with the layer of digital nodes and networks. Over a number of years some researchers have approached this problem by looking at ways of visualizing these networks; such as the work of Carlo Ratti (Ratti *et al.* 2005) with real-time mobile phone networks and the Equator Project with the mapping of Wi-Fi nodes and GPS availability (Dix *et al.* 2005). These go some way to giving people in physical space an understanding of where the spatial presence of wireless and mobile networks overlay the space. But on another level these mappings use false metaphors, and in trying to reduce complexity create instead oversimplified models of how these technologies exist in the physical world. Quite simply these technologies do not occupy space in the same sense that we have come to perceive metric space. They exist as ethereal flows of information, which change temporally and are affected by our physical world with the consequence that signal strength and thus availability can disappear as quickly as it becomes available. Such characteristics cannot be imagined or visualized in the visual terms by which we understand the world around us. In attempting to better integrate these technologies into our physical world there will need to be a shift in both frameworks; the physical structure and layout of the built world will need to be designed around the features of these technologies, and the technologies themselves will need to be adapted so that their presence in space can be better utilized.

These technologies reconfigure Euclidean spatial frameworks framed around spatial proximity and bounded-ness, in a manner which is fundamentally different from the PC internet (Ito 2005). Thus, the previously defined aspects of the 'real world' which can be exploited as part of a spatial model, need to be informed by the aspects of the affordances of mobile and wireless technologies, and can be summarised as follows:

Separation Displacement in layered media spaces is not confined to the physical properties of 'real world' objects, but also extends to include the specific ranges of technologies. For instance Bluetooth, enables interaction within a radius of approximately ten metres, whereas Wi-Fi nodes offer access within a range of up to one hundred meters. As such the definition of interaction in a space of communication flows is structured around spatial nodes of opportunity.

Bounded-ness Regions are not only defined by spatial extents, but also by patterns of informational or social access. Consequently, collectively defining boundaries becomes part of the pattern of communication; for example the common practice of asking for and reporting location at the beginning of a mobile phone call. Boundaries are still an omni-present characteristic of space, but moving in and out of bounded zones can occur much like the flicking of a switch, rather than involving some form of graduated change.

Presence Technologies create a form of shared background space, not based on physical presence. Presence becomes more ambiguous, since previous reliance on the visual to orientate and structure awareness in space is augmented with non-visual presence in technological networked spaces. For instance, a form of co-location becomes possible, where interaction can occur in represented models of the "real world", whilst simultaneously being physically present in the real-world. One of the consequences of this is that actual physical co-presence; or the 'flesh meet' (Ito 2005, 260) is elevated to a higher level of importance.

Linkage The concept of linkage is intensified, and in many ways more subtle and differentiated levels of connectivity frame interaction. Action in network type structures is characterised by a whole array of weak and strong links. Networked infrastructures start to dominate over physical spaces.

Temporality Interactions occur in a "real time", which is de-sequenced and person centred rather than a global time. Stability and permanence are comprehended as particular qualities of the "real world", and fluidity and change are valued. As such time becomes more malleable and capable of division into non-linear segments.

The nature of mobile and wireless communication is still fairly unfamiliar, and the quality of moving physically while keeping the networking connection to everything we do is a realm of then human adventure, on which we know little (Castells 2004, 87). Therefore spatializing these communication technologies and reconnecting them to spatial settings requires new views on the inter-connectedness of location and behaviour.

Spatial Presence of Wireless Nodes: A Case Study

In order to better understand how wireless networks can affect spatial settings a case study was undertaken of an urban space to investigate the nature of the spatial presence of wireless networks. The study was undertaken in a defined one kilometre square area in South-East London called Deptford. The area studied roughly covers a neighbourhood which is home to a number of community based and public arts centres, including theatres and community centres as well as public amenities such as swimming pools, schools and parks. It also has a vibrant street culture, with a twice-weekly street market, and is home to a lively multi-cultural local population. However the main reason the area was chosen for this study is that Deptford is also home to a community-led public wireless mesh of nodes; called 'Boundless'. Established during 2004, Boundless developed from a direct evolutionary line out of the two initiatives which were the first to offer free Wi-Fi access in London; the Clink Street and the Consume network (Bleecker 2004, 7). According to James Stephens, one of the founders of the project: 'Boundless consolidates theory and free network practice pioneered in the wild back then'.[3] It aims to support community development of user-owned and operated local internet access, inter-linking residential, business, educational, cultural and digital media communities. In studying of the area of Deptford that hosts the wireless network, this project seeks to address how the perception and use of public space is affected by the presence of

3 Personal correspondence, 2 November 2007.

these nodes in public space. The study is split into two distinct stages; the first being an investigation of the spatial configuration of the network of wireless nodes and how this affects the social and spatial characteristics of public space. The second stage of the study focuses on a single public wireless node, and how individuals perceive the spatial presence of this technology in its corresponding spatial setting.

Locating the wireless nodes in public space

The first stage of the project involved using a series of mapping techniques to establish the actual location and density of the pattern of wireless nodes. An important factor to consider in studying the presence of such networks is the almost complete lack of any kind of overview mapping or viewable representation of the location and availability of these networks. Individual wireless access points are advertised locally at distinct locations, but the network of nodes is nowhere present to be seen as a whole. A second consideration is the temporal nature of the technology. Depending on a whole number of conditions (including weather and network capacity) the signal strength and availability can vary widely. This is despite the fact that the black box

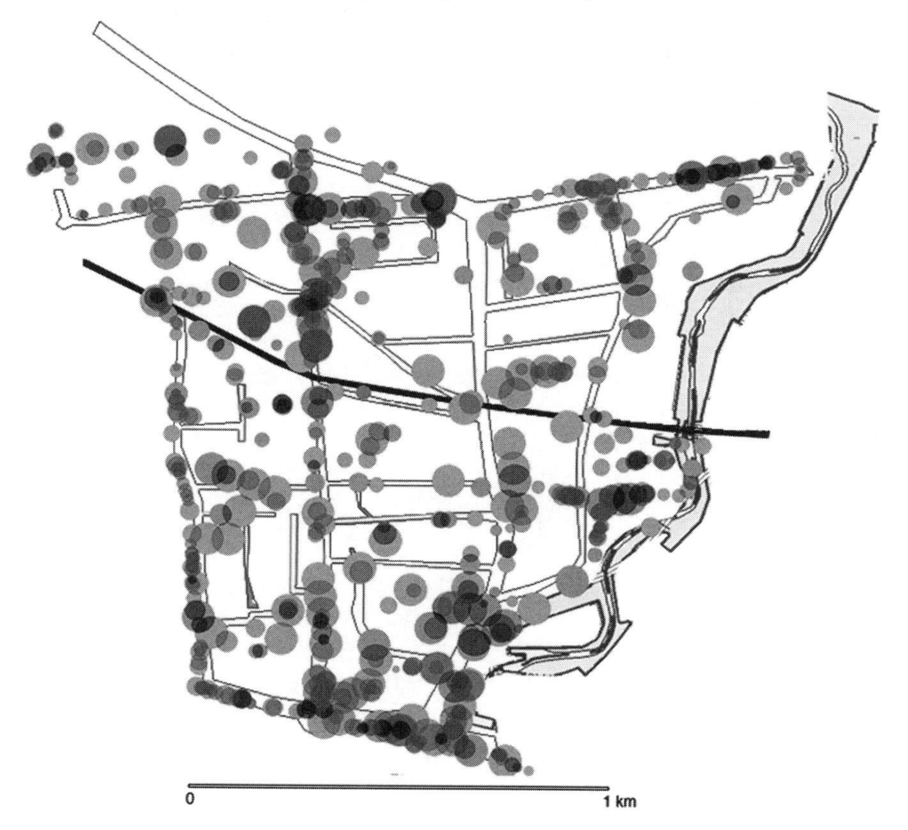

Figure 1.3 Density of Public and Encrypted Wi-Fi Nodes, Credit: Author

emitting the wireless signal has not changed in any way. This means that a pattern of nodes detected on one day may differ from those detected on another day. For this reason the detection was undertaken on three separate occasions to validate the data. In the approximately one kilometer square area considered the detection process was carried out using a method similar to that known as wardriving.[4] However in this case the Wi-Fi nodes were detected whilst walking around the area on foot, rather than by driving, so as to ensure all accessible public space was mapped.

The first outcome to note about the study is the large number of Wi-Fi nodes detected. Over five hundred and eighty-nine public and encrypted nodes were found. The nodes were not spread evenly, but tended to show patterns of clustering based on building occupancy and use. In the most densely covered locations, there was a density of up to one Wi-Fi node per ten square metres (see Fig 1.3).

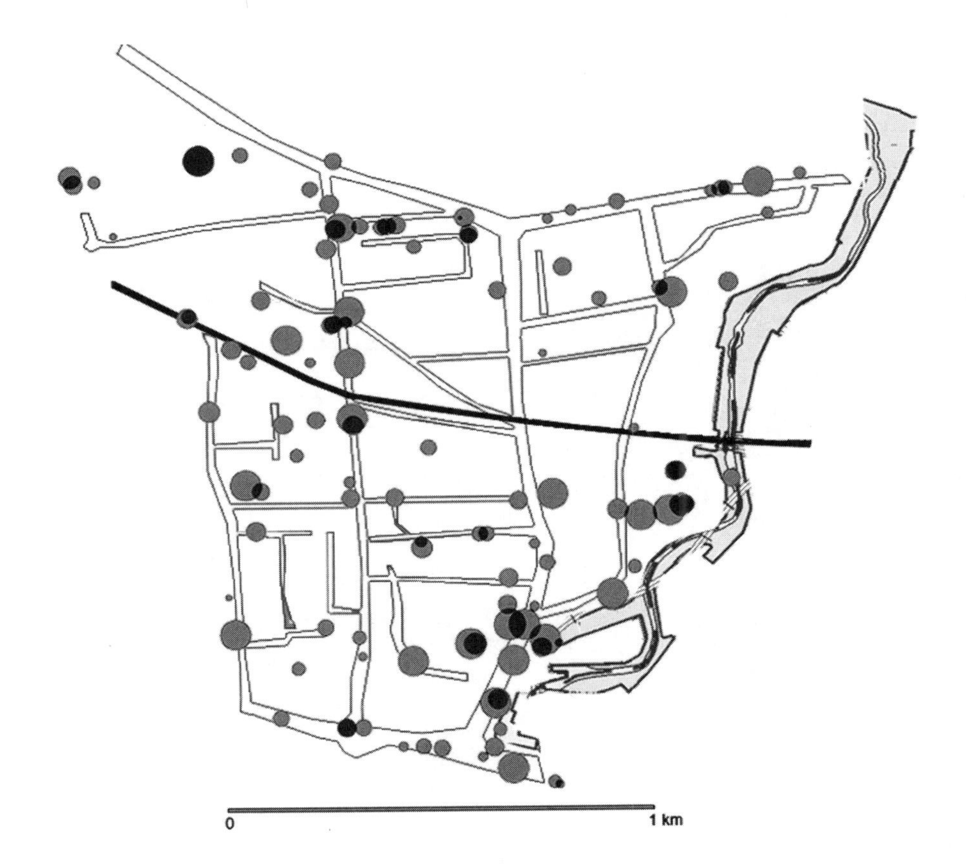

Figure 1.4 Density of Public Wi-Fi Nodes, Credit: Author

4 Wardriving is the practice of detecting and mapping wireless access points.

Of this total, one hundred and nineteen nodes were public or open which represents approximately twenty per cent of the overall number (see Fig 1.4). Thirty-one of the public nodes were those which were part of the Boundless wireless network. A further fourteen were sited in community centres or public amenities such as schools. The remaining seventy-four can be identified as nodes located in private property which have been either deliberately or unknowingly left open by their owners. This means that sixty per cent of the open nodes are not part of any organizational framework, and thus their availability can be considered as conditional on the control of individual persons who may or may not have altruistic intentions. Ten per cent of the public nodes are hosted by centres or organisations that already offer public physical space; theatres, arts organisations, visitor centres and even public houses. Importantly these nodes tend to be located in places where the facilities make it practically possible to access the Wi-Fi such as a warm place to sit and a power supply. They are also

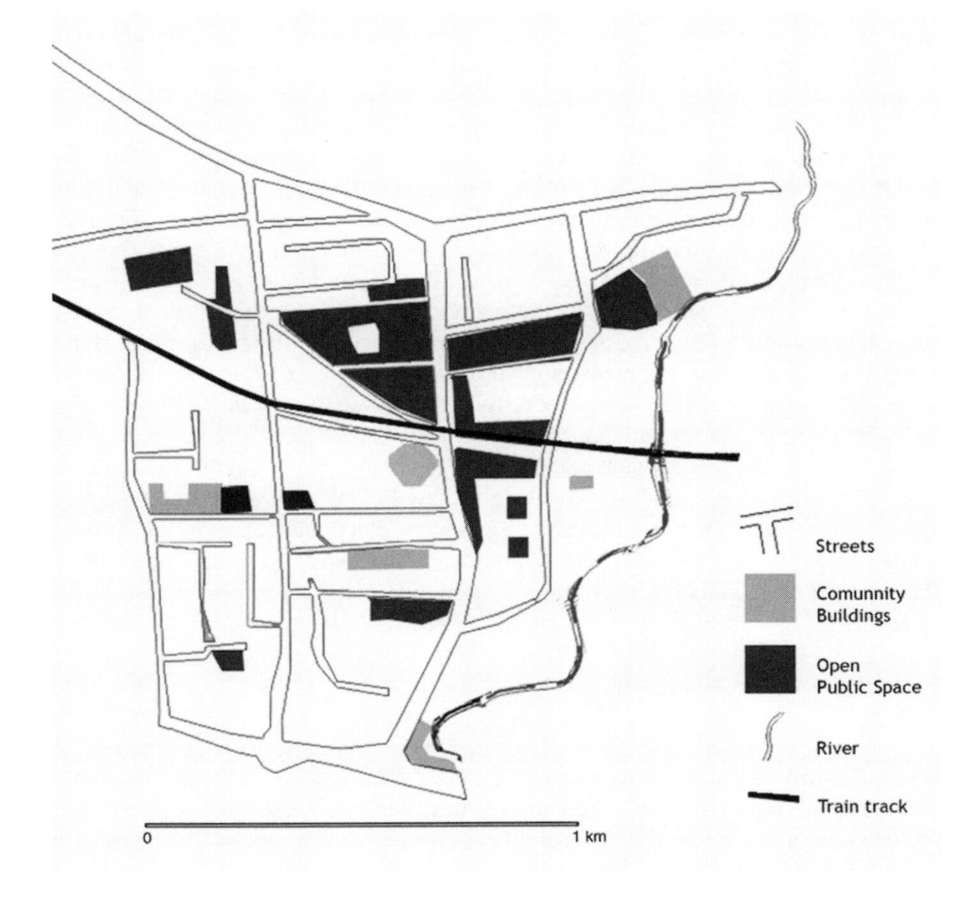

Figure 1.5 Characteristics of Public Space in Area Studies, Credit: Author

literally integrated into an existing social structure and physical location, with the associated guarantee of technical dependability and a degree of social interaction arising out of link to the relevant organization. In fact the key infrastructure is often the people, not the technology, a characteristic highlighted by James Stephens of the Boundless network:

> We have a network of fifty nodes which link together and redistribute any broadband connectivity nodeholders have to offer. (Yet) it's the people who are most the important and who require the most support, attention and encouragement.. To establish a broad range of coverage in an area requires clear line of sight roof to roof connections but the ground level solutions are the hardest to co-ordinate and sustain.[5]

For comparison purposes, the characteristics of the public space for the area studied are illustrated in Fig 1.5.

Aside from these community nodes, the remaining public nodes are often located in private spaces such as houses or offices. The consequence of this is that although the Wi-Fi access may spill out into open public space such as streets and parks, it is not primarily intended to be used in these locations. In fact one of the marked characteristics of the location of the nodes in this study is that none coincide with any form of green open space. The traditional function of green open space as offering a place to spend spare time and for entertainment is thus never coincidental with the opportunity to access the wireless internet for leisure or relaxation purposes. Obviously this is not an intentional quality, but the location of the nodes means that Wi-Fi access is often correlated with busy and functional public spaces.

The main street running through the centre of the studied area is home to twelve (ten per cent of the total) of the public nodes, giving opportunities for the many shops along the street to link their commercial activities in physical space with online access. But there is not one example of crossover between these on and offline worlds either in the street space or any other site within the area studies. This highlights just one example of the predominant paradigm of the Wi-Fi coverage in the area studied; that there is almost no meshing of the use and activities occurring in the urban public space and the corresponding location of public wireless access points. Aside from a comparatively small number of exceptions, the somewhat surprising outcome of this study is that the two worlds operate in almost independent spatial and social spaces.

Perception of a single wireless node in public space

The second stage of the study seeks to understand in more detail the nature of perception of the spatial presence of a single Wi-Fi node. Wireless access is structured around nodes which are the point at which a person actually accesses the network, and are literally devices that have a physical location. Yet in terms of their presence in space they may be viewed through an interface which shows the presence of all available nodes, but are rarely present in the sense that they can be seen with the eye. Most often they are concealed within the physical structure of

5 Personal correspondence, 2 November 2007.

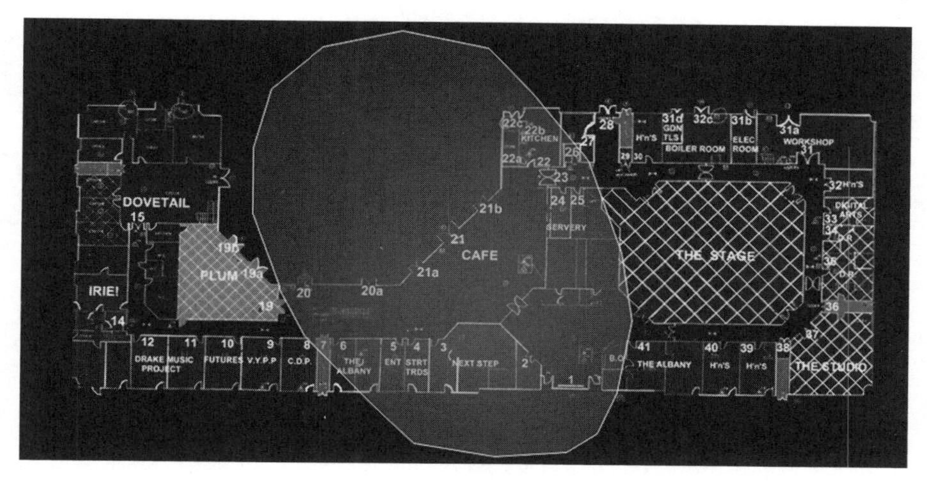

Figure 1.6 Actual Spatial Presence of a Public Wi-Fi Node, Credit: Author

the building, or a merely such anonymous devices in terms of visual appearance that they are not noticed. Yet on order for us to act "sense-ably" in urban public space it is often useful to be able to literally find a node. So, how do people act in the presence of wireless networks in public space if they do not base their action on what they can see? The study was conducted in the café space of the Albany Theatre, a community centre and theatre located in the heart of the Deptford area. The café space is indoors, situated close to the entrance to the building, and has an adjacent outdoor courtyard. The wireless network is available throughout the café space, but

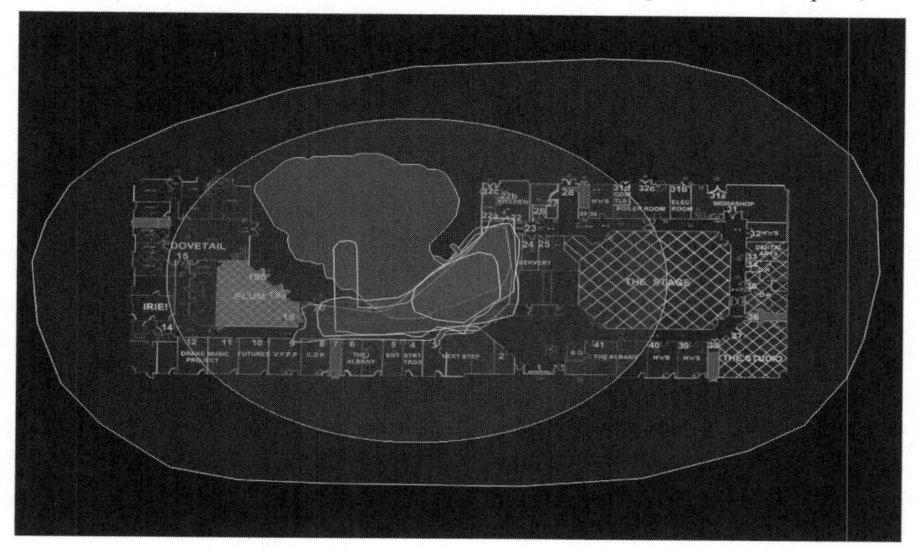

Figure 1.7 Perception of Spatial Presence of a Public Wi-Fi Node, Credit: Author

actually offers access to the internet in a wider area than that confined to the physical walls and boundaries of the building, so that it also extends out into the street space to the front of the building and the courtyard to the rear (see Fig 1.6).

Over a period of three days the usage of the café public internet was observed, and visitors who accessed the internet using a laptop where questioned as to how they perceived the presence of the availability of the wireless internet. Fourteen people where interviewed and asked to draw a shape indicating the extent of the wireless internet availability on a scaled plan drawing of the building. The results (fig 1.7) indicated that people have in general a poor understanding of how the wireless technology is present in the space.

Almost all the participants described availability of wireless access as being confined within the physical territory of the indoor café space. Only two participants indicated the actual condition that the wireless access extended to cover an area not confined to the physical boundaries of the space. For instance one participant reported: 'well I know you can get access out in the foyer, but there's nowhere to sit out there so it's not usable'.

Other participants readily acknowledged that they didn't know where the wireless signal reached to and had simply based their perception on the limits of their experience, with one person explaining that: 'I know I can get it here because that's where I always sit, but I don't know about whether you can get access over there' whereas another participant commented: 'you can maybe get access out in the courtyard, but I've never tried'. When asked to show the location of the router on the plan none of the participants interviewed could estimate its position, apart from one person who had once noticed it being relocated from the ceiling space by the in-house technician.

The interesting outcome of participants' response to the study was that they saw the technology not in terms of where it could reach to in space and possibilities it offered on its own terms. Instead they perceived the technology as having presence only where they could access it, in the sense that it was usable to them. A second aspect of the usage of the wireless network in the Albany is that the flow of people through the space is slow. Instead of people spending a few minutes spent checking some piece of information online and then returning to a task in the local area the actual pattern of behaviour is almost the opposite. According to usage statistics from wireless traffic at the café ,[6] ninety percent of users are regular users making regular and often daily visits. In this way the space is not frequented by people dipping into the world of digital information before continuing on their journey, but rather as a work space or home-from-home offering the added benefit of the social interaction offered in a public space.

6 Obtained during personal exchange with Stuart Calder, Events and Digital Technician, 1st October 2007.

Enacting the Spaces of the Wireless City

The way in which we view it, public space has a visible appearance with salient features that structure it; there are landmarks, visually familiar places, open viewpoints, and closed spaces. These visuo-spatial properties of public space enable and frame patterns of behaviour and activity. For example we tend to meet people at commonly recognizable landmarks; below a clock or on a street corner, and we relax in spaces which often have a physical openness. But as demonstrated in the study of the perception of the spatial presence of a wireless node, wireless technologies are not visible structures in public space. The presence of networks in public space exists in a manner more similar to our concepts of a social network. Our notion of the social network of friends, relations and acquaintances exists as a highly developed framework in the mind of an individual, not as a visuo-spatial mental image, but instead as a network of possible relations connected through threads of weak and strong ties. Thus we see these technologies as connection points with opportunities for accessing information. A person is thus perceived as being separated from another only by a switch to a network connection, not a physical distance in space.

Public space has at its heart the notion of equal access. This used to refer to literal physical and social access; equality between individuals to be present and act a common space. But mobile and wireless technologies reconfigure this concept of access in the sense that access also becomes access to the technology. Yet, in all the imagining of un-tethered access to wireless worlds, there still remains a frustratingly practical issue associated with interacting with a laptop or mobile device delivering the wireless information. Even despite the large number of unencrypted wireless nodes it is not really possible to move seamlessly between them. Moving between points of connection still has boundaries and concentrated centres with transition spaces between. This is for two reasons; firstly most available wireless access technology does not allow for seamless switching between available networks; the user has to take a conscious act to switch nodes. Secondly the physical features of the built and natural world create boundaries for the wireless signal. The signal will not pass through dense physical objects, such as concrete, metal or over natural features such as hills. In this way technology creates shadows or pockets of bounded 'off limits' space; gaps in network availability. The boundary of the space is not just defined by the physical limits of visible space, but limits of access and usability.

Physical line of sight is still often a requirement. In this sense the visual aspect of such technologies still has some relevance. Since these nodes are to all extents and purposes concealed black boxes, the individual has to have an awareness of where they need to position themselves in space in order to get a good signal. This is usually achieved by referencing the strength of signal bar in the mobile device interface, but if a user returns to a place many times, they slowly start to develop an awareness of the presence of the node in that space, even if they aren't able to visually see it. For external public space this can sometimes be a little easier in that wireless aerials are by necessity located on rooftops and are thus often visible against the skyline. In this way rooftops take on a different relevance, and there can be some comparison with the traditional role of the church spire as the highest visible point in a neighbourhood. More and more the placing of wireless antennae supercedes this visual role by occupying a dominant position on the skyline.

As mobile and wireless become ubiquitous in these transitional spaces, those using it still exhibit an often passive engagement with the space itself, but they in contrast they start to literally spend more time in the space. The purposeful interaction with a laptop or a long mobile phone conversation, give the user a legitimate reason to remain in a public space. The increasing ubiquity of wireless networks similarly affects how people engage with urban public space. Increasingly wireless access delivered through high-end mobile devices will start to change patterns of use, as these require less attention to the physical nature of the device, and are typically intended to be accessed for shorter periods. But even still the moment when an individual is required to pay attention to the device is a moment when their visual attention is concentrated on the device itself, and in most cases the individual will also voluntarily or involuntarily physically stop moving. Interacting with wireless networks in public space may simply have the effect of literally slowing down the normal pattern of movements through the spaces, as people's attention is increasingly taken up by acting and reacting to flows of information.

Summary

Communication technologies shape our experience of urban public space. Starting with the telephone we have learned to communicate at a distance, and to adapt to the corresponding changes in the way we inhabit and imagine physical space. With the emergence of mobile and wireless technologies, which are becoming ubiquitous in public space, we are still limited by our essentially metric perception of spatial settings. The complex and rich nature of social interaction in public space is transformed when these interactions are less defined by physical boundaries and frameworks. Spatial concepts such as separation, bounded-ness, linkage, presence and temporality are reconfigured by mobile and wireless technologies so that although the physical setting still influences our actions, many aspects of social connectedness are further elaborated and accentuated. In the case studies the effect of wireless networks on how public space is inhabited and perceived was discussed. They identified that the non-visual presence of such networks mean that people tend not to see such technologies in spatial terms. In terms of the way the technologies are enacted in public space, this results in patterns of behaviour where people slow down their movement through the space and generally. All of these social and practical aspects of acting wirelessly tend to contrast against the idealised image of the person on the move, flicking between flows of information whilst walking through the public spaces of the city. In a sense what we have is a discord between the physical and social possibilities offered by wireless technologies and the reality of the physical and social world. The two domains are operating on different structures, layered one on top of another but in many instances not working as a unified domain. In order to resolve these disparities it will be necessary to rethink some of the ways we act, occupy and also construct our physical world.

References

Adams, P. (1998), Network Topologies and Virtual Place, *Annals of the Association of American Geographers*, 88: 1, 88-106.

Batty, M. et al. (2000), 'Representing and Visualizing Physical, Virtual and Hybrid Information Spaces' in D. Hodge et al. (eds) pp. 133-146.

Batty, M. (1990), Editorial: Invisible Cities. *Environment and Planning B,* 17, 127-130.

Bell, A. (1876), Lab Notebook, March 10th 1876. (Washington: American Library of Congress) also available at http://lcweb2.loc.gov/cgi-bin/ampage, 22.

Brunn, S.D. et al (eds) (2004), *Geography and Technology*. (Netherlands: Kluwer Academic Publishers).

Carter, E. et al. (1993), *Space and Place: Theories of Identity and Location* (London: Lawrence and Wishart).

Castells, M. (2000), *The Rise of the Network Society*. (Oxford, UK: Blackwell Publishers).

Castells, M. (2004), 'Space of flows, Space of Places: Materials for a Theory of Urbanism in the Information Age', in Graham, S. (ed.).

Dix, A. et al (2005), 'Managing multiple spaces' in Turner, P. (eds) Space, Spatiality and Technologies.

Dodge M. et al. (2000), *Mapping Cyberspace* (Routledge: London).

Goffman, E. (1963), *Behavior in Public Places; Notes on the Social Occasion of Gatherings* (New York: The Free Press).

Goffman, E. (1969), *The Presentation of Self in Everyday Life*. (London: The Penguin Press).

Gottmann, J . (1977), Megalopolis and the Antipolis: The Telephone and the Structure of the City in de Sola Pool (ed.)

Graham, S. et al. (1996), *Telecommunications and the City: Electronic Spaces, Urban Places*. (London: Routledge).

Graham, S. (ed.) (2004), *The Cybercities Reader* (London; New York: Routledge).

Harrison, S. et al. (1996), 'Re-place-ing Space: The Roles of Place and Space in Collaborative Systems', in *Proceedings of the ACM CSCW*, 67-76. New York: ACM.

Hodge, D. et al. (eds.) (2000), *Measuring and Representing Accessibility in the Information Age*, (Berlin: Springer Verlag).

Ittelson et al. (1966), *Environmental Psychology*, (New York: Rhinehart and Winston).

Lakoff, G. et al. (1980), *Metaphors we Live By* (Chicago: University of Chicago Press).

Lynch, K. (1960), *The Image of the City* (Cambridge: MIT Press).

Mackenzie, A. (2005), 'Untangling the unwired: Wi-Fi and the cultural inversion of infrastructure', *Space and culture* 8: 3, 269-285.

Meyrowitz, J. (1986), *No Sense of Place: The Impact of the Electronic Media on Social Behaviour* (USA: Oxford University Press Inc).

Mitchell, W. (1995), *City of Bits: Space, Place, and the Infobahn* (Cambridge: MIT).

Ratti C., et al. (2005), 'Mobile Landscapes: Graz in Real Time', *Proceedings of the 3rd Symposium on LBS & TeleCartography*, 28-30 November, Vienna, Austria
de Sola Pool, I. (1977), *The Social Impact of the Telephone*. (Cambridge: MIT Press).
de Sola Pool, I. (1990), *Technologies without Boundaries: On Telecommunications in a Global Age* (Cambridge: Harvard University Press).
Suchman, L. (1987), *Plans and Situated Actions* (Cambridge, England: Cambridge University Press).
Townsend, A. M. (2000), Life in the Real-Time City: Mobile Telephones and Urban Metabolism, *Journal of Urban Technology* 7:2 , 85-104.
Turner, P. (eds) (2005), *Spaces, Spatiality and Technologies*, (Dordrecht: Springer).
Wellman, B. et al. (1988). *Social Structures: A Network Approach*. (Cambridge, England: Cambridge University Press).
White, H. C. (1992), *Identity and Control*. Princeton, (NJ: Princeton University Press).
Zook, M. et al. (2004), 'New Digital Geographies: Information, Communication and Place', in Brunn, S.D., et al. (eds).

Internet-based references

Bleecker, J. *The State of Wireless London* (published online March 31st 2004) http://informal.org.uk/people/julian/publications/the_state_of_wireless_london/
Consume Project, http://www.consume.net.

Chapter 2

Framing, Locality and the Body in Augmented Public Space

Patrick Allen

Introduction

This chapter deals with a range of issues related to the structure and appearance of augmented public space in terms of framing. It also develops key theoretical perspectives concerning the ways that information and media content is superimposed onto the urban environment. In doing so, it analyses the importance of locality on the character of display and argues that in the long run it is the body that is central to the framing of content and so is crucial to our understanding of augmented public space. This is exemplified in the widespread adoption of urban screens in UK city centres which forms a case study, but is not exclusive in its application to urban screens. The issues dealt with are relevant to all forms of augmented public space and in any situation where the built environment coexists with layers of information and media content – the "media layer". As a result of this coexistence, it appears that on the surface the distinction between the physical and the digital is self evident. It is argued throughout that one of the defining characteristics of augmented public spaces is actually to hold this distinction at bay and so prompt the questioning of its transparency and to present the real and the virtual as something of a problematic distinction. Another critical issue is that, on the one hand, there is an experience of such spaces that is both transitory and incomplete (Augé 1995), especially when the intervention of technologies of visual display is taken into account, and on the other hand, there is an experience of the urban space that is clearly fixed and tangible when considering the physical structure of the built environment and especially when aspects of representation in the built environment become fixed in terms of locality. These two opposed senses of space are, however, united via the body as a frame, as a container of all information (Featherstone 2006, 235), which is perceived by subjects and is the locus for the perception of place and location: a frame that is both content and context.

Research into augmented public space, though still to be fully articulated, either in terms of any general theory or of any concrete application, has tended to encompass the following three strategies as different ways of interpreting the increasing integration of architecture, communications technology and various forms of display technology in the urban environment. First, urban planners and architects themselves now employ forms of three-dimensional display, so-called "virtual" or "augmented reality", to extend more conventional ways of constructing plans for urban spaces. An example of this is MIT's 'Augmented Urban Planning Workbench' (Ishii *et al.* 2002). Second, is in terms of the integration of real-world architecture in urban spaces that is itself combined with display technology. This can take the form of big screens in city centres, digital signage and information spaces. Added to this are other forms of information display and media content received by the individual through a variety of mobile communications devices. Third, is the way that ideas concerning the overlapping of media content have been extended to encompass other forms of display and communications technology and interaction that interrupts people's inhabitancy with the urban environment and links them potentially to other spaces and other localities and consequently to forms of public space that have no tangible boundary. This chapter is only concerned with the latter two of these strategies.

Framing Augmented Public Space

There has been a significant impact on the visual appearance and experience of the urban environment as a consequence of the emergence of augmented public space. What follows is an attempt to capture at least some of the reality of this transformation in the inherent structure of urban spaces, where the virtual forms inherent in the interaction with media content are experienced simultaneously with the real structure of the built environment. Such a transformation has occurred primarily through the integration of the built environment with the media layer in that a range of digital display technologies and communications media now coexists with architecture. At the most basic level, one key argument is that, due to the application of display technologies such as large LED screen displays into city centre environments and the pervasive use of personal mobile communications devices, the majority of city centres in the UK now fall into the category of augmented urban spaces rather than this being confined to major international and global cities as suggested elsewhere (Manovich 2006).

Characteristics of augmented public space

The opening scenes of Ridley Scott's film *Bladerunner* (1982) – oft cited for its portrayal of Los Angeles in a starkly post-modern future world – provide a fictional impression of what, if taken to an extreme, augmented spaces might look like in the future. Harvey (1990, 308) discussed this film as both a visual expression of post-modernity and as an emblem of postmodern culture and representation: as a discussion of current ambiguities relating to time and space in contemporary culture as well being useful for its articulation of the components of the urban environment that constitute what is now called augmented public space. Many features of this film contain components of this highly visual and electronic landscape. For example, the entire face of high-rise buildings is taken up by massive digital screens displaying advertising and other forms of media. The striking aspect of these scenes is the sheer scale and saturation of media spaces in the cinematic version of this futuristic urban environment albeit via a somewhat dystopian vision. Yet many of the features of the contemporary built environment already contain features that bear out this vision. Manovich (2006, 219) gives a further impression of these spaces, listing major international economic centres such as, Tokyo, Hong Kong and Seoul. But one could easily add New York's Times Square and major European cities such as London and Berlin to this list. However, the advent of urban screens in many city centres in the UK and the sheer pervasiveness of a wide variety of mobile communications – including cell phones, PDAs and laptop computers – means that augmented spaces have now become a distinctive reality in the majority of city centres in the UK. What has happened with mobile media is a notable overlapping of media content onto the urban environment. A similar phenomenon is to be found in the combination of architecture and forms of display technology, from large LED screens being the most prominent example, to smaller scale digital signage and so forth. Representations that are available in alternative modalities, say, for example, those that are provided by mobile communications devices are one of the many layers of representation implied in the notion of augmented public space – that is, the 'overlaying of physical space with dynamic data' (Manovich 2006, 223).

The overabundance of information and events in this environment has the effect of disassociating the person from the space as a real or authentic experience. Thus, the individual is not only distracted from the living space, but is caught up in a world of private messages which are not connected to any single location or scene. Transformations in the experience of public space form an important historical backdrop to current investigations of augmented public space (Manovich 2006). Such transformations also emphasise the visual nature of the urban environment from the placement of advertising hoardings to the existence of large cinema sized digital screens. These transformations are distinct from what, over the last decade, has been labelled "Augmented Reality" (Haller, Billinghurst and Thomas 2006). This can be expressed as an experienced reality, combined with some form of computer generated simulation or modelling in which technological artefacts are adopted because of their ability to provide adjuncts to the real. In contrast, attention here is given to the physical experience of urban environments and space that is overlaid with dynamic information and media content (Manovich 2006, 219). Augmented public space is dominated by information spaces and for the most part these have been an integral part of the built environment ranging from simple street signage, advertising, pedestrian and traffic control systems, to more sophisticated electronic display mechanisms and the current use of "digital signage" (Latta 2006). A relatively new feature of the urban environment is the large LED screen placed in central locations within the city that provides another layer of information and media content. Add to this the use of personal communications devices in the form of PDAs, mobile telephones and so forth, and you have a rich layering of signs, information and media that is superimposed onto the urban scene.

As a consequence of the superimposing of media onto the built environment there is a collision, therefore, between "non-place" (Augé 1992), in the form of superimposed information media content, and the experience of urban public spaces in their physical form producing an experience that as multi-layered, multimodal and multi-sensory. Rather than connecting the urban environment with its location and surrounding, what these different layers of representation do is connect the space to a continuous city of signs and representations – a set of global messages like the universal language of airport signage or continuity seen in interiors in global hotel chains. These are homogenous spaces that are connected through the consistency of signs and representations. I could, for example, board a train at King's Cross station in London, where a the big screen displays the national news, complete my journey at Bradford, enter Centenary Square and be presented with the news from the same source. This is exactly the way in which the homogenization of public spaces is manifested in "non-place": 'Place and non-place are rather like opposite polarities: the first is never completely erased, the second never totally completed: they are like palimpsests on which the scrambled game of identity and relations is ceaselessly rewritten' (Augé 1992, 79).

Non-places are exemplified by such structures as massive shopping malls, airport terminals, global transport systems and global communication systems and the persistent experience of "always passing through". The ubiquitousness of the many forms of signage and information display on a global level is all part of the saturation of visual messages which is again an important characteristic of augmented public space.

Real space and virtual space

There is a difficulty in distinguishing between real space and virtual spaces. This is so frequently a matter of context and location, whereby, non-places, for instance, exist only as a matter of interpretation or rather their existence only occurs through the representation of contextual cues provided by the environment. Here, the notion of framing is central, since it identifies the ways in which expectations are invoked on the basis of the presentation of content. For example, particular cues are employed, when watching a television news program. These relate to the form of the content or its genre and thus indicate important information about the type of message that is being communicated and, therefore, about the way in which such a text is to be interpreted. This could easily be applied to the analysis of urban screens in the built environment. On one level, they occur as large television screens. Add to this that, in the majority of cases, these screens are frequently used for broadcasting national media content and news output in particular – such that there are specific cues in the text that are given in the sense that they have become transparent. A similar logic can be applied to the space surrounding these screens: all architectural details function in the same way by providing cues about the nature of how they are to be interpreted.

Urban Screens and the Localisation of Content

Currently, Bradford in the UK is an important case where aspirations of economic, environmental and social regeneration are being realised in plans to reorganise and rebuild much of the city centre. An integral part of these aspirations is articulated via an architectural vision that incorporates the display of media content and public access to information technology.

Figure 2.1 Big Screen Bradford, Centenary Square, Bradford, Credit: Author

For example, a large LED screen display has been attached to the side of a key new architectural structure in the City's main public space, Centenary Square (Figure 2.1) which houses bars, restaurants, shops and an art gallery. This space also includes hotspots where inhabitants can connect wirelessly to the internet in addition to accessing media content and communication with others using mobile devices and engage in virtual interactions outside the space. Such an environment integrates all of the trapping of an augmented urban space. An environment where inhabitants are both physically present in the space at the same time as engaging in virtual interactions that occur outside of the physical space.

It is by virtue of their location that urban screens are public spaces and it is this that frames them as integral parts of the urban scene, perhaps as natural a part of the environment now as billboards and advertising hoardings, as well as being important sites for the display of media content to the public. This is why organisations such as the BBC, in partnership with local authorities, are so interested in establishing ways of managing these spaces which offer new opportunities for public participation and community engagement. Participation on such a scale has led to many experiments, such as in Manchester, where recently the use of these spaces has been investigated as sites for generating locally-based content and engaging users as active participants in events.

Struppek (2006) argues that urban screens are themselves sites of engagement with the city which is, in a sense, a self-reflective organism. In many respects, this has been the case ever since signage and billboards have been an integral part of the city in the modern era and where media spaces such as advertising hoardings that in themselves have become such a defining feature in the development of the urban experience through saturation of visual messages. Now these spaces combine with many forms of digital information display. All of which coexist with buildings. On top of this the inclusion of large screen displays into strategic locations in city centres has created sites where dialogue between the city and its participants takes place. This gives a privileged position to urban screens, not because they employ any novel technology but because of their location and potential audience. It is this that provides their novelty and interest. It is the location of the screens within the urban environment that determines the context in which they are used and engaged with. Part of their transformative potential, therefore, is articulated through the dialogue that the city has with its inhabitants and the locality (see also Jewitt and Trigg 2006).

Their positioning, in such prominent locations in city centres, means that the locality – the space in which they are positioned – has a direct impact on their function. There are two immediate consequences of this: first, their position directly affects the type of content that is displayed and its planning, management and organisation, and second, in relation to the way that content is specifically designed to engage with the inhabitants of the space in the immediate vicinity and surrounding space. In addition, their salience is given by their prominence in the locality, their size and position such that what is most significant is their scale and location. In the first instance their locations are selected to ensure that media content, in a variety of forms, can be presented to very large audiences. Thus, it is their site-specific nature that provides their novelty and Anna McCarthy has argued this in relation to the pervasive occurrence of television in public space (McCarthy 2001, 2). But this site specific phenomenon, especially when the nature of the display is taken into account, is equally as applicable to urban screens. What also makes them unique in contemporary urban spaces is that they occur seamlessly within the built environment. They have also become sites through which major players in broadcasting use as a public medium and through their joint administration by both local authorities and organisations such as the BBC can result in tensions between local and national/global representations.

The localisation of content is most certainly one of the most important developments in the strategic use of urban screens by organisations such as local authorities and the BBC and for many signifies an opportunity to tailor their use for a local audience and participation as well as to experiment with new modes of interaction. In addition, specific events have been created around the screen in Bradford city centre, for example, intended to display local creative talent and using the screen as a central focus of interest in contemporary arts events. It is important to recognise that the visual presentation of content is framed to serve the expectations of local audiences through the presentation of cues and through the visual composition that indicates place. News content, for example, is branded to reflect the location and for consumption by a local audience and is literally framed by messages about local news and events.

The site specific nature of the content is worth dwelling upon. Many urban screens are, partially at least, being employed for the purposes of the presentation of artistic content and to encourage the participation of local inhabitants. *The Bradford Grid* is a visual arts project using artists who either live or work in the Bradford region. In essence, it is a long term survey of the region's urban environment producing a substantial archive of both still and video images as well as sound recordings established in 2003 and set to continue on a long-term basis. In short, a map of the region is cut into squares and, on a regular basis, individual squares are chosen at random and each member of the team visits the area represented and records, in their own chosen style, salient aspects of the urban environment. Transformations in the urban scene, particularly relevant as the region itself is undergoing regeneration and major shifts in urban planning, are recorded using a variety of modes, still photography, video, sound recordings and written responses to the urban scene.

The project was commissioned to produce installation work based on the new output from the locality, the immediate vicinity around the screen and displayed as part of a contemporary arts festival. The creative strategy employed in the production of the installation had to make reference to the way that inhabitants use the space in order for it to work and involved constructing a series of short video pieces none of which lasted for more than three minutes. Interestingly, this duration for each episode was based on an estimate of how long it takes for an average user of the space to perambulate through it, including a little dwell-time. There are two critical issues here, therefore, that have an impact on the way that media content is managed on the basis of how aspect the physical space determines how inhabitants use and engage with it. Firstly, the choice of content adds to this notion of the screen acting as a mirror for the space around. Second, the duration of each of the episodes for this commission was based on an assumption of the level of mobility expected of participants and the average length of time that they dwell within the space.

In addition, a soundtrack was used in an attempt to draw passers-by into the work, the assumption here being that visuals are actually less likely than sound to capture the attention of inhabitants of the space. All of these issues in relation to the organisation and management of content for urban screens indicate some important new avenues for research. They all have implications for the way that augmented public space is characterised in terms of locality and how this impacts on their character of display. Framing is central to how we begin to understand the display strategies for media content intended for urban spaces and fundamental to how we form an understanding of augmented space. The compositional characteristics mentioned above lead directly to issues of identity and branding since augmented public space is so often dominated by branding, advertising messages and information displays that contain more subtle forms of branded message.

Whilst the content of Bradford's Big screen – owned by the local authority and managed by the BBC – is dominated by BBC news and current affairs programmes, it also incorporates a strong local flavour by covering local events and promoting events in the district. The Big Screen experience with this emphasis on locality works in opposition to the notion of non-place and interestingly provides the opportunity of transforming the function of similar forms of display technology for uses that engage the participation of the inhabitants of the space. Experiments in artistic content and

Figure 2.2 Extracts from the "Bradford Central" Installation for Bradford's Big Screen as part of the STIR Contemporary Arts Festival, Credit: Author

community engagement are an example of this. On a theoretical level there is also an important juxtaposition that deserves to be fleshed out. On the one hand, there is a tendency towards the fragmentation of experience that is characteristic of how subjects are located and participate in and interact with augmented public space and in forms of virtual environment. On the other hand, urban screens, as an integral part of the media layer in augmented public space, has a tendency to position the inhabitants of real spaces through their locality and engagement with the immediate environment, space and surroundings. Both have the potential to subvert the global and homogenized world of non-place. This is the case, especially if much of the content is displayed in a creative and community context. There is also an overlap between the real physical space of the city and the virtual spaces of the media that are superimposed upon the built environment.

Framing and Locality

The previous section explored how framing governs the organisation of content in relation its orientation to the audience and ways that the branding of information is framed for a local context. In this way the actual compositional and spatial structures of content and its branding function simultaneously to serve the requirements of the local context. Thus, the character of display in augmented space, exemplified in the spatial properties in the design of content indicates place and location. In addition, both aspects of framing connect the character of public display in more general terms with features of the surrounding space and can, therefore, be translated onto other features of augmented public spaces. Framing can also be related to the actual physical structure of urban spaces, indicating, for example, entry points and navigation paths through the space, or suggesting the extent of boundaries and borders.

Framing binds aspects of the represented spaces with those of the social world. The closer an element is to another, the likelier it is that both are seen as being related in some way. They are connected in some way purely by their proximity to one another. It is here that spatial properties are bound up with forms of social action. This position is taken initially from 'social semiotics' (Hodge and Kress 1988) whereby the closeness and distance of represented objects maps directly on to those of the social world. This notion of proxemics, first developed by Hall (1966) and further applied to the analysis of multimodal texts by Hodge and Kress (1988) has found recent application to the built environment (O'Toole 2004; Alias 2004). Its application to augmented space might be fruitful, whereby the relative position of bodies in space indicates specific forms of social relationship. For example, this notion of framing has been applied to the navigational properties of both printed and online texts and it is not difficult to see its potential in explaining aspects of augmented public space. Proximity is the way that content elements in a multimodal text are grouped together. These form distinctive zones where related informational items can be found and in so doing assist navigation as is the case with the spatial qualities of the visual design of a website, for example. Both the building upon which an urban screen is placed, and the space in which the building is located can be considered as a form of multimodal text and, as a consequence, can be described in terms of the framing of elements within a composition. In many forms of multimodal text grouping can occur on the basis of the organisation of content to serve navigation and facilitates specific modes of access to the text.

As city centres become saturated with the layering of dynamic information and media content and inhabited by users of communications devices, the more these technologies are either integrated with the built environment or encourage users to be dislocated and disassociated from it. This gives rise to another paradox. For example, in particular forms of virtual interaction inhabitants are present in spaces outside of their physical location. This will inevitably mean that the fundamental nature of space in urban environments as unified or fixed entities is brought into question and also questions the relevance of describing these spaces in terms of their physical boundaries. Thus, the urban environment, largely as a consequence of its integration with new technologies, is clearly shifting notions of mobility, from forms that are fixed within a specific set of physical relationships within a bounded space to much more transitory and boundless ways of treating such concepts as location and place. And in terms of framing, once you have drawn a boundary around an urban space as a physical entity the position of entry points can be established, for example, informally by the users of the space, or, indeed, by planners. The problem is, with augmented space, once these boundaries become less distinct, so too will the manner in which inhabitants enter into it – its entry points.

Not only have architecture and the planned aspects of the built environment within urban spaces become less stable as physical structures, but the saturation of display technologies has also meant that there is a further blurring of the distinction between the physical environment and digital display technologies and this leads to the questioning of the distinction between virtual and real. This is especially the case with regard to the media layer where display technologies, including urban screens, digital signage in all its forms, as well as everyday use of computers and miniature screen technologies, not to mention the potential for virtual interaction, are all now superimposed onto the experience of urban public space. This gives rise to a further blurring of the distinction between "the real" and "the virtual". Such an ambiguity can function at many different levels. Where there was once a real physical boundary to urban spaces there is now a variety of possible boundaries dependant upon how each individual uses the space and upon the nature of the media with which they interact. Clearly, the physical structure of urban spaces is shifting significantly in terms of framing. A city-dweller might walk into a central square and be quite clear about its external boundaries on a physical level but in an augmented space where do these boundaries actually occur and at which point does a person actually enter the space?

Framing and the Body

One of the most critical questions to come out of this discussion rests on where to locate the body within augmented public space. One way might be in relation to an inhabitant's interaction with content, for example, displayed on urban screens. Here there are similarities to what Manovich (2001) has described in relation the interaction inherent in the engagement with new media artifacts as 'cinematic vision' (Manovich 2001, 103) but with one fundamental difference. For him, the body is usually fixed at a single location. In the new context of augmented space, the body is usually highly mobile, both in terms of its transit through the space, but also in relation to any virtual interaction taking place outside the space.

The body itself acts as an interface both on a sensory level in terms of its reception of information from the environment and in terms of receiving information from many technological artifacts, from personal stereos to mobile phones, all of which, can be argued, augment the body in some way. The consumption of media, therefore, becomes part of the augmentation of the body through the use of these artifacts. Featherstone stresses the 'importance of the body as a framer of information and this has become more urgent with digitized media' (Featherstone 2006, 235). Thus, framing and the framing of information ultimately becomes a question of the body and its location in space. Like a mirror, the body is both the site of representation and is a representation itself. This runs almost counter to much recent debate about the virtual whereby it is assumed that the greater the immersion into virtual spaces, the greater the disembodiment and disassociation from the real and real space. Far from the latter, the body still has to be the site of consumption and therefore of interpretation: the point at which affect, mobility and action can occur (Massumi 2002).

There are two sets of current theoretical issues that impact on how to conceptualize augmented public spaces. Firstly, navigation can be considered in a particular way, namely in relation to how the body is positioned in space and also in relation to the space that is occupied by it. Secondly, the distinction between the real and the virtual through an examination of recent theories of human action can easily be applied to navigation and the body and incorporated into framing as a theoretical strategy. How is it possible to make sense of the uneasy distinction between the virtual and the real – digital/analogue – in terms of locality, the body, and the experience of urban spaces? At this level of abstraction, the argument seems to rest largely on how we position or locate the body in space, rather than on any consideration of the nature of technology *per se* or its use in urban spaces. The body has a sense of itself; it inhabits real space and is positioned at a specific location at any one time. It also becomes an individual element in itself within the environment that it inhabits or navigates from point to point within space. It consumes and interacts with representations, as part of the structure of experience, for example. Yet the body itself must be represented in some way, as an appearance or an image (Featherstone 2006).

It was Massumi (2002) who captured the essence of this distinction with regard to the virtual and the relation between the virtual and the body (2002, 30). He indicated that the two were inseparable and, moreover, questioned the basis upon which the body is moved to action (p.5). This idea can be applied to navigation and human mobility in general, in either real or virtual spaces (p.134). This is important because the whole notion of navigation is predicated on the assumption that the body exists, or is always located, within real space. This is always the case even if a participant is engaged in some form of interaction in an implied space elsewhere. Even in virtual spaces there is a body that has taken some form of action and this action has occurred in real space. Further, it requires that there also exists the intention to move to another location, another point in space: another point on the map, so to speak.

Thus, another important set of abstractions comes to light. The argument here is that there is no clear boundary between the virtual and the real and in fact the two different forms of representation are interdependent and is an important part of this distinction. Massumi argues convincingly that there are many representations, such as TV images or paintings, to take just two examples, which contain qualities of what has come to constitute the virtual: 'Digital technologies have a connection to the potential and the virtual *only through the analogue*' (p. 138). This ambiguity would lead us, therefore, to question any clear distinction between them as Massumi continues: 'If all emergent form brings its fringe of virtuality with it, then no particular medium of expression has a monopoly on the virtual. Every medium, however "low" technologically, really produces its own virtuality' (p. 175).

Thus, committing all representations in some way to the virtual means that many of the visual representations found in the urban landscape can easily be labelled as such. In theoretical terms there is a link between the issues relating to the real and the virtual, the position of the body in space, and modes of representation in augmented public space.

Conclusions

In augmented public space an inhabitant's trajectory is always transitory. On a simple level the engagement with the space is always one of just passing through. To capture the spirit of place, it is necessary to move beyond the branded messages prevalent in so much of what we call augmented public space to a sense of locality that reflects the reality of the urban environment and its inhabitants. The case study on urban screens at the centre of this chapter bears this out. The management of urban screens in the UK where important curatorial partnerships have tended to work to manage the presentation of content has meant that, in many cases, these spaces can be used for the display of content fixed to the locality and as an important avenue for the display of artistic content and a way of engaging the participation of the inhabitants of augmented public space. In theoretical terms, the understanding of augmented public space from the point of view of the breaking down of the physical boundaries of space and place, or the blurring of the distinction between the real and the virtual can both be resolved by focusing on the body and its location in real space.

References

Alias, S. (2004), 'A Semiotic Study of Singapore's Orchard Road and Marriot Hotel', in *Multimodal Discourse Analysis: Systemic Functional Perspectives* (London: Continuum).

Augé, M. (1995), *Nono-places: Introduction to an Anthropology of Supermodernity* (London: Verso).

Couldry, N. and McCarthy, A. (2003), *Media Space: Scale and Culture in a Media Age* (London: Routledge).

Featherstone, M. (2006), 'Body Image/Body without Image', in *Theory, Culture and Society*, 23 (2-3) pp. 233-236.

Hall, E. (1966), *The Hidden Dimension*, (New York:Doubleday).

Haller, M., Billinghurst, M., Thomas, B. (2006) *Emerging Technologies of Augmented Reality: Interfaces and Design* (New York: Idea Group Publishing).

Hansen, M. (2004), *New Philosophy for New Media* (Cambridge, MA: The MIT Press).

Harvey, D. (1990), *The Condition of Postmodernity: An Enquiry into the Origins of Cultural Change* (London: Blackwell).

Hodge, R. and Kress, G. (1988), *Social Semiotics* (Oxford: Polity Press).

Ishii, H., Underkoffler, J., Chack, D., Piper, B., Ben-Joseph, E., Yeung, L., Kanji, Z. (2002), 'Augmented Urban Planning Workbench: Overlaying Drawings, Physical Models and Digital Simulation', in *Proceedings of IEEE and ACM ISMAR 2002*.

Jewitt, C. (2006), 'Screens and the Social Landscape', in *Visual Communication*, vol. 5 issues 2, pp. 131-140.

Latta, J. (2006), 'Power Signs', in *The Wave Report*, <http://www. Wave-report.com/conference_reports/2006/PowerSigns2006.html>, accessed 21st December 2006.

Manovich, L. (2001), *The Language of New Media* (Cambridge MA: The MIT Press).

Manovich, L. (2006), 'The Poetics of Augmented Space', in *Visual Communication*, vol. 5 issue 2, pp. 219-240.

Massumi, B. (2002), *Parables of the Virtual* (London: Duke University Press).

McCarthy, A. (2001), *Ambient Television*, (London: Duke University Press).

O'Toole, M. (2004), 'Opera Lundentes: the Sydney Opera House at work and play', in *Muiltimodal Discourse Analysis: Sytemic Functional Perspectives* (London: Continuum).

Struppek, M. (2006), 'The Potential of Urban Screens', in *Visual Communication*, vol. 5 issues 2, pp. 173-188.

Chapter 3

Mobile Networks, Urban Places and Emotional Spaces

Heesang Lee

Introduction

Recently, one of the most important characteristics of electronic landscapes has been that they are supported by personal, mobile or wearable technologies such as mobile phones, digital cameras, MP3 players, wireless game players and so on. Among the technological devices, the mobile phone may be the one that has the most significant effects on our everyday lives. As a result of its development as 1G, 2G and 3G multimedia, the mobile phone produces new kinds of mediascapes and bodyscapes in the so-called information city and network society.

When it comes to the network society, as a new kind of society emerging through the development of information and communication technologies, this has been explained mainly with regard to the global space of flows mediated by global cities at a macro-level (Castells 1996; Sassen 1991). However, our everyday lives in ordinary cities are somewhat far from the 'macro-network society' or "global network society" (Webster 2000). This chapter looks at the 'micro-network society' fabricated and facilitated by mundane technologies such as mobile phones and people's practices in their everyday lives. As Wittel (2002, 52) states, 'it is worthwhile translating this macro-sociology of a network society into a micro-sociology of the information age'. One of the most outstanding landscapes of the micro-network society underlined here, is that ordinary human bodies which are connected to electronic machines act as nodes in techno-social networks such as mobile networks.

Machines, cities and bodies have been evolving together for a long time, and the recent development of information and communication technologies has transformed both cities and bodies into new forms through the implosion of humans and machines and the explosion of the body in the world. As Grosz (1992, 250-1) puts it, 'the city is an active force in constituting bodies, and always leaves its traces on the subject's corporeality. It follows that, corresponding to the dramatic transformation of the city as a result of the information revolution will have direct effects of the inscription of bodies'. Alongside the development of state-of-the-art technologies such as info-technologies, bio-technologies or nano-technologies, contemporary cities are undergoing "cyborg urbanization" (Gandy 2005; Chatzis 2001; Graham and Marvin 2001) in which the boundaries between human and machine spaces are increasingly blurred and human bodies are evolving into "human-machine hybrids" called "post-humans" or "cyborgs" (Haraway 1991; Stelarc 1998; Hayles 1999).

In this context, concerned with the relations between machines, cities and bodies, this paper explores how mobile networks are related with the physical space of the city and the psychological space of the body. For this, I surveyed mobile landscapes produced by Korean university students in their 20s, who are more active in mobile phone use than any other social class in Korea, within Daegu which is the third largest city in Korea. This was done mainly through semi-structured interviews during 2003. The paper is organized into four main sections. First, it provides a theoretical review of the ways in which mobile networks transform urban spaces and human bodies. Secondly, it explains the generation of mobile networks through technological and institutional changes in Korea. Thirdly, it looks at the socio-spatial scales and time-space landscapes of mobile networks in relation to mobile users' motions and practices in their everyday lives. Finally, it attends to the ways in which mobile networks involve paradoxical emotional spaces in relation to mobile users' emotions and desires to be dis/connected with those networks.

Mobiles, Spaces and Bodies

Mobile networks and spaces

Although the mobile phone is having very significant effects on our everyday lives or urban spaces, there have been very few studies of the mobile phone in urban, social or cultural studies. It is perhaps not surprising at a time when even the traditional telephone's implications for cities and societies have not been fully explored. Recently, however, some researchers have begun to draw their attention to the landscapes created by mobile phones (Kopomaa 2000; Townsend 2000; Sussex Technology Group 2001; Laurier 2001; Brown et al. 2001; Henderson et al. 2002; Green 2002; Fortunati 2002; Katz and Askhus 2002; Weilenmann 2003; Ito 2003; Yoon 2003; Rafael 2003; Goodman 2003; Ling 2004; Carey 2004; Williams and Williams 2005). The following pages present a brief overview of their research, particularly focusing on the ways in which mobile phones produce the blurred boundaries between absent and present spaces and bodies-with-mobiles as nodes in mobile networks.

One of the outstanding effects of mobile phones is that they tend to blur the boundaries between binary realms. For example, the boundary-blurring process between public and private spaces by mobile networks has been underlined by many researchers (see Sussex Technology Group 2001; Henderson et al. 2002; Green, 2002; Sheller 2004). As Green (2002, 287) puts it

> a kind of spatial and temporal "boundary rearrangement" becomes possible. ... This involves both the case of "public" activities and responsibilities (as in the case of work) that become embedded in the temporal rhythms of the home, as well as "private" relationships becoming integrated into the public sphere in mobile relations.

That is, whether people exist in public spaces or private spaces is not determined by whether the people are physically in public spaces or private spaces but is dependent on whether their mobile phones are on or off.

More importantly, mobile networks shatter the boundaries between absent and present spaces. With outstanding insights, Gergen (2002) stressed that we need to distinguish between the potential of mobile phones and that of other technologies in regard to the separation and integration of the present and the absent. The development and proliferation of our major communication technologies of the past century have expanded the dimension of "absent presence" which results in 'the erosion of face-to-face community, a coherent and centered sense of self, moral bearings, depth of relationship, and the uprooting of meaning form material context' (Gergen 2000, 236). However, the mobile phone makes "absent presence" tenuous in local communities, resulting in the integration of "the absent" which other technologies have facilitated so far, and "the present" which the technologies have eroded. As a result, 'with the cell phone, one's community of intimates more effectively sustains one's identity as a singular and coherent being' (Gergen 2000, 238).

In a similar vein, Fortunati (2002) argued that the mobile phone enables 'present absence' through which the binary opposition presence/absence is undermined. What is important is that this process restructures the sense of belonging to places. 'This shift makes it possible to suffer less from nostalgia, a tormenting feeling which frequently accompanies immigration, mobility, tourism and so on, and which is connected to the sense of loss of one's own relationship with a place' (Fortunati 2002, 520). However, Fortunati concluded that as the sense of belonging to one single place is translated into the sense of belonging to many places or an unlimited space, people come to suffer from a sense of uncertainty, insecurity and confusion.

In the recent special issue on "absence" and "presence" in *Society and Space* (2004, Vol.22), Urry (2004), Sheller (2004) and Licoppe (2004) explained complex network spaces produced by mobile machines such as mobile phones in terms of 'intermittent connections and mobility', 'fluid and contingent social structures' and 'connected presence and social relationships' respectively. Here, the mobile phone can be perceived as a key technology transforming the existing Euclidean frame of time-space into fluid, multiple and complex time-spaces where spatial and temporal boundaries are reordered and rearranged.

In addition, the techno-social networks of mobile phones can be characterized as individualized and decentralized spatial networks and accelerated and speeded up temporal networks. Townsend (2000) suggested that the arrival of mass mobile communications in the city results in the acceleration of urban metabolism through decentralized and complex information networks, producing the 'extension of the body' and the 'real-time city'. In the real-time city, mobile phones tend to make people move according to indeterminate and uncertain time-space coordinates at which people get together and meet each other through incessant mobile communications. Ling and Yttri (2002; see also Ling 2004) called this 'hyper-coordination' or 'micro-coordination'.

In this sense, Carey (2004, 136) said that 'we're no longer required to make prior arrangements – activities can keep their chosen sites for demonstration (physical and virtual) secret until the last minute, thus stealing a march on the authorities'. Furthermore, Kopomaa (2004, 271) argued that 'increased use of the phone and spontaneous phone conversations in public enliven the street scene. … Non-places – with the ideal type being the traveller's space – are transformed to loci of here-

and-now by mobile phone users'. Carey (2004) and Kopomaa (2004) called the city based on such mobile communications an 'indeterminate city' and a 'condensed city' respectively. This effect of mobile phones on time-space coordinates is reflected in conversations on mobile phones. As Laurier (2001) observed, (especially, for those who live their lives as nomadic worker) "geographical locating" appears often in conversations on mobile phones and thus at the heart of the conversations are temporal and spatial orderings on a spatio-temporal context which is mutually accomplished between users. While Laurier said that the most frequent question is "where are you?" Weilenmann (2003) said that it is "what are you doing?". According to Weilenmann, the frequent question sometimes causes a location to be given as part of the answer which shows how location, activity, and availability are strongly related.

Mobile networks and bodies

In the present generation of mobile communication a new pattern of individualized and decentralized electronic networks seems to have appeared. We can think of this spatial structure of mobile networks as an important aspect of the micro-network society in which ordinary bodies themselves become nodes and the (individual or collective) temporal and spatial coordinates of everyday life come to be fluid and floating. After all, people come to be "bodies-with-mobiles" as nodes in mobile networks in the micro-network society. As Sheller (2004, 49) puts it, 'persons themselves are not simply stationary nodes in a network, but are flexible constellations of identities-on-the-move'. Human bodies become not only biological entities but also electronic nodes combined with mobile phones acting as not only technological objects but also prosthetic parts of their bodies. In such bodies-with-mobiles, the mobile phone can be thought of not just as objects but also as "organs". As Callon and Law (2004, 9) put it, 'technologies such as cell phones are not best thought of as extensions to the body. Instead they are organs, integrated into the body'.

As the mobile comes to be more and more directly and closely connected to the human body, the technological machine does not exist as a pure object any longer but rather as a "quasi-object", and likewise the human body cannot exist as a pure subject any longer but rather as a "quasi-subject" in terms of Latour's (1993) "actor-networks". In this sense, the emergence of bodies-with-mobiles implies that the mobile phone, as an organ existing not "inside" but "outside" the human body, can transform the human body into a "cyborg" in Haraway's (1991) terms. Furthermore, the 'bodies-with-mobiles' can be seen as de-territorialized bodies or "bodies-without-organs" in Deleuze and Guattari's (1987) terms. That is, as Lash (2001, 108) puts it, 'technological forms of life, whether natural or social are like Deleuze and Guattari's "body without organs". As they open, they externalize their organs, and open up to flows of information and communication'.

In addition, the development of bodies-with-mobiles provides the possibility of the production of a kind of "wearable" or "ubiquitous" computing space in which human-machine hybrid networks blur the boundaries between human and machine spaces, leading to "cyborg urbanization". Through the connections of body and media and of body and space 'space becomes wearable when affect becomes the operator of spacing or the production of space through bodily experience' (Hansen 2002, 321). The mobile phone can be explained as one of the technological devices that are capable of constructing a new kind of technological environment. It brings about major changes in "the geography of calculation" where 'from being centred and stable entities located at definite sites, through the medium of wireless computing, computing is moving out to inhabit all parts of the environment and users are able to be mobile' (Thrift 2004, 182).

The mobile phone can also be described as bringing into being the world of "local intelligence" where 'everyday spaces become saturated with computational capacities, thereby transforming more and more spaces into computationally active environments able to communicate within and with each other' (Thrift and French 2002, 315). Particularly, in these mobile spaces of calculation and intelligence, the mobile phone can play a role as a kind of GPS (Geographical Positioning System) in that mobile networks make it possible to identify the locations of bodies-with-mobiles on the move. This point can be found in the fact that communications on mobile phones are characterized by "geographical locating" in the sense that the most common and frequent question found in conversations on mobile phones is "where are you?" (Laurier 2001). In a sense, this relates to the networks of control.

In such a new technological environment, we can think about how mobile networks produce "relative" and "relational" networks between bodies and spaces. First, mobile phones enable human bodies to be extensible and divisible. Reviewing studies of the relations of practice and structure in the context of Hägerstrand's time-geography (mainly, studies by Pred, Thrift and Carlstein in the 1980s) and studies of the relations of space and technology in the context of McLuhan' media theory (mainly, studies by Abler, Gould and Jannelle in the 1960s to 1980s), Adams (1995) argues the possibility of "personal extensibility" like "amoeba" in space-time through electronic media. For example, 'when a person in city A telephones a person in city B, he or she is partly present in city A and partly present in the virtual space of the phone call' (Adams 1995, 270). In a similar vein, Townsend (2000) says that 'the mobile phone is more and more becoming perceived as an extension of the body'. As Fortunati (2002, 518) puts it, 'with the spread of the mobile, that is, the phenomenology of the presence of individuals in social spaces also changes, in that individuals apparently present in a given place are actually only half-present'.

Time-geographers have seized on the idea that 'one individual cannot exist in two places at one time and therefore has to allocate his path in time-space' or 'all individuals are indivisible – never being able to be at more than one place at a time' (Adams 1995, 271 and 269). However, the mobile phone makes the argument invalid. Furthermore, 'the ability to be in one room or building, linked to a computer accessing data elsewhere, while phoning someone who is at another point, begins to unravel any simple time-geography based on physical presence' (Crang 2000, 306). Thus, 'the points of a personal network need not be unitary Cartesian individuals' (Bridge 1997, 622). The mobile phone makes it possible for the body to be extensible and divisible into off-line and on-line spaces or into co-present and tele-present spaces. As a result, the body can be located at different points, and different spaces can be located at the same point at the same time.

Second, the mobile phone makes spaces around the body "multiple" and "eversible". Let us suppose a person communicating with their friends at point C on their mobile phone in the bus or underground moving from point A to point B. How many spaces are they related to? First, they exist in the moving space of the bus or underground. Second, they lie in the transit space between point A and point B. Third, they exist partly in point C through mobile networks. In addition, one's mobile phone contains others' mobile phone numbers at different locations which could be instantly and directly connected through mobile networks. These multiple networks can be seen as invisible, potential and virtual social networks contained in the small chip of the mobile phone, forming the "liminial" spaces between absent and present spaces in everyday life (Shields 1992).

Such multiple spaces are eversible spaces where "inside" and "outside" spaces can be easily changed. In Euclidean or Newtonian absolute space, physically close spaces can be regarded as inside spaces, and remote spaces as outside spaces. However, mobile networks can easily reverse or dislocate the two spaces' positions. As Gergen (2002, 238) puts it, 'cell phone conversation typically establishes an "inside space" ("we who are conversing") vs. an "outside space" constituted by those within earshot but prevented from participating'. In the sense, mobile spaces are like Möbius spaces with no boundaries between inside and outside spaces.

The Generation of Mobile Networks in South Korea

In Korea, mobile communication services began in the 1980s (car-phones in 1984 and 1G mobile phones: analogue cellular phones in 1988). Until the mid-1990s mobile phone ownership was only affordable for a small group of people. Furthermore, most mobile phones were imported from foreign countries, mainly the USA. To create the stable supply and demand for mobile phones, it was necessary to develop new mobile technologies that were not dependent on foreign technologies, and which furthermore could be exported into foreign markets.

In this situation, in 1989, the government set up a project for the development of digital mobile communication systems, and in 1995, IT institutes and companies in Korea succeeded in developing digital mobile phones using CDMA (Code Division Multiple Access) technology, which was possible only in theory but uncertain in practice at that time. As a result, 2G mobile phones such as digital cellular mobiles (in 1996) and PCS (Personal Communications Services) (in 1997) began to be used in Korea, and furthermore 3G mobile phones (CDMA-2000) began to be introduced in the early 2000s. These technological changes were important in making Korea one of the most wired countries in the world.

The changes in mobile technologies were combined with changes in the structure of mobile service markets from monopoly to competition systems through the government's institutional regulations. In fact, such regulatory changes have taken place in almost all communication service markets in Korea since the mid-1990s. In the case of mobile services, until 1994, the public company KMT (Korea Mobile Telecom) had acted as the main and only mobile carrier. However, in 1994, according to the government's policies for facilitating competition systems in mobile service markets, the public mobile carrier KMT was privatized into SK Telecom which is now the largest mobile carrier and another private mobile carrier (Shinsegi Telecom) was allowed to launch mobile services. With the development of CDMA-based mobile technologies, the two companies began digital cellular phone services in 1996, and three new telecom companies (Korea Telecom, Hansol Telecom and LG Telecom) selected as new mobile carriers (PCS) started their mobile services in 1997 (Table 3.1).

Table 3.1 Change in the Market Share of Mobile Carriers

	Cellular		PCS		
	SK Telecom	Shinsegi Telecom	KT	Hansol Telecom	LG Telecom
1995	100	-	-	-	-
1996	90.9	9.1	-	-	-
1997	66.9	16.5	5.1	6.1	5.4
1998	42.7	15.3	16.8	10.1	15.1
1999	43.1	13.8	18.2	11.7	13.2
2000	40.8	13.1	19.7	11.7	14.7
2001	52.3	(merged into SKT)	33.0	(merged into KT)	14.7
2002	53.3		31.9		14.8

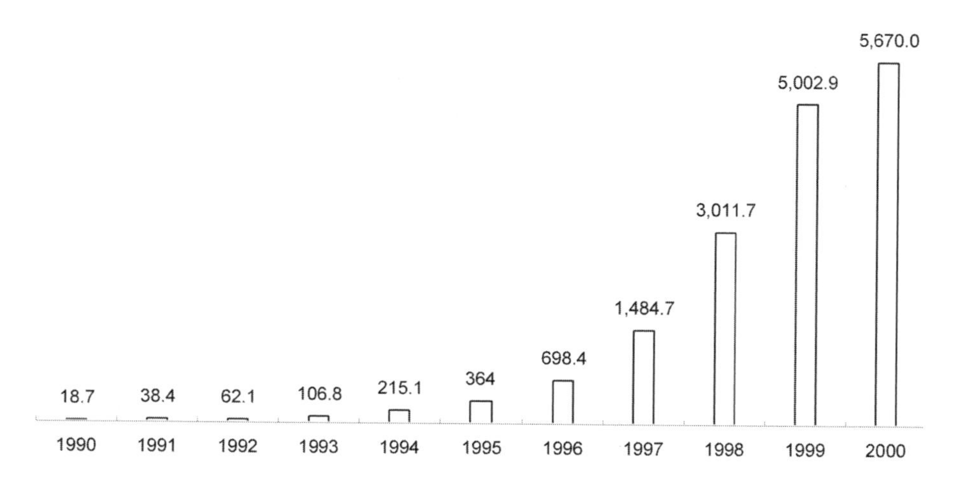

Figure 3.1 Change in the Number of Mobile Subscribers per 10,000 Persons
Source: Ministry of Information and Communication, Korea (2001, 79)

As a result of these technological and institutional changes there has been a tremendous increase in mobile devices, services and users since the late 1990s. From 1996 to 1998, the subscribers increased by 100 per cent per year, and by 1999, about 50 per cent of the population (more than 20 million people) were using mobile phones (Figure 3.1). Mobile users began to surpass pager users and fixed telephone subscribers in 1999.

Mobile Networks and Urban Places

Mobile-mediated and place-based networks

Many argue that increases in human mobility have detached people from their original territories, resulting in the demise of their local communities. For example, Bauman (2001, 38) argues that 'the cellular telephone, offering independence even from wired networks and sockets, delivered the final blow to the claim physical proximity might have had on spiritual togetherness'. However, I suggest here that mobile phones, which can be seen to facilitate and accelerate people's mobility, are highly bound to their local places. In order to address this question, I investigated the socio-spatial scales of mobile networks, comparing them with e-mail networks.

Both mobile and e-mail networks can be seen not only as technological tools for personal communications, but also as social networks. Table 3.2 shows, in the case of university students, how many phone numbers were booked in their mobile phones and how many e-mail addresses were booked in their e-mail accounts. First, it was found that mobile phones were being used more than e-mails. More importantly and surprisingly, participants in the study tended to say the same things about the differences between mobile phone and e-mail communications. They tended to use their mobile phones for communicating with those whom they met frequently in their everyday lives, while they tended to use their e-mails for communicating with those whom they did not meet often in their everyday lives due to social, temporal and spatial distances, in addition they used e-mails with those whom they had to communicate with because of public reasons or affairs. It is helpful to explain this in more detail in order to understand the socio-spatial scales of mobile networks in relation to e-mail networks.

In general, the mobile phone users interviewed for this research had the numbers of intimate friends and family stored in the phone books of their mobile phones. These were people that they met in their homes, universities or other locales, and were people that they kept in close touch with in their everyday lives. Of course, this does not mean that all people whose numbers were stored in their mobile phones were close associates. In fact, the interviewees' mobile phones also contained the numbers of people that they did not call on a frequent basis. This is because on first meeting a new person they had the habit of storing that person's number in their phone book, even though they knew that the number might not be used. They seemed to tend to get a kind of pleasure by filling their mobile phones with others' phone numbers, whether the invisible, potential and virtual social networks were

Table 3.2 Number of People Booked in Mobile Phones and E-mail Accounts

Interviewee' name (sex)	In mobile phone	In e-mail account	Overlapped
Ji-Wook Jang (M)	229	31	24
Kyoung-Wan Seong (M)	136	89	75
Hae-Min Kang (F)	84	0	0
Eun-Kyoung Jeon (F)	94	20	10
Myoung-Sook Lim (F)	83	9	2
Min-Jeong Baek (F)	93	1	1
Jeong-Min Kim (F)	130	25	20
Kyoung-Mi Kim (F)	113	25	20
Soo-Jeong Lee (F)	164	53	53
Jung-Lim Heo (F)	230	41	16
Hyo-Jin Park (F)	200	25	25
Ki-Heon Park (M)	172	30	13
Seong-Ik Jang (M)	30	0	0
Jong-Ki Moon (M)	100	10	10
Chang-Soo Cha (M)	77	17	9
Chang-Jun Lee (M)	11	0	0
Hyeon-Seok Jeong (M)	51	6	6
In-kyoung Kim (F)	188	98	43
Young-Dong Cho (M)	168	17	16
Jin-Geol Shin (M)	138	27	21
Dae-Yeol Bae (M)	22	4	2
Min-Gwak (M)	178	10	2
So-Jeong Heo (F)	164	29	29
Da-Jin Jeong (F)	134	32	13

important or not. Furthermore, they tended to interpret having the stranger's number as a sign of friendship and thought it as a part of the social ritual of politeness.

The interviewees said that the email addresses stored in their email accounts tended to be those of people that they rarely met in everyday life. These can be divided into some types. (a) Those whom they once met intimately or were acquainted with in the same city or other cities: for example, ex-girl/boy friends or their primary or secondary schoolmates. (b) Those whom they could not meet easily because they lived in different cities or countries: for example, friends based in foreign countries. (c) Those whom they needed to communicate with for public reasons: for example, classmates with whom they had to communicate about shared tasks, or university professors or lecturers to whom they were required to submit their papers or assignments by e-mail. (d) Those whom they came to know in the on-line world: for example, friends made in on-line communities or on-line chat rooms. Most of these people were not those whom they had strong or private links with at the moment, and their communications with them were already broken off or were being practiced on a temporary or momentary basis.

Finally, people in the overlapping part between mobile phones and e-mail accounts were somewhat ambiguous. In fact they were often closer to the user than those described in the first and second cases. Many said that people in this group were those with whom they had originally or previously kept in touch through e-mail communications, whether they first met them in the off-line world or the on-line world, but with whom they were now keeping in touch through mobile communications after purchasing their mobile phones or as they developed a closer relationship. In this case, e-mail communication tended to be replaced by mobile communication. However, some said that people in this group were those whom they already had been acquainted with in the off-line world and had always kept in touch with through face-to-face or mobile communication. They occasionally got in touch with these people via e-mails for private or public reasons. For example, they used e-mails when they needed to say something special which was hard to say through face-to-face or mobile communications for private reasons, or because they sent and received files by e-mail for private or public reasons. In this case, e-mail communications did not replace mobile communications, but supplemented face-to-face or mobile communications.

In short, they tended to use mobile phones for communicating with those whom they met socially intimately, spatially closely and temporally frequently. In terms of Granovetter's (1973) social networks, whereas e-mail networks relate to 'weak links' formed at larger spatial scales, mobile networks relate to 'strong links' at smaller spatial scales. The localized socio-spatial scale of mobile networks represents 'network enclosure' (Bridge 1997, 616) with relatively dense networks, and indicates that they are associated with a strong sense of place. We can know that social, temporal and spatial networks in the off-line world have effects on the ways of on-line communications though mobile phone or e-mail networks and that the ways of on-line communications reflect social interactions, relations or networks in the off-line world. The two on-line worlds of mobile phones and e-mails correspond to two off-line worlds of "now-and-here" and "then-or-there" respectively. In other words, while "co-presence" in the off-line world is related to "tele-presence" through

Table 3.3 Time-diary of Electronic Media Use

Media			Counterpart	
Time	Media	Location	Who	Where
Friday 31 October 2003				
07:45-07:47	Mobile	Home	Friend	Local
09:30-09:30	Mobile	University	Friend	Local
12:50-12:52	Mobile	University	Friend	Local
18:00-20:40	CATV	Bar	-	-
20:45-20:52	Mobile	Street	Friend	Non-local
23:30-00:50	Internet	Home	-	-
Saturday 1 November 2003				
00:56-00:58	Mobile	Home	Friend	Local
01:13-01:17	Mobile	Home	Friend	Local
08:21-08:22	Mobile	Street	Friend	Local
09:34-09:35	Mobile	Home	Friend	Local
09:46-09:48	Mobile	Taxi	Friend	Local
10:01-10:02	Mobile	Bus	-	-
13:07-13:10	Mobile	Coffee shop	Friend	Local
13:28-13:29	Mobile	Street	Friend	Local
14:08-14:10	Mobile	Coffee shop	Friend	Local
15:53-15:55	Mobile	Coffee shop	Friend	Local
16:56-16:57	Mobile	Photo studio	Friend	Local
17:30-17:32	Mobile	Street	Friend	Local
17:41-17:42	Mobile	Street	Friend	Local
19:18-19:26	Mobile	Bus	Friend	Local
19:31-19:35	Mobile	Bus	Friend	Local
20:43-21:01	Mobile	Home	Friend	Local
21:13-21:14	Mobile	Street	Friend	Local
21:23-21:28	Mobile	Street	Friend	Local
21:28-21:29	Mobile	Home	Friend	Local
22:00-23:00	Internet	Home	-	-
23:06-23:10	Mobile	Home	Friend	Local

mobile communications, "absence" in the off-line world is related to "tele-presence" through e-mail communications.

The point that mobile communications tend to be based on local places is evident in the case of a female interviewee. When comparing the contacts she had stored in her mobile phone and e-mail account, she said of the effects of the mobile phone on her social networks:

> Those who are in my hand phone [mobile phone] are socially and geographically close people I usually meet often. … Those who are in my e-mail account are otherwise, though they are also in my hand phone. They are people I hardly meet or people I send and receive files by e-mail. Though I booked them in my e-mail account, I don't get in touch with them often (student, female aged 21).

In the time-diary where she recorded her behaviors related to the use of various electronic media such as the mobile phone, the Internet and cable TV for two days (Table 3.3), we can see how the mobile phone produced personal or urban time-space fabric: (a) The mobile phone was overwhelmingly used more often than any other media, increasing real-time interactions in the city. (b) The mobile tended to be used locally. (c) The mobile tended to be used regardless of day and night; private and public spaces; and fixed and moving spaces. (d) Finally, the mobile phone tended to be used at weekends rather than weekdays, and this implies that the mobile phone was used for private activities rather than public activities. These mobile landscapes signify that the mobile phone can provide alternative socio-spatial networks in the urban mediascape of 'absent presence' (Gergen 2002).

Floating time-space coordinates

The mobile phone enables us to access others instantly and directly, and thus entails the socio-temporal networks of "flexibility" in two ways. First, the mobile phone also enables "always-accessible" networks, extending time in which we can access others. Just as 'much of what many people now think of as "social life" could not be undertaken without the flexibilities of the car and its availability 24 hours a day' (Urry 1998), so 'one aspect of temporal location significant for user (and for service providers) is the "anytime, anywhere" availability provided by mobile devices' (Green 2002, 287). Second, the mobile phone enables "immediately-accessible" networks, reducing time we spend in accessing others. While the always-accessible networks relates to the extension of time, the immediately-accessible networks to the reduction of time. The mobile phone is a typical device to reduce the time taken in the intermediary zone. This aspect of mobile phones can be compared to that of private cars (Bauman and May 2001, 40). Just as private cars are used for people to increase temporally accessibility to destinations, so are mobile phones. The socio-temporal networks of flexibility can lead to the socio-temporal networks of "uncertainty" in terms of "floating time-space coordinates".

The mobile phone enables their users to call and be called instantly and directly, making a 'real-time lifestyle' (Townsend 2000) or 'just-in-time lifestyle' (Sussex Technology Group 2001) where 'the old schedule of minutes, hours, days, and weeks becomes shattered into a constant stream of negotiations, reconfiguration, and rescheduling' (Townsend 2000). The emergence of this "on-the-spot lifestyle" in turn produces two paradoxical temporalities. On the one hand, it results in speeded up temporalities, making users move according to accelerated rhythms. As Kopomaa (2004, 269) notes, 'the mobile phone challenges its user to engage in real-time participation, which brooks no delays. The mobile phone side-steps anticipatory social arrangement and allows for spontaneous forms of real-time interaction. It offers a tool for a social and practical control of the urban environment'. On the other hand, it results in loose temporalities, making users free from strict schedules. Again, Kopomaa (2004, 269) states, 'mobile phones change our notion of time as something linear and mechanical, distinct from rhythm of nature and always divisible into smaller parts. … Continuous availability means more flexible working hours; the mobile phone rearranges the division of time into work and leisure, previously dictated by the clock'.

In these urban temporalities, the mobile phone produces floating time-space coordinates in that it can unexpectedly change locations, directions and connections in time-space prisms in terms of time-geography. For instance, the mobile phone makes it possible to easily arrange prompt or impromptu events and to easily change or cancel already arranged or anticipated events, making uncertain the time-spaces of events even until the last moment. In interviews with mobile users, I found that mobile phones made them have frequent meetings, they tended not to decide previously or precisely the time-spaces of appointments, and they tended to constantly change the time-spaces of appointments even until some minutes before they actually met each other.

> [Thanks to the hand phone] appointment times and venues tend to remain undecided often, and even already decided times and venues tend to be changed easily (student, female aged 20).

> Certainly, I have become more and more insensitive to appointment times, since the hand phone made it possible to immediately get in touch with others (student, male aged 23).

Ling and Yttri (2002) call these kinds of mobile communications 'micro-coordination', and Ito (2003) also describes these aspects as one of the important landscapes of mobile communications. In this context, Carey (2004, 136) claims that the mobile phone engenders the 'indeterminate city': 'by virtue of the cellular phone, meeting places have become indeterminate; fluid territories rather than precise spots'. As such, time-space coordinates according to which mobile users move, meet each other and get together in urban space always remain uncertain, changeable, fluid and floating.

Mobile Networks and Emotional Spaces

Social networks and desires to be connected

Mobile phones tend to make their users always ready to call and be called, and thus desire to be connected to mobile networks. It seems that people are more addicted to the mobile phone than any other technological device, maybe even more than their attachment to the Internet. That is

> The mobile can play the part of a technological injunction (you will never ignore my demands upon you, you will never be free of my intrusion) as much as a technological conjunction (you will never be outside the network of always-immediately-available presence, you will always be able to reach me) (Sussex Technology Group 2001, 220).

This is possibly because mobile networks involve social networks. Human communication through the phone relates to the 'true psychological need for "social belonging"' (Lull 2000, 106). To be sure, the mobile phone is a tool which can satisfy such a psychological and social need in contemporary life through social networks, increasing 'connected presence' (Licoppe 2004) or 'present absence' (Fortunati 2002) and decreasing 'absent presence' (Gergen 2002). For many people, mobile phones are 'significant for social capital because they are accessible to unprecedented numbers of people' (Goodman 2003, 5), and living without the mobile phone means a kind of exclusion from social networks. For example, some studies of mobile culture in Korea found that '(adult) users of mobile phones were more active in getting together with their colleagues' (Kim 2002, 71), and also that 'young people's cultural practices via the mobile phone did not contradict that which was valued by adults' (Yoon 2003, 342).

However, when invisible, potential and virtual social networks in the mobile phone could not be unfolded and realized for different reasons (for example, when other people do not call, the user leaves his/her mobile phone somewhere else, the power of the mobile phone runs out, the mobile phone is missing and so on), that is, the desire to be connected to mobile networks could not be satisfied, the user comes to have painful emotions (such as anxiety, irritation, frustration, disappointment, depression and so on). These cases are shown well in the example below.

> When I leave my hand phone at home sometimes, I become nearly irritated and almost crazy. However, when I come back home and find that no text message is left for me in my hand phone during the day, it makes me more disappointed and depressed. … And, I become really nervous when the battery of the hand phone is gone and I could not remember the phone numbers of my friends booked in my hand phone, thus I could not call them (student, female aged 20).

Because the number of calls or messages can be thought of as the popularity or social currency of the user (see Ling and Yttri 2002, 161; Yoon 2003, 338), receiving very few calls or messages can make this person feel very disappointed. In addition, the mobile phone tends to make users oblivious of others' phone numbers for they usually tend to store these in their mobile phone's memories without keeping them in mind noting them elsewhere. This mobile-referring habit or mobile-induced amnesia makes users more frustrated when their mobile phone is missing. For the social networks stored in the mobile phone come to disappear and be lost and users come to be isolated from their social networks.

Control networks and desires to be disconnected

Mobile phones make their users desire not only to be connected, but also to be disconnected from mobile networks. This is because the mobile phone can be a tool for remote controls with unpredictable and interruptive networks, for example, in the social networks between employers and employees and between boyfriends and girlfriends (Sussex Technology Group 2001) and between parents and their children (Green 2002, 288-9). In this sense, concerning such a contradictory and paradoxical function of the mobile phone, Green (2002, 291) stresses the duality of time-space in relation to social relationships by mobile phones: 'on the one hand, social space and time are 'extended', and on the other, they remain locally continuous. Communities are being formed in highly contradictory ways, which reflect new disjuncture, as well as new continuities, in the relationship between space, time, and location'.

The mobile phone can be used to forbid users from even being apart from their social networks and relations, forming somewhat homogenous and standardized social spaces by restricting the user to existing social networks (Gergen 2002). Furthermore, just as 'visual representation of the moving body by GPS introduces the possibility of subjective mapping – or plotting the personal' (Parks 2001, 212), so too the mobile phone enables vocal representations of the moving body. Actually, GPS services through the mobile phone are provided in Korea, and the police use the mobile phone as a means for the networks of surveillance like CCTV in order to trace criminals.

As such, while the mobile phone can be seen as a means of mobility and freedom, it can also be viewed as a means of control and surveillance. This is the reason why people want to turn off or throw away their mobile phones. A male interviewee explained that he deleted voice and text messages other people sent him and did not store females' mobile numbers in his mobile phone because his girlfriend got jealous about them. Similarly, a female interviewee said that she found out that her boyfriend was cheating on her by checking the text messages stored in his mobile phone. Another female interviewee said that she sometimes turned off her mobile phone to avoid calls from her home, and another male interviewee said about the interruptive and embarrassing effects of the mobile phone on his everyday life:

> The hand phone gives me many chances to meeting people. … It gives me on the one hand convenience, and on the other hand restrictions. … When I would like to be alone at home, if someone calls me, then I have to go out. Sometimes I think it would be better to get away from the hand phone. Of course, if I have no hand phone, then I would feel anxious. However, this also is because I have the hand phone now. It is not something I can escape from. Even though I want to be severed from it, I cannot do so (student, male aged 24).

After all, human bodies linked to mobile phones have ambivalent desires: to be connected to mobile networks as social networks on the one hand, and disconnected from mobile networks as control networks on the other hand. The point signifies that mobile phones produce emotional spaces in the everyday lives and lived spaces of their users.

Conclusions

This chapter has pointed out that mobile networks are embedded in urban places and are related with emotional spaces. Producing the socio-temporal landscapes of floating time-space coordinates according to which mobile users shift in the city, mobile networks involve the socio-spatial scale of mobile networks which are highly localized for people tend to use their mobile phones for communications with those who are socially, temporally and spatially close to them in their everyday lives.

In such an urban mediascape, mobile networks entail social networks to which their users desire to be connected, fabricating the "rhizome-city" in terms of Deleuze and Guattari's (1987). At the same time, they involve control networks from which the users desire to be disconnected, producing Foucault's (1977) space of power (closed, disconnected and centralized spaces) or more exactly Deleuze's (1997) space of control (open, connected and decentralized spaces). This point means that ambivalent desires appear in human bodies acting as nodes in mobile networks and thus paradoxical emotional spaces are shaped in urban places through mobile networks.

In general, alongside the development of discourses on the emergence of the so-called "knowledge-based" society or economy, and more recently on the construction of "ubiquitous" computing spaces or cities, information and communication technologies have been explained in relation to the production of intelligent spaces where people can access information or knowledge wherever and whenever they want to do so. However, we need to draw attention to the fact that the technosocial networks that mobile phones produce are not only "intelligent" spaces, but also "emotional" spaces where mobile users have various negative feelings or discontents when their desires to be dis/connected with the techno-social networks of mobile phones cannot be satisfied.

Finally, it should be argued that contemporary cities need to be seen as cyborg spaces in the sense that human bodies and electronic machines are imploded and physical and electronic spaces are deterritorialized into each other. Increasingly, human bodies are operated and urban spaces are produced through electronic machines and networks such as mobile phones. Such urban spaces can be regarded as in-between spaces where the boundaries between various binary categories such as natural/cultural, social/technological, human/machine, actual/virtual, global/local, public/private, inside/outside, present/absent and so on are blurred. In other words, the city comes to be more and more hybrid and fragmented.

Acknowledgments

I would like to thank Dr. Alessandro Aurigi and my ex-supervisor Dr. Mike Crang for their helpful comments and encouragement.

References

Adams, P.C. (1995), 'A reconsideration of personal boundaries in space-time', *Annals of the Association of American Geographers*, 85:2, 267-285.

Bauman, Z. (2001), *The Individualized Society* (Cambridge: Polity Press).

Bauman, Z. and May, T. (2001), *Thinking Sociologically* (Oxford: Blackwell).

Bridge, G. (1997), 'Mapping the terrain of time-space compression: power networks in everyday life', *Environment and Planning D: Society and Space* 15, 611-626.

Brown, B., Green, N. and Harper, R. (eds) (2001), *Wireless World: Social and Interactive Aspects of the Mobile Age* (London: Springer-Verlag).

Callon, M. and Law, J. (2004), 'Introduction: absence – presence, circulation, and encountering in complex space', *Environment and Planning D: Society and Space* 22, 3-11.

Carey, Z. (2004), 'Generation txt: the telephone hits the street', in Graham, S. (ed.), *The Cybercities Reader* (London, Routledge), 133-137.

Castells, M. (1996), *The Rise of the Network Society* (Oxford: Blackwell).

Chatzis, K. (2001), 'Cyborg urbanization', *International Journal of Urban and Regional Research*, 25:4, 906-911.

Crang, M. (2000), 'Urban morphology and the shaping of the transmissible city', *City* 4:3, 303-315.

Deleuze, G. (1997), 'Postscript on the societies of control', in Leach, N. (ed.), *Rethinking Architecture: A Reader in Cultural Theory* (London: Routledge), 309-313.

Deleuze, G. and Guattari, F. (1987), *A Thousand Plateaus: Capitalism and Schizophrenia* (Minneapolis: University of Minnesota press).

Fortunati, L. (2002), 'The mobile phone: towards new categories and social relations', *Information, Communication and Society* 5:4, 513-528.

Foucault, M. (1977), *Discipline and Punish: The Birth of the Prison* (London: Allen Lane).

Gandy, M. (2005), 'Cyborg urbanization: complexity and monstrosity in the contemporary city', *International Journal of Urban and Regional Research* 29:1, 26-49.

Gergen, K.J. (2002), 'The challenge of absent presence', in Katz, J. and Aakhus, M. (eds), *Perpetual Contact: Mobile Communication, Private Talk, Public Performance* (Cambridge: Cambridge University Press), 227-241.

Goodman, J. (2003), '*Mobile telephones and social capital in Poland:* a case study with Vodafone Group', Part of Deliverable 12 of the project Digital Europe: E-business and Sustainable Development, Forum for the Future.

Graham, S. and Marvin, S. (2001), *Splintering Urbanism: Networked Infrastructures, Technological Mobilities and the Urban Condition* (London: Routledge).

Granovetter, M. (1973), 'The strength of weak ties', *American Journal of Sociology* 78:6, 1360-1380.

Green, N. (2002), 'On the move: technology, mobility, and the mediation of social time and space', *The Information Society* 18, 281-292.

Grosz, E. (1992), 'Bodies-cities', in Colomina, B. (ed.), *Sexuality and Space* (New York: Princeton Architectural Press), 241-253.

Hansen, M. (2002), 'Wearable space', *Configurations* 10:2, 321-370.

Haraway, D. (1991), *Simians, Cyborgs, and Women: The Reinvention of Nature* (London: Routledge).

Hayles, N.K. (1999), *How We Became Posthuman: Virtual Bodies in Cybernetics, Literature, and Informatics* (Chicago: University of Chicago Press).

Henderson, S., Taylor, R. and Thomson, R. (2002), 'In touch: young people, communication and technologies', *Information, Communication and Society* 5:4, 494-512.

Ito, M. (2003), 'Mobiles and the appropriation of place', *Receiver* 8, (Vodafone Group).

Katz, J. and Aakhus, M. (eds) (2002), *Perpetual Contact: Mobile Communication, Private Talk, Public Performance* (Cambridge: Cambridge University Press).

Kim, S.D. (2002), 'Korea: personal meanings', in Katz, J. and Aakhus, M. (eds), *Perpetual Contact: Mobile Communication, Private Talk, Public Performance* (Cambridge: Cambridge University Press), 63-79.

Kopomaa, T. (2000), *The City of Your Pocket: Birth of the Mobile Information Society* (Helsinki: Gaudeamus).

Kopomaa, T. (2004), 'Speaking mobile: intensified everyday life, condensed city', in Graham, S. (ed.), *The Cybercities Reader* (London: Routledge), 267-272.

Lash, S. (2001), 'Technological forms of life', *Theory, Culture & Society* 18:1, 105-120.

Latour, B. (1993), *We Have Never Been Modern* (Cambridge, MA: Harvard University Press).

Laurier, E. (2001), 'Why people say where they are during mobile phone calls', *Environment and Planning D: Space and Society* 19, 485-504.

Licoppe, C. (2004), ''Connected' presence: the emergence of a new repertoire for managing social relationships in a changing communication technoscape', *Environment and Planning D: Society and Space* 22, 135-156.

Ling, R. (2004), *The Mobile Connection: The Cell Phone's Impact on Society* (San Francisco, CA: Morgan Kaufmann Publishers).

Ling, R. and Yttri, B. (2002), 'Hyper-coordination via mobile phones in Norway', in Katz, J. and Aakhus, M. (eds), *Perpetual Contact: Mobile Communication, Private Talk, Public Performance* (Cambridge: Cambridge University Press), 139-169.

Lull, J. (2000), *Media, Communication, Culture: A Global Approach* (Cambridge: Polity).

Ministry of Information and Communication, Korea (2001), *Korean Information and Communication's Twentieth Century History*.

Ministry of Information and Communication, Korea (2003), *The Informatization Strategy of Korea*.

Nie, N.H., Hillygus, D.S. and Erbring, L. (2001), 'Internet use, interpersonal relations, and sociability', in Wellman, B. and Haythornwaite, C. (eds), *The Internet in Everyday Life* (Oxford Blackwell), 215-243.

Parks, L. (2001), 'Cultural geographies in practice – plotting the personal: global positioning satellites and interactive media', *Ecumene* 8:2, 209-222.

Rafael, V. (2003), 'The cell phone and the crowd: messianic politics in the contemporary Philippines', *Public Culture* 15(3), 399-425.

Sassen, S. (1991), *The Global City: New York, London and Tokyo* (Princeton: Princeton University Press).

Sheller, M. (2004), 'Mobile publics: beyond the network perspective', *Environment and Planning D: Society and Space* 22, 39-52.

Shields, R. (1992), 'A truant proximity: presence and absence in the space of modernity', *Environment and Planning D: Society and Space* 10, 181-198.

Stelarc (1998), 'From psycho-body to cyber-systems: images as post-human entities', in Dixon, J.B. and Cassidy, E.J. (eds), *Virtual Futures: Cyberotics, Technology and Post-human Pragmatism* (London: Routledge), 116-123.

Sussex Technology Group (2001), 'In the company of strangers: mobile phones and the conception of space', in Munt, S.R. (ed.), *Technospaces: Inside the New Media* (London: Continuum), 205-223.

Thrift, N. (2004), 'Remembering the technological unconscious by foregrounding knowledges of position', *Environment and Planning D: Society and Space* 22, 175-190.

Thrift, N. and French, S. (2002), 'The automatic production of space', *Transactions of the Institute of British Geographers* 27, 309-335.

Townsend, A.M. (2000), 'Life in the real-time city: mobile telephones and urban metabolism', *Journal of Urban Technology* 7:2, 85-104.

Urry, J. (1998), 'Automobility, car culture and weightless travel', Department of Sociology, Lancaster University.

Urry, J. (2004), 'Connections', *Environment and Planning D: Society and Space* 22, 27-37.

Webster, F. (2000), 'Information, capitalism and uncertainty', *Information, Communication and Society* 3:1, 69-90.

Weilenmann, A. (2003), '"I can't talk now. I'm in a fitting room": formulating availability and location in mobile-phone conversations', *Environment and Planning* 35, 1589-1605.

Williams, S. and Williams, L. (2005), 'Space invaders: the negotiation of teenage boundaries through the mobile phone', *The Sociological Review* 53:2, 314-331.

Wittel, A. (2001), 'Toward a network sociology', *Theory, Culture and Society* 18:6, 51-76.

Yoon, K. (2003), 'Retraditionalizing the mobile: young people's sociality and mobile phone use in Seoul, South Korea', *European Journal of Cultural Studies* 6:3, 327-343.

Chapter 4

Epigraphy and the Public Library

Malcolm McCullough

Preface: Who has the Right to Mark up the City?

To augment urban space with information is not so new an act. Just about any city in history has left us inscriptions (figure 4.1). Neither printed nor broadcast, these are an earlier form of "locative media". Most of what has endured are the markings that were intended to do so: proclamations carved in stone.

Not just anyone made these. Most lasting inscriptions came from positions of authority, which of course had the most means to leave lasting traces. More permanent markings, of the sort that last the ages, tended toward the rule of law and the immortality of kings. Even the milestones along Roman roads express authority. Titles, treaties, and laws were set in stone to end all dispute. Although some private shrines and burials have also survived, and instances of graffiti as well, scholars generally assume that common signage was more ephemeral, and the more enduring monuments were to public aspirations. Other media may well have been considered ambient, for example the daily dressing of temples with offerings, but there is little evidence for writing anywhere beyond the foreground. Because these inscriptions came against a background otherwise nearly devoid of text media, (no words printed on clothing for example), their power may be difficult to estimate for someone in today's media-saturated world.

Some cultures were more given to public inscription than others, and not just because they were more authoritative. Different dispositions exist with respect to markup itself. For example, while the Greek alphabet was well suited to carving, Japanese pictographs were less so, and worked better on large painted banners. Because Islamic culture discourages figurative art, more of its sacred sites were extensively lettered; Christian churches instructed too, but through symbolic and decorative form. The industrial age juxtaposed Beaux-Arts and Bauhaus, the one prominently inscribed in the public spirit, the other against ornament and toward a machine poetics for the age of private comforts. The automobile era infamously crowded its cultural landscape with billboards. Ephemeral and irreverent advertising made most acts of lasting inscription seem pompous.

For the last quarter of the twentieth century in architectural theory, (gross summary being necessary in a preface) academics mistook the city for a text. This was perhaps an inevitable reaction at the threshold of experiential saturation by "media," in environments where the signs are more prominent than the things they describe. The more that cultural communication conflated the electronic with the physical, and the more that culture itself became the raw material being processed and consumed,

Figure 4.1 Inscription of the Augustus Gate at Ephesus, Credit: www.livius.org, Marco Prins and Jona Lendering

the easier it became to interpret everything as signs. Following sharp critique of broadcast cultural politics by the Frankfurt School (radio was the first truly ambient medium, and propaganda was its first major abuse), postmodern critics exposed consumerist excess through its communications. Critics often found architecture, even where not overtly inscribed, to be forcibly serving an age of disinformation.

In today's augmented city, the main legacy of that cultural era is a situationist one. Semiotics and literary criticism have long since hit their limits in architecture and urbanism. According to explanations in embodiment, cognitive learning, organizational memory, and haptic orientation – all concepts also dear to human-computer interface designers – the city bears many texts but is itself much more, and in non-linguistic ways.

What has carried forth from the critical project is an emphasis on context. Interaction designers express this as activity theory, which explains how experts play situations, and how ability rises from habitual engagement with objects and settings. Similarly in art, intervention in context reveals the world's working; it is no accident that many more telling examples of pervasive computing are art installations.

Territorial markings naturally predate all this thinking, and have always been a form of identity politics. Archaeologists find ancient graffiti. Contemporary graffiti artists remain the masters of testing where an act becomes transgressive. Marks too obviously amateurish or destructive are removed; only those most on the edge between art and crime are protected.

But now in the augmented city a new domain emerges between the authority of broadcast and the defiance of graffiti. As a consequence of rapid technological development in mobile communications, geographic positioning, tagging (radio frequency identity, RFID), and remote sensing (not only video surveillance); and of accelerating social patterns of sharing photos, videos, and links, the experience of "media" moves beyond the desktop or television screen, beyond passive viewing, and out into the city. The expression "urban markup" is a shorthand for the participatory aspects of mobile, embedded, and "locative" media. The latter include many applications of geographic positioning systems (GPS) but also technologies of tagging, sensing, and urban screens. Urban markup turns the privileged reader into an active tagger, an embodied interpreter, and at some level, and with some unstudied degree of access and duration, also a cultural producer.

Participation drives the so-called "Web 2.0," which is a belief system about the social internet. Cultural communication becomes many-to-many on the net, not one-to-many as it was with television, or with corporate attempts to dominate the web. Although there is no particular technological leap to this, as there was with the introduction of web browsing in the 1990s, web 2.0 emphasizes bottom-up development. It is more granular, casually programmable, and continually evolving than the existing media-moguls' approach to the web. Participatory practices of collecting, tagging, and posting become social, to the point where they challenge the authority of broadcast media. The active participant in social media differs enough from a passive subject of the broadcast era to remake some much more major cultural expectations.

Change is perhaps more evident in domains closer to previous conceptions of "media," most notably the video-sharing service *YouTube*, which has transformed viewing habits well beyond the usual early adopters of new technology, or the photo-sharing service *FlickR*, in which the social aspect of bottom-up communications has emerged most distinctly. If this is a bubble economy (2.0) it is one inflated by social groups, not media companies. Instead of consuming the work of a few "content producers" that has been pushed down the channels mainly for the benefit of the channels' owners, unlimited numbers of participants post, share, and organize themselves around their own subject matters. The former establishment, uncomprehending, amusingly refers to this practice with a favorite new oxymoron: "consumer-produced content."

Not all may be personalized. Not all is sociomedia or identity politics. You cannot go placing your own stop signs on the streets, for instance. That right to mark up the city is very specialized. That is even so for more temporary road markings. Public works crews, and no others, may arrange orange cones, insert colored flags, and spray bright arrows onto the pavement.

2. Keyword: Epigraphy (and beyond)

Does the augmented city form a pattern of enduring inscriptions that is worthy of later cultural interpretation? How does urban markup leave something for posterity?

To help ask such questions, let us blow some dust off a discipline that was more prominent long ago.

Epigraphy, study of intentionally durable writings – as opposed to accidentally surviving papyrus for example – became a prominent feature of academic institutions in the nineteenth century. This generally took residence in departments of classical studies, which then played a very central role in universities, where just about everyone studied some Greek and Latin.

Regard for antiquity gained momentum as the industrial age gave its scholars not only more historical perspective, but also more mobility and means. Overseas discoveries accelerated. Patriarchies were guiltlessly plundered. Heinrich Schleiman excavated Troy in the 1870s, and it is difficult to imagine that happening in any other era. Expanding cultural contact fueled runaway eclecticism and historicism in architectural style, but recast neoclassicism from its formerly regal role into a new civic eminence. In the public monumental architecture of the age, such as the museums still popular today, the art of durable public inscriptions reached its broadest audience.

Epigraphers decipher what archaeologists have found, for the benefit of stories that historians construct. "Rosetta Stone," the artifact whose decoding remains one of the field's most famous events, as it opened the interpretation of Minoan scripts, which prefigure the Greek culture so central to Epigraphy – itself a Greek word – became a figure of speech any seminal act of decipherment. Will the augmented city provide some future its own Rosetta Stones? The epigrapher's challenge goes beyond unfamiliar alphabets in deceased languages, however. Because inscriptions are necessarily short, they '[tended] to omit pertinent information that [was] already known by the intended audience' (McLean 2002, 2). The epigrapher must somehow reconstruct that cultural context. Among those named in one recent epigraphic classification for ancient Greece are official decrees, funerary inscriptions, milestones and boundary markers, treaties, dedications, stonecutters' and stonemasons' marks, sacred sites and shrines, and of course graffiti (p.181).

Although a separate branch of classical study, Numismatics exists for the study of coinage, as that medium has also survived. There is no discipline, little physical evidence, and surprisingly little mention in surviving literary texts, of everyday urban markup. It is difficult to imagine any past city without its signs and banners, both sacred and profane, but it is also difficult to study one.

Historians of material culture reveal past practices of everyday life without much mention of ambient information. For example there is little evidence of urban markup reported in Braudel's vast treatise on the rise of mercantile Europe. Urban technological historians such as Mumford or Hughes focus on transport factories or power more than saturation in communications. Early environmentalists lamented the contextually oblivious contention of commercial signage, especially in spaces that gave free reign to the automobile. Neo-Luddites such as Postman and Rozsak protested the social and psychological cost of the early internet era. Yet so far, information pollution lacks its Rachel Carson. To protest is to contribute to the problem.

The history of pollution is a mostly short and recent one. Toxins and nuisances have always existed; city waters have been eternally foul. Only in recent decades has pollution been subject to systematic regulation, however; only in the last two centuries (and not in all places) has it been recognized as addressable. John Snow's linkage of cholera to well water occurred in London in 1854. US clean air and water acts were passed in the early 1970s. Environmental impact studies become usual only then. Although this began with persistent material nuisances, especially those with biochemical effects, the idea of pollution soon carried over to information effects. Most cities regulate noise pollution, which was already an industrial phenomenon, and increasing numbers of them also address light pollution.[1] The state of Vermont first banned billboards on highways in the 1960s. Recently the city of São Paulo, (hardly known for its pastoral scenery) banned billboards in its urban core. The city of New York prohibits phone conversations while driving. Set against the onscreen techno-euphoria of the 1990s were such books as *Data Smog, The Cult of Information,* and *The Age of Missing Information.*[2] Meanwhile if there is any place that you can tell the difference between floods of raw information and the architecture of selected information, it is the library.

3. Scene: The Library and its Discontents

From the standpoint of ambient information, reconsider that traditional monument to the authority of writings, the library. For this our chosen example is the Boston Public Library, the first publicly supported municipal library in the United States

	Experience	Editorial	Urbanism
Epigraphic	*Public*	*Public*	*Facade*
Bibliographic	*Individual*	*Public*	*Reading room*
Hypertext	*Social*	*Varies*	*Ambient*
Graffiti	*Territorial*	*Contention*	*Transgression*

Figure 4.2 Epigraph, Old and New, Among Different Forms of Writing, Credit: Author

1 Advocacy against light pollution: www.darksky.org

2 Warnings on information glut, even before the web: David Shenk, 1997, *Data Smog—Surviving the Information Glut*, San Francisco: Harper Collins. Theodore Roszak, 1986, *The Cult of Information-- the folklore of computers and the true art of thinking,* New York: Pantheon. Bill McKibben, 1992, *The Age of Missing Information*, New York: Random House.

(1848), one of America's most celebrated neoclassical buildings, designed by McKim, Mead, and White (1888-95).

Of all building types, the library may best illustrate the emerging role of urban markup. This is because it demonstrates the multiple nature of literacy (Gardner 2000). More specifically, it interrelates different forms of writing in ways important for the augmented city. Consider these in turn (figure 4.2).

In the abstract, the library has long been the best model for a literacy-center. This meant something different when information was scarce than it does now with information superabundant. To some critics, the age of Google dooms libraries to immediate obsolescence. When all work has been scanned, when enough new information is produced annually to surpass the volume of the largest library by at least factor thirty, and the best information architecture is available from the contexts of the office or the comforts of home, when search engines crawl all libraries' and indeed all servers' holdings as one; is there a there, there, at the public library anymore? To others, the lack of stability, credibility, or guidance in that flood of data makes libraries more important than ever, no longer just for searching and retrieving information, but instead for making sense of it. The Seattle Public Library, which has become synonymous with innovation in the field, uses a series of "living room" spaces in combination with personal device services to help give context and reputation (rather than mere popularity tagging) to the elements of its holdings (Kenney 2005). This library is incidentally also the leader in RFID tagging especially for automated sorting, a clear instance of the augmented city. The building design by Rem Koolhaas, standing across a century and much else from Charles McKim's design for Boston, does lack lettering, however.

While this is not the moment to unpack the history of libraries, nor to join debate on the future of the book, it is worth noting that any institution confers identity on its members. Among libraries this was one thing in a monastery, or a club such as Boston's fine Athenaeum, or a university research library, and another when made "free to the people," as Carnegie's libraries were inscribed. The public library movement was a social welfare project. It sought to redress some social inequities of the industrial age by applying some of that era's unprecedented wealth toward a project of universal literacy. Of course this meant books. Among all forms of writing, the one that has held sway longest and most authoritatively, at least in the modern west, is bibliographic.

Mass access to print has been the standard background story to electronic media studies. The historic trajectory of text from rarity to superabundance was a favorite story of ubiquitous computing pioneer Mark Weiser. Text was rare before printing, and one had to go to a specialist – often a cleric – to use it. Printing liberated not only the location but also the purpose of text, and that story has too often been told. Less recited in media histories, printing has transformed text into something found in small disposable pieces everywhere. Gutenberg has long since become a cliché; Aldus, who invented the portable, tradable book is more interesting; Victor Hugo, whose *ceci tuera cela* ("this will kill that") argument has become canonical in architectural education. As seen through the widely-used interpretation *The book and the building* (Levine 1984), the rise of industrial mass media in print, (under which Hugo, or his English counterpart Dickens pioneered the serial novel in the penny newspaper), soon undermined architecture's preeminence at instructing the public. In other words, epigraphic literacy fell to bibliographic literacy.

Here it is worth expanding the former term. Epigraphic writing communicates publicly, for all to see, whereas bibliographic writing does so privately, one reader at a time. Even a book may be read in an epigraphic manner, to be glossed by one person visibly before a group. Indeed this was the norm, whether among medieval scholastics or rural schoolchildren, wherever just one copy of a book was at hand (or where just one person knew how to read it). Epigraphic literacy is recently rampant in PowerPoint as well. And while the latter exists so that any fool can deliver the corporate party line, all these forms involve a process that is more two-way than proclamations cut in stone. The key moment in epigraphic communication is assent.

The key moment in a library, by contrast, is cross-reference. Librarians mainly practice information architecture (Rosenfeld and Morville 1998).[3] This is the basis for a third kind of writing, which has regularly been expected to kill the bibliographic kind, namely hypertext. As also quite often told in media studies, the brilliance of hypertext is its power to accelerate the associative trail of cross references—no more riding up and down the open stack in a tiny elevator. Library lineage is evident in that first widely-usable instance of hypertext, modeled on the now-forgotten card-catalogue, *Hypercard.*

The key moment in publishing, however, is editing; and this is just what the hyperlinked web so lacks. Whereas once the right to publish text was fairly exclusive, and the mere fact that something was in inscribed in print lent it some authority, the web and electronic prepress opened the field to vastly more participants, (as noted above.) This of course drives the amount of information up, and the quality any given piece of information down, *ad absurdem.* Finding and having more information is no longer of much benefit unless one also has ways of knowing its reliability and reputation.

As any movie critic or bookseller could have foretold, one of the difficulties of the social internet is that popularity is not the same thing as quality. Mob rule is no better in culture than in politics. When on the web the items that show first as search results are the ones with the most tags, links, or recommendations from an entire populace of participants, the voice of an informed editor or critic is drowned. When popularity is ephemeral, recently important items disappear as well. Better systems for sense-making, reputation, information architecture, and just plain search all must meet such challenges, and they do so in increasingly sophisticated ways. Among the strategies used to interpret the rising flood of information, the one of most interest here is context. While that context may itself be informational, it is commonly acknowledged to have physical, organizational, and situational embodiment as well (Dourish 2003).[4] Dematerialization has its limits. In Boston, for one, staggering real estate prices are sufficient evidence of that.

So there is more to learning than sitting home alone with Google, even for casual foraging. The library is no mere portal. Although many of its electronic resources are available from anywhere online, to use them is a more discerning experience than casual search of the full web. As bodies of information architecture, they feel curated.

3 At the University of Michigan, the information school was formerly library science.

4 Phenomenology of interactivity.

Meanwhile most city libraries have seen an increase in foot traffic among those from across the digital divide, who come in for the internet access that they otherwise lack. Library strategists realize that social networks have become the main opportunity and challenge (Albanese 2006). Information architects understand the importance of streaming media (some of that written). Our concern is just the physical context. There the traditional reference services merge best with the electronic overabundance. The more information that exists, the more a trusted guide helps make the best of it. Events and classes add to the sense-making, whether for specialized research, in, say, early American newspapers, or for just learning English. BPL web resources are inscribed (but not in stone) that "books are just beginning."

Despite Victor Hugo's dictum, in this case the very building still instructs. This begins from its relationship to Copley Square, which gives visual perspective on it as an object and symbolic power to it as a public edifice (figure 4.3). It continues to the top of the facades, which are prominent inscribed in the Roman manner: "THE PUBLIC LIBRARY OF THE CITY OF BOSTON • BUILT BY THE PEOPLE AND DEDICATED TO THE ADVANCEMENT OF LEARNING • A.D. MDCCCLXXXVIII". Adjoining facades proclaim "MDCCCLII • FOUNDED THROUGH THE MUNIFICENCE AND PUBLIC SPIRIT OF CITIZENS" and on

Figure 4.3 The Copley Square Façade of Boston Public Library, Credit: Daniel Schwen, Wikimedia Commons

Boylston Street, "THE COMMONWEALTH REQUIRES THE EDUCATION OF THE PEOPLE AS THE SAFEGUARD OF ORDER AND LIBERTY".

Next the sequence of arrival and entry creates cultural context passively through form and painting, and without recourse to an excess of way-finding signs that one might need in, say, a hospital. Most libraries struggle with the usability and appeal of their service desks, which casual users fear. Boston's counterpart in Pittsburgh (Carnegie Libraries' origin) recently combined moves in physical architecture, information architecture, and information design to improve access, for instance; another clear case of augmentation.[5] Boston's building clearly divides popular lending from research services, for instance. Finally, the grand reading room at the top reminds that among architectural types the library is traditionally the one most based on ambient daylight.

Meanwhile at the newsstands outside, a quick survey of the goods reveals clearly perceptible differences across market segment and intellectual spectrum. Scanning a page of search engine results does not afford such obvious differentiation. The best prospects for useful markup may yet be on the street.

4. Claim: Between the Epigraphic and the Everyday

There exists a new middle ground of urban markup that must become subject matter for cultural production, research, and scholarship. Neither organized "media" as the twentieth century knew them, nor random graffiti as all the ages have witnessed, new practices of mapping, tagging, linking, and sharing expand both possibilities and participation in urban inscription. These inscriptions have greater duration and access than the casual signage that must always have been a part of urban experience, for they involve distinct codes and they leave a trail. They have greater freedom and extent than the short official wordings of permanent physical inscriptions. They involve a wider or at least different population than graffiti. They serve a wider, or at least a different set of purposes than advertising or street signs. They present a new range of subject matter for cultural interpretation. They may stretch literacy in different ways that the usual tension of print versus internet. Do they form any lasting cultural body of work?

Evidence for this claim has been thin, but increasing, mainly so far in art installations. In other words, despite what libraries, our representative case, have been doing about augmentation with tagging, way-finding, and social networks, they have done fairly little with massively multi-user markup.

Such a position seems warranted when new practices are advocated by those who do not have a financial stake in the spread of technology, when they interrelate with existing cultural patterns rather than declaring those obsolete, and when they appear to favor increased participation. But of course there are objections. The foremost caution from librarians concerns the limits of bottom up architecture. Folksonomies make them shudder. More broadly the usual lament is that information technology usage in any sort of critical manner is just a narrow social niche; more people use unquestioningly; more still hardly connect at all. Above all, and like ambient pollution, the costs of ephemeral, unedited information glut demand ever more

5 Maya Design, Inc, 2005, "Carnegie Library | Dynamic Information Environment," <http://www.maya.com/web/what/clients/what_client_clp_dyninfo.mtml>.

vigilant concern. Therefore the usual academic response to pervasive computing has been dread. At least with radio or television you could turn it off and walk away.

To qualify the claim in light of these objections requires a new cultural sphere of ambient information. What has changed in media studies, or more specifically "media urbanism," is the prospect for a useful middle ground. Between the expression of authority – whether oppressive or benign – and expression of denial or transgression (another well-established ambient information medium: graffiti), does there now arise a body of urban markup that is culturally responsible? Where is an environmental history of information? What is an epigraphy of ambiently embedded electronic records?

5. Agenda: Toward A History of Ambient Information

More is different. This is a fundamental principle of emergence. Increase the scale, velocity, of volume of a system, and new patterns arise. Now as much more urban markup occurs, what patterns can be identified? What are the essential traits? A strange mix of everyday commercial notices, personal memory tokens, ad-hoc social networks, mass news feeds, ubiquitous reference resources, and of course defiant graffiti increasingly floods and blurs perceptual experience.

Throughout the history of the city, density has been a key catalyst. As Braudel has observed, districts form when the volumes of apparent competitors provides critical mass for a greater differentiation of production support services. The latter have been characterized by Sassen, in an earlier stage of networking, as the most characteristic indicator of global cities; indeed entire neighborhoods and sometimes entire cities became devoted to production support for global capital markets. Those kinds of information services and capital channels rarely penetrated for enough into the emerging middle classes of developing nations however. This is the most citied argument in the economic geography of the networked city, as expressed best by Joseph Stiglitz.[6] Better agendas in ambient information therefore address the "bottom of the pyramid," or the "long tail," to name two recently popular theories. Production, not consumption is the key to augmented city geographies; and the emerging networked class, not the deprived, is the social pattern most worth watching. As the anthropologists Jyri Engestrom and Ulla Maaria Mutanen have advocated, one immediately practical prospect for the augmented city is for groups to develop their 'own logo' tagging systems (Engestrom 2006). When group or personal authorship is casually indicated without recourse to registered trademarks, uniform product codes (UPC), or stock keeping units (SKU) that adds to the long tail of microtransactions that exist outside the scope of the global supply chain.

Then, fed by commercial prosperity, other cultural practices aggregate. The augmented city is one whose cultural information layer becomes denser. Urban

6 Three standards from three decades on the economic history of the networked city: Fernand Braudel, 1983, *The Wheels of Commerce, (Civilization and Capitalism, 15th-18th Century Volume II)*, New York: Harper Collins. Saskia Sassen, 1994, *Cities in a World Economy*, Thousand Oaks CA: Pine Forge Press. Joseph Stiglitz, 2003, *Globalization and its Discontents*, New York: Norton.

markup is the belief (and so far not much else) that new density comes not only through additional channels of universal broadcast, or personal portable collections, but also through new means of local inscription (figure 4.4). For a new epigraphy, some of those inscriptions must endure, and be readable by whatever means become current. For ambient information, this markup must be plentiful, and in the background, not contending for attention the way advertising does. For a history of all this, more durable media must record selective interpretations of more ephemeral media, and do so in a cultural commons. And in that history, the belief that the good life is an urban life may be inscribed.

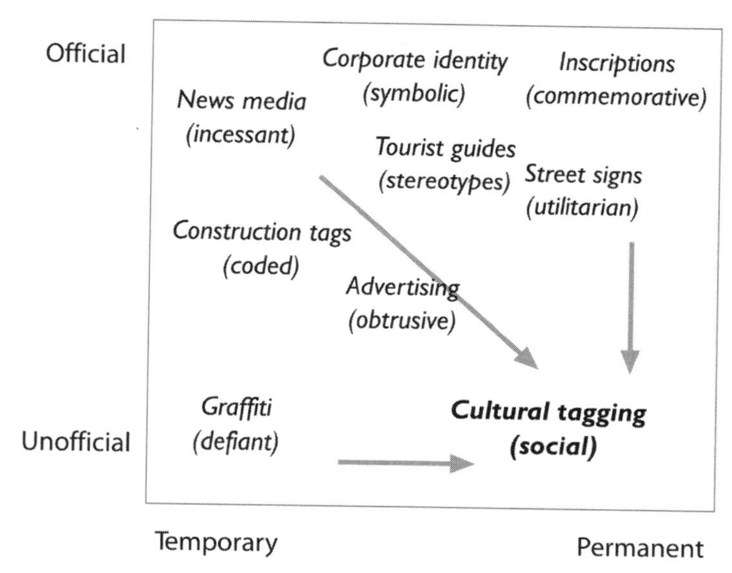

Figure 4.4 The Copley Square Façade of Boston Public Library, Credit: Daniel Schwen, Wikimedia Commons

References

Albanese, A.R. (2006), 'Google is Not the Net: Social networks are surging and present the real service challenge—and opportunity—for libraries', *Library Journal*, September 15 2006, <http://www.libraryjournal.com/article/CA6370224.html>.

Dourish, P. (2003), *Where the Action Is*, (Cambridge MA: MIT Press).

Engestrom, J. (2006), 'Own Logo', <www.zengestrom.com/blog/tagging/>

Gardner, H. (2000), *Intelligence Reframed: Multiple Intelligences for the 21st Century*, (New York: Basic Books).

Kenney, B. (2005), 'By discarding every preconception about a public library building, they created the first 21st-century library', *Library Journal*, August 15 2005, <http://www.libraryjournal.com/article/CA633326.html>.

Levine, N. (1984), 'The Book and the Building', in Robin Middleton (ed.), *The Beaux-Arts and Nineteenth Century French Architecture*, (London: Thames and Hudson).

McLean, B. (2002), *An Introduction to Greek Epigraphy*, (Ann Arbor: University of Michigan Press).

Rosenfeld, L. and Morville, P. (1998), *Information Architecture for the World Wide Web*, (Sebastopol CA: O'Reilly).

Chapter 5

Impacts of Social Computing on the Architecture of Urban Spaces

Marcus Foth and Paul Sanders

Introduction

After the home and the workplace, public spaces are the most prominent building blocks of a city. They act as "social catalysts", places where urban residents and members of neighbourhood communities meet to create and maintain social ties and friendships and engage in discussion and debate. They are paramount in establishing the identity and culture of a city and a sense of cohesion and belonging.

The emergence and uptake of new media and networked information and communication technology have added a range of online public spaces that provide opportunities for city dwellers to meet collectively, such as chat rooms, discussion forums, community networks, digital cities and massive multi-user online games, as well as peer-to-peer through email, instant messengers and SMS (short message service). Early pessimistic voices interpreted these forms of interaction as alarming expressions of increasing "individualism" and "privatisation of leisure time" that provide evidence for the disappearance of traditional forms of civic engagement and community values and for a strong decline of social capital in society (Putnam 2000).

However, online participation in public spaces can facilitate new connections to work, education, civic participation and a healthy social fabric. Watters rightly argues that

> social capital comes from much more fluid and informal (yet potentially quite close and intricate) connections between people. [...], social capital could as easily accrue among a tight group of friends yet still have an effect on the community at large. (Watters 2003, 116)

The internet and mobile phones provide means for city residents to connect with each other and to negotiate face-to-face meetings and social gatherings that take place somewhere in the city. Hence, physical place is increasingly important not despite but because of the range of social ties, bridging links and local interactions that occur online (Fallows 2004; Horrigan 2001), an effect that has been termed "glocalization" (Robertson 1995; Wellman 2002). These connections are created and maintained in both virtual and physical urban spaces and city residents traverse these worlds seamlessly as they are increasingly interwoven. Thus, it is time to depart from simple binary oppositions and compartmentalised dichotomies such as "physical place" vs. "cyberspace" or "online" vs. "offline" (DiMaggio *et al.* 2001; Lovink 2005) and embrace the complex hybrid nature of urban spaces.

The role that urban neighbourhoods play in this new era has changed. The premise that a strong place will ensure a strong community needs to be revisited. Previously, neighbourhoods were marked by central public places that provided traditional meeting spots such as the market place or town square. These locations were used to meet with friends and peers. Mobile communications technology such as the mobile phone and SMS, and ubiquitous communications technology which can be accessed anywhere, such as wireless local area networks, are now enabling users to negotiate meeting places and venues on-the-fly anywhere and anytime.

Studying urban public space requires a cross-disciplinary approach with contributions from three main areas, that is, the people dimension (urban sociology, community development, communication studies), the place dimension (urban design, town planning, architecture), and the technology dimension (community informatics, interaction design, computer science). So far, much interest has focused on how new technology enables new forms of public space that digitally augment city life and lead to so-called "digital cities", and how these spaces are designed, developed, maintained, used and administered effectively as well as what impact they have on the life of city residents. Surprisingly, the reverse direction, that is, the impact of new technology facilitating social networks and peer-to-peer interaction on the design and architecture of physical urban spaces, has not been met with the level of attention necessary to invoke a truly cross-disciplinary exchange that goes both ways.

The impact of new technology on residential architecture can be divided into three areas:

1. New technology enables innovation in production and construction. New materials and construction processes allow architects to design buildings that realise unique forms and shapes that had been impossible before. Prominent building works by Frank Gehry, Norman Foster and others are examples of this.
2. New technology is being implemented into buildings in a range of styles. The integration of universal ducts and wires to "future-proof" the home and to provide local area connectivity, especially in master-planned community sites, is becoming a standard in new buildings, alongside electricity, gas and water. On the other end of the scale, interactive artistic experiments such as the Aegis Hyposurface (dECOi Architects, Paris, & RMIT, Melbourne) provide an artistic outlook of what the future of digital augmentation and integration may hold.
3. New technology is being used by urban residents for personalised networking to form social formations that are different from conventional images of "community", "neighbourhood" and "urban village". Their use of new technology enables a fluid, swarming social behaviour that has implications on residential architecture and the design of urban space.

This paper is about the latter point, the new social formations as they emerge in three inner-city apartment buildings in Australia and the implications for the residential architecture of the public spaces in those buildings. We briefly outline the relevance and significance of this topic that is established two-fold by (a) an ongoing trend towards more and more compact cities in the light of urban renewal, and (b) findings from urban sociology that describe the emergence of networked individualism. These two notions are now discussed in turn.

Urban Renewal in an Australian Context

Australia is one of the most urbanised countries in the world in terms of the high proportion of urban dwellers among its total population. Approximately two-thirds of the total population reside in major cities (Australian Bureau of Statistics 2004). South East Queensland (SEQ), a region of approximately 75 km radius around the City of Brisbane, is one of the most pressured given its long history of low density urban sprawl and now its status as the second highest growth region in the world after Phoenix in the US. Current projections for the region are 3,709,000 by 2026, an increase of around 1.05 million people, or almost 50,000 each year on average (Queensland Government 2005). The management of this growth has been the subject of a strategic regional plan, developed under the auspices of the Office of Urban Management of the Queensland Government. This document provides some statistical data as the backdrop rationale for urban renewal in SEQ towards higher densification of inner-city areas.

> SEQ has experienced high and sustained population growth since the 1980s, growing at an average of 55,300 persons each year between 1986 and 2004. The estimated resident population of the region in 2004 was 2,666,600. (Queensland Government 2005, 14)

> The projected population increase, combined with the continuing trend towards smaller households, will require an estimated 575,000 new dwellings in the region by 2026. There will also be a greater demand for a diversity of housing forms to match the needs of changing household structures, particularly an increase in one- and two-person households. (Queensland Government 2005, 16)

The Queensland Government as well as local government representatives are aware that the continuation of the low density urban sprawl in the SEQ region is not sustainable. A range of implications have been proposed in the Regional Plan such as the implementation of policies to ensure that new developments are contained within the existing urban footprint of the region, protecting areas of urban landscape and rural production, and delivering more compact and higher density residential solutions. A further complexion relating to residential trends that have been identified here is the proliferation of large detached dwellings on small lots. Statistics indicate that family sizes in these large homes are decreasing with a tendency to single and couple occupancies and a related strong demand for one and two bedroom units.

At the same time recent economic trends in Australia have seen a rapid escalation in real estate value to a point where entry level residential accommodation in inner-city areas is becoming unattainable for the average income earner. More and more households with limited resources are excluded from high amenity areas in the inner city and gravitate to areas offering relatively low housing costs in city fringes and new greenfield estates (Healy & Birrell 2004). Issues of affordability and density in residential accommodation further impact on strategies for urban zoning as well as future typologies in the design and delivery of adequate residential stock.

These trends that are similar in other urban and residential areas elsewhere in the world have global economic relevance and reflect a changing role of cities internationally. In Australia, compact city policies are being developed and implemented in all capitals to deal with population pressures and urban expansion. The strategies proposed in these policy documents open up new research questions around issues of governance and sustainability (Gleeson *et al.* 2004). They require a re-interpretation of what archetypical concepts, such as "neighbourhood community", "urban village", "smart growth" and "new urbanism" (De Villiers 1997; Walmsley 2000), mean in practice. Randolph rightly argues that

> the language of community has come back with vengeance in policy areas that ignored it for many years. Cities are becoming, perhaps more than ever before, collections of distinctive communities and neighbourhoods, all the more differentiated as the cities grow in size and complexity. As the city expands, people remain focused on their small part of it. (Randolph 2004, 483)

Mixed-use residential apartment complexes are "a small part of it", yet arguably one of the most prominent components of urban densification and thus play a crucial role in urban renewal. Apartment buildings provide the immediate surroundings in which location-based interactions with other residents could occur and "communicative ecologies" (Foth & Hearn 2007) and social networks could emerge. However, their architectural design and layout (beyond issues of market demand, scope and scale) is rarely informed by societal developments and sociological insights and has hitherto been guided more by the functional requirements of the individual resident and by rental and investment returns than by the resident community at large and their need for public space and interaction. These conditions are being aggravated by prevailing attitudes of developers who confuse "planning for community" with "master-planning community" (Gleeson 2004; Ziller 2004).

There are few exceptions. In Brisbane, the architectural practice of Donovan Hill acknowledges the essential commodity of public space in the pursuit of sustainable environments in residential design. Private residences are construed as fragments of cities, the design for the components of the houses are set around a plaza or courtyard, a focal public place within a private realm. Not surprisingly, Donovan Hill's designs for multi-unit developments embrace this theme of public place. Their design for an eight townhouse development in Terrace Street, New Farm, establishes a large lawn space as common garden, from which all units relate.

The context of urban renewal in SEQ as outlined suggests that innovative models of housing will need to be considered in addressing the impending pressures on the availability of residential accommodation. Solutions that yield higher densities will be sought, and opportunities to inform residential architecture through advanced understandings of social networks and communicative ecologies will be essential in order to create public space that accommodates the needs of urban residents and their new social formations.

New Social Formations in the Urban Mediascape

Since the advent of modern means of transportation and global communication, the importance of door-to-door and place-to-place neighbourhood ties, which (apart from family and kinship ties) used to provide the closest and most convenient way to socialise, has been diminished by friends and peers other than neighbours who fulfil social needs in various person-to-person and role-to-role relationships (Wellman 2001). The portfolio of sociability (Castells 2001) of urban residents, that is, the result of maintaining a range of individual social ties with selected friends through the internet, mobile phones and other media, tend to be place-independent. Nevertheless, the frequency of contact with the nodes in our portfolio is mostly dependent on the nodes' proximity to our locality. We remain what Baker & Ward (2002, 221) describe as 'physically-instantiated and geographically-centred individuals and citizens'.

The hybrid nature of maintaining a portfolio of sociability that is at the same time both "individualistic" in the sense of social control and private ownership, and "networked" in the sense of being connected to a personalised set of friends and peers, has led to the term 'networked individualism' (Wellman 2002). Watters' (2003) detailed description of 'urban tribes' illustrates how the theoretical concept of networked individualism applies in practice in an urban context.

Networked individualism introduces challenges to conventional understandings of "place" and "public places". It opens up opportunities for architecture, city planning and urban studies to re-conceptualise their understanding of community and neighbourhood planning in the light of opportunities presented by new media and network ICTs (cf. Castells 2004; Florida 2003; Graham 2004; Mitchell 2003; Oldenburg 2001; Walmsley 2000). The contemporary interpretation of community is shifting from "village" and "neighbourhood" to "social network" and "urban tribe". However, such a re-conceptualisation has not been achieved yet in all relevant areas due to a lack of theoretical and practical understandings of the freedom and constraints and the social and cultural meanings that urban dwellers derive from their use of location-based ICTs.

Neighbourhood identity and a sense of belonging is derived less and less from the bricks and mortar of the built environment itself and more and more from a combination of the usage of the built environment – especially the "third place" (Oldenburg 2001; cf. Soukup 2006), such as cafés, bars, parks, etc. – and the transitory meaning residents associate with these places. It could be any decent café that a group of friends decide to meet at. The decision to use this particular café as today's meeting place bestows meaning on this place, and frequent use will raise its identity as a favourite meeting place – yet, tomorrow, it could be another favourite café across the street, as long as it is conveniently located within the proximity of group members and fulfils their needs and expectations. A public place cannot invoke meaning or a sense of belonging per se. The culture of place making involves humans adding layers of shared experiences. The agora of the group's interaction can be quite motile but remains essentially face-to-face and place-based, either within the neighbourhood, suburb or city. ICT plays a role in preparing the meeting, and possibly during or after the meeting to prepare the next gathering.

New light has recently been shed on the location preferences and decisions of citizens in the context of diversity and creativity (Florida 2003). Early results indicate that people prefer to settle in open, accepting and permeable cities. That said, an online community network (Day 2002) might contribute to a city's permeability by affording personalised networking and by offering a choice of residents to socialise with on the basis of self-selected criteria such as age, interest, family status, profession, nationality, etc. However, the new emerging social formations and communicative ecologies which are at the same time networked and individualistic have implications not only for systems architecture of online urban space but also for the residential architecture of physical urban space.

Public Space in Residential Apartment Buildings

One of the significant common denominators in well functioning residential architecture is the provision of social spaces, interstitial places that offer opportunities for interaction and exchange. The cloistered monastical courtyards provided inhabitants with a public place of relief from the humble quarters of the private cells. In another context the Public Houses ("pubs") marking the street corners of nineteenth century British mass terraced housing, provided the scale of a lounge environment for social gatherings spaces, as private living rooms were modest and inadequately sized for group interaction. In the mass housing solutions of the twentieth century, the street was replaced by the access corridor in high-rise developments, mostly void of places to dwell, providing mere circulation. As these corridors became devices of internalised access, the mounting disfunctionality increased in the face of developers' slim profit margins.

The modernist residential tower blocks and vertical real estate mostly fail to recognise the model established in Le Corbusier's *Unité d'Habitation* in Marseilles, France, completed in 1952, that of an elevated podium (allowing the landscaping to flow beneath the structure), the allocation of public amenities on mid block floors (shops, laundry, etc.), and recreation facilities (pool, playground, crèches) on the roof.[1] The need to optimise the return on real estate investment focuses the attention of today's developers of apartment buildings on the apartments themselves; for they are sold according to size and location. Public space may add value, but also increases body corporate fees and maintenance requirements. It is thus not surprising that public space in residential apartment complexes appears all too often to be an afterthought and a way to fill gaps.

In the following section, we introduce a case study that examines the public spaces of three inner-city apartment buildings. The design and usage of these spaces is analysed with a view to better understand the articulation of physical urban

1 For an extraordinary example of a postmodern version of Le Corbusier's residential apartment complex, see Andrew Maynard's Corb 2.0 at <http://www.andrewmaynard.com.au/corb.htm>.

spaces. The combination of the theoretical understanding of social interaction and the empirically illustrated understanding of physical urban spaces is necessary to help inform the design of ICT to augment urban spaces.

Case study of three inner-city apartment buildings

Our case study research comprises three different inner-city residential apartment complexes in metropolitan Australia. To protect the privacy of residents, the sites will be referred to as “Alpha”, “Melba” and “Sigma”. Research methods that have been employed are situated within an action research framework (Foth 2006c; Hearn & Foth 2005) and include mostly qualitative and ethnographic methods such as surveys, focus groups, participant and site observation and interviews.

Research on Alpha started in late 2002. Melba and Sigma have been added to the case study at the end of 2004 to control for certain demographic factors and to enable a more comparative analysis. Opened in 2000, Alpha is an apartment complex for international students who are about 17 to 24 years of age and study at nearby tertiary institutions. They come from a variety of national and cultural backgrounds. The majority of tenants only stays short-term, that is, for one or two semesters of study. About a fifth of tenants come to Australia to study a full degree program which usually lasts three to four years. Alpha contains 94 one, two and three bedroom units with a total of approximately 160 tenants.

Melba was built in the mid 1990s and is the home of mostly working singles and couples in their Twenties and Thirties. It contains 39 two and three bedroom units with a total of approximately 90 residents, mostly tenants and some owner-occupiers. Length of residence at Melba is medium to long-term. Sigma is the largest site which was completed in the early 1980s. It consists of three high-rise buildings, a low-rise two story building and 48 townhouses. There are 156 apartments and approximately 300 residents in total with the majority being owner-occupiers and some tenants. Residents are mostly couples and families in their Forties and Fifties working in diverse occupations with some retirees. Length of residence at Sigma is usually long-term. Unlike Alpha where every tenant is an international student, there is no pre-existing underlying common link at Melba or Sigma other than living in the one complex.

Interaction between residents and public spaces

The public spaces at Alpha, Melba and Sigma are examples of contemporary residential architecture. In this study we are interested in analysing how the use of digital information and communication technology and resulting social behaviour impacts on the purpose of public space and how it is used and seen by the residents of our case study sites. Each apartment or unit at all three sites includes one or more bathrooms and a kitchen, so there is no need for residents to leave their unit and use shared facilities which is common in shared accommodation and college-style dormitories and which could stimulate the initiation of interaction with neighbours.

Alpha consists of two six-storey buildings which are linked through a gateway on each level. There is a reception and lobby area on the ground floor, a laundry room and a common room with a pool table and ping-pong table on Level 1, an outdoor swimming pool on Level 3, as well as two barbecue sites. Melba consists of three three-storey apartment buildings which are built along the corner of two streets. Seven separate entrances give access to a cluster of about six apartments each. The only underlying link is the common underground car park through which all residents have to traverse in order to get to the courtyard pool and barbecue area on the inside of the building (Figure 5.1). Sigma is a gated multi-building complex with its own private road infrastructure. There is a swimming pool and a lap pool at Sigma, a tennis court, as well as a barbecue site. In relation to its size, public spaces at Sigma are sparse.

Figure 5.1 Entrance to the Pool Area via the Car Park at Melba, Credit: Authors

The number and size of public spaces also depend on the size and layout of the apartments themselves. The smaller an apartment is, the less social space it offers for entertainment and other purposes, especially in shared accommodation. Public spaces can make up for this lack by offering break-out areas. Collective ownership of public spaces also enables residents to access and use facilities which would be too large, too expensive or too inconvenient to maintain on their own such as pools, gyms or tennis courts.

> A gym would be fantastic – none of the units are large enough to cater for basic gym equipment and the gyms nearby are quite expensive. (Resident at Melba)

> I rarely use the shared facilities. (Resident at Melba)

> Advantages: Don't have to maintain the public areas (more time on our hands). If feeling sociable there are generally people around. Security, there is always someone around. (Resident at Melba)

However, collective ownership does not mean collective use. Most public spaces are meant to be "public" in relation to access, but "private" in relation to use. Yet most of them do not offer the adequate level of privacy that residents desire. The barbecue area at Sigma (Figure 5.2) as well as the combined outdoor pool and barbecue area at Melba are surrounded by apartments and open to the gaze of spectators. The lack of privacy of these panoptic spaces make many residents feel uncomfortable and awkward.

> More interesting space around [the] barbeque – more landscaping etc. Currently very open and not a terribly interesting place to bbq. Would love it to be a place you want to go, and enjoy eating a meal rather than feeling like everyone is watching you. (Resident at Sigma)

> More privacy, most people can see what is going on. (Resident at Melba)

Public spaces also give residents the opportunity to invite a number of friends and visitors over who cannot be accommodated in the private space of an apartment. Thus, public spaces offer three distinct types of use: single use, collective/ shared use by residents, and individual use by residents with friends. Policies and rules may need to be in place to govern access and to allow residents to book a space for private use. However, it is difficult to negotiate priorities between exclusive use by individuals or groups since it depends on the social attitude of residents and group sizes.

> I usually wait until the other residents have finished because that provides me with the privacy that I need. (Resident at Alpha)

> Depends on the number of them, and again my mood. It is overwhelming at times meeting tons of new people. Though sometimes it is nice. Smaller groups are more aproachable. (Resident at Alpha)

Figure 5.2 Public Barbecue Space at Sigma, Credit: Authors

> Depends on the groups, kids can deter me as they tend to be a little annoying, large groups deter me as I feel they would be better to have the space for themselves, I haven't been disappointed with any people, generally when I want to use the areas it is reasonably quiet. (Resident at Melba)

> Depends on how many are there. More likely to wait until they have gone if it's busy. (Resident at Sigma)

Interaction between residents and their friends

One of Watters' (2003) findings about the social behaviour of urban tribes describes their apparent invisibility to external observers. Urban tribes, or similar social formations, do not appear as one coherent entity to the public. They are private networks that integrate seamlessly into the social fabric of urban life. Members of an urban tribe may not even be aware of their membership or of the extent of the network. The interactions between the nodes of these social networks take place in both physical and virtual spaces. They traverse cyberspace (email, instant messengers, mobile phones) and the 'third' space (cafés, parks, bars) with ease. However, in any case, interaction usually remains private and peer-to-peer, whether it is mediated online or direct face-to-face interaction. Hence, the preferred social spaces of urban tribes are private spaces (someone's home) or private places in public spaces (cafés, bars, internet). Even if groups of friends meet up in a large public space such as a night club or discothèque, they exchange SMS to form private clusters that gravitate towards each other through an invisible bond.

The design of public space needs to acknowledge and accommodate this behaviour. Yet, most public spaces are designed to cater more for a collective many-to-many than a private peer-to-peer form of interaction. Although the choice between private spaces and private places in public spaces depends on situational circumstances and personal choice, the public spaces of their apartment buildings are considered not to be desirable meeting places in any case for residents to meet and socialise with their friends and peers.

> I meet my friends a lot and it is usually away from [Alpha] probably [in a nearby park] or in the city. I don't like socialising at anyone's house even if it is my own house. (Resident at Alpha)

> I generally have more fun at home or at another person's home than at a café, pool etc. (Resident at Melba)

> I am far away from my established group of friends who are back home [...]. I used to see them daily at University, in the halls etc. I am slowly making new friends here. My flatmates and I hang out with a few others we've met. Usually the meeting place has been a restaurant or other such location. (Resident at Alpha)

> I meet people all over the place, home might be the stop before heading out, sometimes we stay at our place or head to our friends. In general there is no preference, but if we are home it is mostly in our unit not in the public areas. (Resident at Melba)

Interaction amongst residents

Although it is easier than ever before to communicate and interact with others, forms of urban alienation remain, and ironically, residents who are socially well-connected otherwise can live in an apartment for years without any interaction with their neighbours or even knowing who lives next to them. We believe that this situation is acceptable as long as it is due to personal choice and not due to a lack of opportunity for local engagement and participation.

Approaches towards neighbourhood development that try to provide such opportunities are mostly based on a utopian objective to try and establish a collective community spirit. They are afflicted with difficulties, because it is impossible to 'make everyone love everyone else'. Physical proximity does not ensure neighbourliness (Arnold *et al.* 2003; Foth 2006a). Hence, approaches to encourage and support interaction amongst residents have to be based on voluntary action and choice to cater for different lifestyles and social needs.

> It would be nice to know my neighbors. (Resident at Alpha)

> I'm not really interested in meeting others to any great extent. (Resident at Melba)

Nevertheless, no resident who participated in our study rejects the assumption that there are residents who share their interests or are at least socially compatible with whom they do not normally interact on a daily basis. If these residents could be easily identified, they may transgress the status of "neighbour" and become new acquaintances and maybe even friends. How can the residential architecture and design of public space stimulate, encourage and support social interaction and networking between residents? We suggest three pathways based on our study's empirical findings (Foth 2006a; 2006b; Foth & Hearn 2007) which we will discuss in turn.

Serendipity 'Bumping into someone' has been reported as the most common form of interaction between residents. These kinds of serendipitous encounters take place in the elevator, at the pool, in the car park, whilst taking out the garbage or walking the dogs. Yet, depending on individual personalities and social preferences, such concurrences may remain without consequence unless people already know each other.

> I feel people are generally sociable to all residents, they will generally say hi, but a more lengthy chat usually occurs between those groups that know each other. (Resident at Melba)

> Most people are reasonably friendly. It is hard to determine who is a resident and who is just visiting most of the time. Generally most people are reasonably friendly. I would say I would most likely chat to a familiar face rather than a new one unless it was obvious they were just moving in. (Resident at Sigma)

On the other hand, residents of a proactive nature may take the opportunity of repeat serendipitous encounters to get to know other residents and to explore possible new frontiers of their existing social networks on the basis of shared demographics or interests.

> Depends on my mood and their body language, if they look friendly such as smile at me and make eye contact... or if they avoid eye contact, you know they don't want to talk, but I am always up to meeting new people. (Resident at Alpha)

> Mostly everyone tries hard not to talk to each other unless they are constantly bumping into the same person and it becomes awkward not to talk. I have managed to become good friends with a once [Sigma] resident, just because we were similar ages, have similar interests and often ended up in the lift together and started chatting. (Resident at Sigma)

The design of public space in residential apartment buildings substantially influences the likelihood, frequency, and intensity of serendipitous encounters. The only public space at Melba where serendipitous encounters happen on a regular basis is the underground car park, however, informal chats are awkward because the environment is dark and uninviting, and residents usually rush between their car and the entrance to their staircase. The absence of paths and pedestrian walk ways in Sigma's site layout favours access by car and makes it difficult for residents to casually visit each other by foot. Alpha's common room on Level 3 has been equipped with board games, a ping-pong and a pool table, but the overall impression of this large and

Figure 5.3 Commons Room at Alpha, Credit: Authors

clinically white room is not very welcoming and conducive to socialise with other residents (Figure 5.3).

Socio-cultural Animation The public barbecue sites at Alpha and Melba have been successfully used in the past to invite all residents to get together for a community barbecue. Although not every resident shows up, most residents that attend such organised events welcome the opportunity to gain a better awareness of who lives in the complex and meet old friends and new acquaintances.

> It is easier to break the ice when someone else does it for you or it is less confrontational. (Resident at Sigma)

> Group meetings are a bit daunting especially when the people who usually attend these things all know each other. (Resident at Sigma)

> If people want to interact they can and it doesn't force those people who wish to go about their existence in the unit as they wish. Also add a bit of alcohol and people tend to loosen up a bit. (Resident at Melba)

These and other acts of socio-cultural animation (Foth 2006d) allow residents to take the initiative to organise collective action. They may take various forms from community barbecues, donation appeals or landscape rejuvenation programs to the establishment of residential community associations (Foth & Brereton 2004). The location and facilitation of such activities requires appropriate public spaces – both physical and virtual – that cater for mixed-use and that offer a heterogeneous fit-out to suit a variety of technical and social needs. Audience sizes change and it is essential that these spaces can be re-appropriated and re-purposed for different contexts and circumstances.

Digital Augmentation Residents at Alpha have broadband access to the internet through a local area network with Ethernet sockets in every bedroom. Most residents at Melba and Sigma have dial-up or broadband internet access at home. These favourable conditions allow residents to explore the potential to develop and install a community network system as a virtual outlet for social interaction to complement existing physical public spaces (Foth, 2006a; 2006b; Gaved & Foth 2006; Foth & Hearn 2007).

Theories of networked individualism and social networks do not only have an impact on the residential architecture of physical urban spaces, but also on the systems architecture of virtual urban spaces. These notions introduce the conceptual context for design scenarios and open up a new set of challenges to create ways to enable, enhance, augment or facilitate existing or emerging social networks between urban residents. Networked interaction for sociability in place describes the more private space occupied by a 'society of friendships', that is, social networks of friends who live within relative proximity to each other. The documentation and dissemination of current activities (e.g., through the use of text or multimedia messages), coupled with simultaneous coordination of the next event, culminates in a shift in the nature of communication itself. Unlike internet chat or even e-mail, the driving force behind the interaction is not about back-and-forth interaction with someone else. Mobile phone applications facilitate a more subtle form of interaction, where communication is mediated through the creation, circulation and consumption of virtual presence (Satchell 2006). However, proximity enables them to gather face-to-face and interact offline. They see each other primarily as 'friends who live closeby' and not as 'neighbours' (Foth 2006a). One of our key goals is thus to find appropriate means to afford residents a seamless, selective and voluntary pathway to transition from "neighbour" to "friend" and to link these new nodes with their existing social networks. This echoes Hornecker *et al.* who examine opportunity spaces where 'there is no urgent problem to be solved, but much potential to augment and enhance practice in new ways' (2006, 47). Neighbourhoods can be such opportunity spaces insofar as they provide residents with opportunities to communicate, interact and socialise with each other. Our analysis of the physical articulation of lived city spaces seeks to inform how new media and ICT can be designed to realise such opportunities and enhance established ways of communicating, interacting and socialising in urban places.

We are currently working on a range of initiatives which operate at the intersection of residential community engagement and digital augmentation of urban spaces. The following two examples demonstrate the cross-disciplinarity and broad appeal of digital augmentation initiatives in an urban context.

Targeting the specific domain of public inner-city places, we are developing a mobile system we call *CityFlocks* (Bilandzic & Foth 2007), which enables urban residents to leave digital annotations with ratings, recommendations or comments on any place or physical object in the city. Thus, *CityFlocks* turns residents into in-situ amateur journalists for visitors or other residents who have questions or need navigational aid related to any place in the city. Based on the outcome of previous studies, *CityFlocks* uses two different design alternatives, one following a direct, the other an indirect social navigation approach. We evaluate how these different design approaches influence the success of participants using a mobile system to socially navigate and find particular places at our case study site. Based on the results of the field study, we analyse how existing design principles for social navigation can be applied, combined and improved in the context of mobile systems to augment urban spaces and to harness the collective intelligence of urban residents towards an effective and efficient navigation tool. We hope the outcomes will provide valuable input to the design of future community driven, mobile information systems.

The *Social Patchwork* project (Klaebe & Foth 2006) explores the use of narrative and new media in community engagement and urban planning processes. The "History Lines" component is part of a suite of engagement tools under the Social Patchwork umbrella that seeks to illustrate residential history and migrational churn. It brings a cross section of new residents together to trace and map where they have lived in the course of their lives. When the longitude and latitude coordinates are collated and augmented with short personal narratives, overlapping and common lines become visible. The stories at these intersections in time and space stimulate interest and offer opportunities for further personalised networking. We see the Social Patchwork project as an experiment to test how urban computing can be used to augment a social network of storytelling, themed around community history and place making.

Conclusions

Good design in housing remains scarce, however innovations in the infrastructure of social space have emerged. The Dutch architect Herman Hertzberger has established principles in social residential projects that targets circulation spaces (staircase, landings and balcony corridors) as opportunities for incidental exchange. On enlarged stairway landings, seating is provided, a simple gesture that allows for resting on the assent to an apartment, a place to meet. Similarly, external corridors are articulated with protrusion outside apartment front doors that also encourage engagement through the opportunity to appropriate a balcony space, although part of the public domain is cared for as if private. For examples, see Hertzberger (2000) and Lüchinger (1987). These simple gestures inform how, with a dimension in design thinking beyond the mere functional minimum, the in-between spaces within a residential development can become more than just circulation.

To Mitchell (2003) designing flexible, permeable, informal public spaces is key in establishing a positive social space as demonstrated at Steven Holl's polemical Simmons Hall Undergraduate Residence, MIT campus, Cambridge, USA (Amelar 2003; Ryan 2004). Holl's philosophy of an architectural porosity enables the building to incorporate a cavernous series of volumes cutting through various stories. These vertical shafts are aligned with group lounges and study spaces. The network of spaces allows for a multiplicity of social events. The buildings plan is based on the traditional central corridor spine, however the departure from the conventional monotonous circulation system through the augmentation of public meeting spaces demonstrates a viable model for residential developments.

> The diagram of Simmons Hall, and its physical exploration, is as if Le Corbusier's economic section of stacked maisonettes for his Unités d'Habitation has mutated with surprisingly spatial, almost surreal incidental volumes. The student rooms, typically paired about small threshold spaces and shared bathrooms, are aligned between floor slabs to either side of the central corridor – a new sort of internal street – whereas the multi-height communal rooms punch through this straightjacket, morphing vertically – in the case of upper rooms – towards fantastical roof lights clear to the sky. (Ryan 2004, 37)
>
> These 'internal streets', inter-dispersed with places for social gathering, recall the earlier models of terraced housing and street corner public houses.

The fact that urban environments in the network society are characterised by fast-paced technological change and a swarming social behaviour of its inhabitants requires a cross-disciplinary exchange between urban sociology, computer science, architecture and urban design disciplines to inform urban planning and public policy making. Design considerations around privacy, exclusivity, permeability and flexibility have to be re-thought in a new light alongside traditional values of access, scale, scope, form and function. If the modern city is to become a dynamic conglomeration of livable "urban villages", a variety of network effects need to be investigated further. In the process of urban renewal, apartment buildings are becoming an essential component of the physical fabric of urban spaces. They provide an integral part of the environment inhabited by social networks. Their significance in the design and development of public spaces that become the new agora of urban dwellers opens up exciting opportunities for future research and innovation.

Acknowledgements

Dr Marcus Foth is the recipient of an Australian Postdoctoral Fellowship supported under the Australian Research Council's Discovery funding scheme (DP0663854). The authors would like to thank Dianne Smith, Robbie Spence, Fiorella De Cindio, Alessandro Aurigi, the organisers and participants of the Digital Cities 4 workshop at the 2nd International Conference on Communities and Technologies 2005, Milan, Italy, and the anonymous reviewers for valuable comments on earlier versions of this chapter.

References

Amelar, S. (2003), Steven Holl experiments with constructed "porosity" in his design for SIMMONS HALL, an undergraduate dorm set in the scientific realm of MIT. *Architectural Record, 191*(5), 204-215.

Arnold, M., Gibbs, M.R., and Wright, P. (2003), "Intranets and Local Community: 'Yes, an intranet is all very well, but do we still get free beer and a barbeque?'" In M. Huysman, E. Wenger & V. Wulf (Eds.), *Proceedings of the First International Conference on Communities and Technologies* (pp. 185-204). (Amsterdam: Kluwer Academic Publishers).

Australian Bureau of Statistics. (2004), *Year Book Australia: Population. Article: How many people live in Australia's remote areas?* (No. 1301.0). (Canberra, ACT: Australian Bureau of Statistics).

Baker, P.M.A., and Ward, A.C. (2002), Bridging Temporal and Spatial 'Gaps': The role of information and communication technologies in defining communities. *Information, Communication & Society, 5*(2), 207-224.

Bilandzic, M., and Foth, M. (2007, Sep 3-5), 'CityFlocks: A Mobile System for Social Navigation in Urban Public Places', paper presented at the Locative Media Summer Conference, University of Siegen, Germany.

Castells, M. (2001), 'Virtual Communities or Network Society?', in *The Internet Galaxy: Reflections on the Internet, Business, and Society* (pp. 116-136). (Oxford: Oxford University Press).

Castells, M. (2004), 'Space of Flows, Space of Places: Materials for a Theory of Urbanism in the Information Age', in S. Graham (Ed.), *The Cybercities Reader* (pp. 82-93). (London: Routledge).

Day, P. (2002), 'Designing Democratic Community Networks: Involving Communities through Civil Participation', in M. Tanabe, P. van den Besselaar &

T. Ishida (Eds.), *Digital Cities II: Second Kyoto Workshop on Digital Cities* (Vol. LNCS 2362, pp. 86-100). (Heidelberg, Germany: Springer).

De Villiers, P. (1997), 'New Urbanism: A critical review', *Australian Planner, 34*(1), 30-34.

DiMaggio, P., Hargittai, E., Neuman, W.R., and Robinson, J.P. (2001), 'Social Implications of the Internet', *Annual Review of Sociology, 27*, 307-336.

Fallows, D. (2004), *The Internet and Daily Life*. (Washington, DC: Pew Internet & American Life Project).

Florida, R.L. (2003), 'Cities and the Creative Class', *City and Community, 2*(1), 3-19.

Foth, M. (2006a), 'Analyzing the Factors Influencing the Successful Design and Uptake of Interactive Systems to Support Social Networks in Urban Neighborhoods', *International Journal of Technology and Human Interaction, 2*(2), 65-79.

Foth, M. (2006b), 'Facilitating Social Networking in Inner-City Neighborhoods', *IEEE Computer, 39*(9), 44-50. <http://eprints.qut.edu.au/archive/00004750/>

Foth, M. (2006c), 'Network Action Research', *Action Research, 4*(2), 205-226.

Foth, M. (2006d), 'Sociocultural Animation', in S. Marshall, W. Taylor & X. Yu (Eds.), *Encyclopedia of Developing Regional Communities with Information and Communication Technology* (pp. 640-645). (Hershey, PA: Idea Group Reference).

Foth, M., and Hearn, G. (2007), 'Networked Individualism of Urban Residents: Discovering the Communicative Ecology in Inner-City Apartment Complexes', *Information, Communication & Society, 10*(5).

Gaved, M.B., and Foth, M. (2006), 'More Than Wires, Pipes and Ducts: Some Lessons from Grassroots Initiated Networked Communities and Master-Planned Neighbourhoods' in R. Meersman, Z. Tari & P. Herrero (eds), *Proceedings OTM (OnTheMove) Workshops 2006* (Lecture Notes in Computer Science No. 4277, pp. 171-180) (Heidelberg, Germany: Springer).

Gleeson, B. (2004), 'Deprogramming Planning: Collaboration and Inclusion in New Urban Development' *Urban Policy and Research, 22*(3), 315-322.

Gleeson, B., Darbas, T., and Lawson, S. (2004), 'Governance, Sustainability and Recent Australian Metropolitan Strategies: A Socio-theoretic Analysis', *Urban Policy and Research, 22*(4), 345-366.

Graham, S. (ed.) (2004), *The Cybercities Reader* (London: Routledge).

Healy, E. & Birrell, B. (2004). *Housing and Community in the Compact City* (Positioning Paper). (Melbourne, VIC: Australian Housing and Urban Research Institute).

Hearn, G., and Foth, M. (2005), 'Action Research in the Design of New Media and ICT Systems', in K. Kwansah-Aidoo (Ed.), *Topical Issues in Communications and Media Research* (pp. 79-94). (New York, NY: Nova Science).

Hertzberger, H. (2000), *Space and the Architect: Lessons in Architecture 2* (J. Kirkpatrick, Trans.). (Rotterdam: 010 Publishers).

Hornecker, E., Halloran, J., Fitzpatrick, G., Weal, M., Millard, D., Michaelides, D., et al. (2006), 'UbiComp in Opportunity Spaces: Challenges for Participatory

Design', paper presented at the Participatory Design Conference (PDC), Trento, Italy.

Horrigan, J.B. (2001), *Cities Online: Urban Development and the Internet* (Washington, DC: Pew Internet & American Life Project).

Klaebe, H. and Foth, M. (2006), 'Capturing Community Memory with Oral History and New Media: The Sharing Stories Project', paper presented at the 3rd international Community Informatics Research Network (CIRN) Conference, Prato, Italy.

Lovink, G. (2005), 'The Importance of Going Public', in S. Lehmann (Ed.), *Absolutely Public. Crossover: Art and Architecture* (pp. 46-48). (Melbourne: Images Publishing).

Lüchinger, A. (ed.). (1987), *Herman Hertzberger: Bauten und Projekte [Buildings and Projects] 1959-1986*. (Den Haag, NL: Arch-Edition).

Mitchell, W.J. (2003), *Me++: The Cyborg Self and the Networked City* (Cambridge, MA: MIT Press).

Oldenburg, R. (2001), *Celebrating the Third Place* (New York: Marlowe & Co)

Putnam, R.D. (2000), *Bowling Alone: the Collapse and Revival of American Community* (New York: Simon & Schuster).

Queensland Government (2005), *South East Queensland Regional Plan 2005 - 2026*. (Brisbane, QLD: Office of Urban Management, Department of Local Government, Planning, Sport and Recreation).

Randolph, B. (2004), 'The Changing Australian City: New Patterns, New Policies and New Research Needs' *Urban Policy and Research, 22*(4), 481-493.

Robertson, R. (1995), 'Glocalization: Time-Space and Homogeneity-Heterogeneity', in M. Featherstone, S. Lash & R. Robertson (Eds.), *Global Modernities* (pp. 25-44). (London: Sage).

Ryan, R. (2004), 'Kinetic Monolith', *The Architectural Review, 215*(1283), 36-41.

Satchell, C. (2006), 'Contextualising Mobile Presence with Digital Images', paper presented at the 2nd International Workshop on Pervasive Image Capturing and Sharing. (UbiComp, Orange County, CA).

Soukup, C. (2006), 'Computer-mediated communication as a virtual third place: building Oldenburg's great good places on the world wide web', *New Media & Society, 8*(3), 421-440.

Walmsley, D.J. (2000), 'Community, Place and Cyberspace', *Australian Geographer, 31*(1), 5-19.

Watters, E. (2003), 'How Tribes Connect A City', in *Urban Tribes: Are Friends the New Family?* (pp. 95-118). (London: Bloomsbury).

Wellman, B. (2001), 'Physical Place and Cyberplace: The Rise of Personalized Networking', *International Journal of Urban and Regional Research, 25*(2), 227-252.

Wellman, B. (2002), 'Little Boxes, Glocalization, and Networked Individualism', in M. Tanabe, P. van den Besselaar & T. Ishida (Eds.), *Digital Cities II: Second Kyoto Workshop on Digital Cities* (Vol. LNCS 2362, pp. 10-25). (Heidelberg, Germany: Springer).

Ziller, A. (2004), 'The Community is Not a Place and Why it Matters - Case Study: Green Square', *Urban Policy and Research, 22*(4), 465-479.

Chapter 6

Towards Spatial Protocol: The Topologies of the Pervasive Surveillance Society

David Murakami Wood

Introduction

In the early years of the Twenty-first Century, a potentially fundamental transformation of the relationships between humans and technology has begun. A conference for technology developers in 2007 expressed this transformation most clearly: 'We are evolving towards an age of convergence in identification technologies where everything that can compute has an IP address, every thing static has an RFID and every individual has a biometric identifier'.[1] A society of pervasive computing is a pervasive surveillance society because it must 'give instantaneous access to any "thing", including tools, books, and people, transforming them into *surveillable things*' (Arraya 1995, 233).

The social sciences are still only just beginning to come to terms with this pervasive surveillance society, and all need to think more deeply about the recombination of the material, virtual, social and spatial. This has been expressed most clearly in the work of actor-network theorists (Latour 2005). Surveillance studies have not neglected the technological, although much of this engagement has been naturalistic if not techno-centric. However the key concept of the 'new surveillance' (Marx 2002) is a mode of ordering (Law 1992): a triple story of new technologies, digitisation of bureaucracy and new thinking about society and risk. It tells a story: the embedding of a particular politics within sociotechnical systems of sorting and categorisation that re-inscribe those categories back on society in increasingly less socially negotiable ways (Lianos 2001; Ball 2002; Norris 2003). This ordering has been thought of in primarily social terms, but across several disciplines some scholars have argued that it is important that this is also a spatial ordering (Curry 1996; Graham 1998; Thrift and French 2002; Graham and Wood 2003; Burrows and Ellison 2004; Donaldson and Wood 2004; Burrows and Gane 2006; Murakami Wood and Graham 2006). This has also been recognised by some in the field of urban design like Krogh and Gronback (2001) and in particular Dana Cuff (2003) who argues that these systems: 'challenge some of our fundamental ideas about the subjectivity, visibility, space, and the distinction

---

1 Call for Papers, 5th IEEE Workshop on Automatic Identification Advanced Technologies, Alghero, Italy, 2007.

between public and private. Together these challenges reformulate our conception of the civic realm' (p.43).

In geography, in cultural studies and media theory, and particularly within the area known as digital criticism, some particularly instructive developments have been occurring. A special issues of *Cultural Studies* in 2004 laid out some of the terrain, and Anne Galloway (2004) argued that pervasive computing 'can be seen to problematize our understandings of spatialization, temporalization and to varying extents, embodiment' (p.397). Amongst digital critics, writers like Geert Lovinck (2002) and Alex Galloway (2004) have seen that the new recombination of the virtual and the digital might mean not so much the transformation of the virtual by the physical but in fact a new kind of spatiality entirely.

This chapter will therefore examine the emerging spatiality of a society increasingly reliant on technologically-mediated forms of surveillance: a pervasive surveillance society. In doing so it ties Geography, Media Theory and Surveillance Studies more closely together. It argues that the ubiquity but invisibility of digital technologies produced whole sets of hidden geographies at scales which are at once more minute but more dispersed, and need to be exposed. This can be seen as part of a movement away from the consideration of space as a topographical phenomenon to space as topological (Mol and Law 1994).

Space is always in process but it is not only worth consideration because it contains the potential to transform into something else: there are 'rules' of process, apparatus, or modes of ordering. This piece argues that transformations necessitate what Foucault called a new "diagram", and the discovery of new "rules" by which a pervasive surveillance society operates. I draw on the recent work by digital critic Alex Galloway (2004), who has argued that such a diagram can be found in the examination of the ways in which distributed computer communication architectures function, and the "rules" (or what he calls "management style") embedded in "protocol". As such architectures increasingly interact and merge with conventional spaces to produce the new topologies in the pervasive surveillance society, I argue that we can see the emergence of *spatial protocols*. These are highly restrictive and controlling rules embedded within the materiality of urban space, which produce all kinds of new liberatory and repressive possibilities. The key lesson is that it is in the production of these spatial protocols that the politics of a pervasive surveillance society may be found.

The Path to the Pervasive Surveillance Society

Foucault (1977) argued that in order for the modern subject to be produced, a kind of productive ordering was necessary: the classification and arrangement of all kinds of properties and entities to maximise their usefulness. Classification and ordering therefore underlies almost everything about modernity (Law 1992). Foucault considered the spatial evolution of modernity through various "diagrams of power": the military camp; the normalizing judgement found within institutions such as schools and factories; and most famously, Bentham's plans for the Panopticon. However he argued further that in the late Nineteenth Century, new kinds of diagrams emerged with the rise of psychology, psychiatry and scientific medicine and 'the steep rise in the use of these mechanisms of normalisation and the wide-ranging powers which, through the proliferation of new disciplines, they bring with them' (Foucault 1977, 306; see also: Murakami Wood, 2007).

Gilles Deleuze (1992 [1990]) extended this temporal trajectory through his brief statement on "societies of control". Whereas discipline is analogical and moulding, control is digital and modulating. The control diagram is the "code", the numerical language that marks access to information. Instead of dealing with the mass/individual dualism, individuals have become "dividuals", and masses "banks" of data. These dividuals exist both as the physical body of the modern subject and as multiple subjects in databases ("data subjects") which are now often considered more important in terms of social identity than bodily selves (van der Ploeg 1999; 2002; Lyon 2001; Graham and Wood 2003). Thus we now live in a 'surveillance society' (Lyon 1994) under 'digital rule' (Jones 2001).

Deleuze and Guattari (1987) advocated understanding the post-modern society as one of 'rhizomes', networks which send up shoots from anywhere, and 'assemblages', heterogeneous objects brought together and working as an entity, giving greater if temporary permanence to flows. Haggerty and Ericson (2000) adapted this in the concept of the 'surveillant assemblage', claiming that in late modernity 'we are witnessing a rhizomatic levelling of the hierarchy of surveillance, such that groups which were previously exempt from routine surveillance are now increasingly being monitored' (p.606). Hinting at some of the issues around pervasive surveillance, they claim that this results in the progressive 'disappearance of disappearance', with the anonymity previously afforded by the city increasingly difficult to keep. Latour (2005) has described the current order as oligoptic, that is made up of multiple surveillant actants with very detailed specific knowledge of very confined areas. Our new diagram does not have to be a physical space, and indeed this chapter argues that in a world where physical and virtual are recombined, it is actually a translation of virtual spatiality.

The recombination of physical and virtual space is achieved through pervasive computing. Pervasive Computing is essentially the ubiquitous distribution and "vanishing" of digital computers (Weiser 1991) into the background of everyday infrastructure, objects and even living things (including the human body). It encompasses wearable devices, distributed systems and all kinds of "context-aware" computers. Within the computing community it is most often referred to as Ubiquitous Computing (or UbiComp), a term generally credited to Mark Weiser of Xerox PARC laboratories, from the late 1980s, although his Scientific American piece of 1991 is titled Pervasive Computing's 'year zero'. Other terminologies also exist: the European Union's Information Society Technology Advisory Group (ISTAG) refers to Ambient Computing or Ambient Intelligence (AmI) (EU 1999), terms adapted from the (European) Philips corporation, which have a more comforting feel of a warm bath than the somewhat more insidious connotations of pervasiveness. Ambient Intelligence is however also an attempt to be explicitly comprehensive and includes UbiComp as well as Ubiquitous Communications (UbiCom) and User Adaptive Interfaces (UAI).

Both hardware and software have complex histories for which there is no room to do justice to here (see: Chabert *et al.* 1999; Stephenson 1999; Ifrah 2000) nor the social history of pervasive computing itself (for a summary see: Anne Galloway 2004; for an accessible introduction to the wider subject, see: Waldrop 2003). There are also other illustrative descriptions of the technical processes, for example Bohn *et al.* (2004). They involve the decreasing size and increasing power of computing, the ability of systems to be distributed, and even mobile, but still connected, locatable, verifiable and addressable in space, eventually sinking into the infrastructure.

However, perhaps the key development and the one most neglected by social scientists is the "development of standard protocols". In order to enable computers to communicate, standard protocols are required. The standardization of TCP/IP, and internet languages like XML, sound and video like MPEG are essential for distributed networks to function. The development of these standards is extremely complex and involves all sorts of wider politics. Indeed occasionally it is in debates of these sorts that the new politics of technological standards occasionally emerges into the public consciousness, although at a very high level. For example this can be noticed in recent debates between the USA and the UN about the governance of the Internet and particular the governance of domain name assignment. It is in the development of these protocols that a new "system of the world" is evolving, what Alex Galloway (2004) defines as 'a distributed management system that allows control to exist within a heterogeneous material milieu' (p.8). However, social scientist rarely engage with the discussion of these protocols, even if they have got to grips with the idea of "code" or "coding".

Surveillance Studies as Critique of Pervasive Computing

It appears that out of socio-technical construction something has emerged that potentially no longer needs the social from which it derived its creation, despite the fact that it may impact on, and interact with, the social itself. In this context, Michalis Lianos has been attempting to develop a theory of a new post-Foucauldian institutional social control, a control that happens almost incidentally and unintentionally because of the requirements of institutional efficiency (Lianos 2001, 2003). This takes place through what he calls 'Automated Socio-Technical Environments' (ASTEs), spaces in which some, if not all decisions on actions taken by humans are pre-determined by particular technologies, which simplify what were once opportunities for rich and complex social negotiations into binary choices, a claim which has strong resonances with Thrift and French (2002).

In surveillance studies, the digital has been in many ways the key to study of the 'new surveillance' (Marx 2002). Databases were the subject of early work in the field by James Rule (1973) and others, but it was Roger Clarke who invented the term 'dataveillance' in 1988. Oscar Gandy (1993) attempted to apply Foucauldian conceptualisations to dataveillance in his consideration of the 'Panoptic Sort,' which was adapted by Lyon (2001) to talk about social sorting. Dana Cuff (2003), Graham and Wood (2003) and Graham (2005) have attempted to ground this further with the latter talking of 'software-sorted geographies.'

Both Dana Cuff (2003) and Anne Galloway (2004) are right to identify Weiser's discourse which privileges the idea of the computer becoming more human, as utopian. Although perhaps it is more appropriate to identify this utopianism as a heroic phase of enthused discovery within the development process: one can also see this reflected in the transhumanist or posthumanist movements and the idea of the biological-technological 'singularity' (see e.g.: Kurzweil 2006). However one should not see the more dystopian visions expressed here as simple a technological determinist vision of "the machines taking over", because the machines are simply the visible part of a collectif that for the most part is "black-boxed" or punctualized out of the view of ordinary social interaction and politics (Latour 1999). As with most of these processes, so long as they function they are not seen: failure or the threat of failure is when they are forced to become visible. Cuff argues that pervasive computing 'can be both everywhere and nowhere... that it acts intelligently yet fallibly; and its failure is complex... and that intelligent systems operate spatially, yet they are invisible' (2003, 43). Thrift and French (2003) use the example of the Y2K "crisis" as an example. Thousands of computer-dependent processes were suddenly "discovered" and for many of the processes governing their maintenance were highly obscure.

However there remains a sense in which we are now stuck. Deleuze's (1995) suggestion of code as the new diagram of power has attracted much attention from the humanities and social sciences, however in a great deal of the writing, code appears as just a kind of metaphor, something akin to poetry, or free floating items which infiltrate themselves into daily life through a rather unspecific set of processes, with the exception of writers already considered. And of course, philosophers of technology have developed writing on the ethics of the technologies on which they are working, particularly on the question of privacy. In a very important but neglected piece back in 1995, Agustin Arraya identified many of these problems with the pervasive computing.

Firstly, he argued that there was a loss of 'otherness' of things. This was vital, he argued, because it prevented the possibility of loss and made a society of pervasive computing devices, a society or pervasive surveillance:

> When the surveillance mechanism fades into the background and we are no longer able to experience it, things in general – not just this manual or that tool or those employees – would have been transformed becoming for us surveillable things, whether we effectively subject them to surveillance or not. A fundamental category that governs our dealings with the world would have been deeply altered. (Arraya 1995, 234)

The next transformation is that our environments become responsive and even anticipatory. In early utopian scenarios this was often portrayed as enabling. However, Graham has warned of the militarized possibilities associated with anticipatory surveillance. More generally however, according to Arraya, this transforms the world into an artefact. When this world can be simulated, reproduced and carried around in some form, this process reaches its peak (c.f.: Graham 1998). Arraya argues that pervasive computing is essentially 'an attempt at obliterating the otherness of the world' but which happens 'in such a way that we are no longer aware of the obliteration' (Arraya 1995, 235). The point that a society of pervasive networks allows nothing to be forgotten has been taken up since (Blanchette and Johnson 2004; Dogde and Kitchen 2005; Bannon, 2006), but actually code is being overwritten and obscured all the time. The dangers of being without an "archive" of coded objects, without any knowledge of origins, addressability, alteration and repairability, seem equally dangerous (Thrift and Graham 2007), especially faced with the consequences of failed heuristic processes – of objects making judgements about the living.

Arraya also produced a strong critique of the place of the human within pervasive computing scenarios, arguing that 'what is striking about most of these scenarios is the *marginal* and irrelevant character of the needs referred to in them and of the envisaged enhancements of the activities'. In a sense Lianos (2001) answered this point in defining the institutional sociality, the rules of flow as privileged. And it does seem clear that in the early Twenty-first Century, many developers appear to be pragmatists enrolled into the programs of action of wider corporate capitalism, who refuse moral or political engagement. An explicit example is the attempt by Fleisch (2004) to identify opportunities and risks to business form pervasive computing. He argues that inefficiency is the main problem for business and that this is predominantly due to a lack of integration between real and virtual, caused by "media breaks", that is the translation process between software, paper and so on. Pervasive computing could remedy this because '[it] has the potential to reduce the cost of integrating the physical world with information systems' (p.818). Some of his identified risks look very familiar to social scientists who study technology, but problems like 'technocratic approach' are quickly dismissed, and 'dependency on technology' is to be solved by 'self-checking routines' and 'interfaces to employees', where the technology remains the centre and the answer is always better technology. Even when he argues, quite provocatively, that 'Pervasive Computing enhances the communication capabilities of objects. It thus can evoke a new dimension of emotions in people confronted with these' (p.828), his answer is not like Nigel Thrift (2004) to explore such emotional transformations, but simply to propose 'special advertising,' to overcome or take advantage of them. His stance assumes an entirely accepted rational choice explanation for human social behaviour and that any resistance to the plans of technological developers is simply a matter of cost and convenience: 'We have learned from history that consumers and employees always opt for convenience at the right price. Thus whenever a technology makes life a little easier people are likely to adopt it' (p.827).

The case of RFID

To take the argument forward it is worth considering one example in a little more depth. There are many one could chose, from the massive increase in geodemographic and geolocational systems (Burrows and Ellison 2004) to the artistic, sensory and enabling environments introduced by Anne Galloway (2004). I will consider Radio Frequency Identification (RFID) tags.

There are two types of RFID chips. Passive tags respond only to an active instrument or reader of some kind. Active RFID chips emit a continuous or periodic limited range radio signal that can be picked up by receivers. Until recently the use of RFID has been restricted to large shipping containers (ports being a major area of vulnerability to smuggling, illegal immigration and terrorist attack, but very difficult to police effectively by traditional methods), as well as consumer goods. Electronic tags have also become increasingly common in the UK and USA for those with judicial restrictions that tie them to a particular area (sometimes as limited as an individual house) for all or some hours of the day. These communicate either constantly or at regular intervals with a receiver, increasingly satellite Global Positioning Systems (GPS) rather than simple radio monitoring (Nellis 2005). However, recently, a notable change has occurred: the implantation of RFID chips into living beings. Race horses and pets were the first groups to be targeted in this way. For pets, RFID chips containing information about immunisation records and ownership have gradually replaced quarantine requirements in the EU since February 2000 through the PETS scheme, which has since been extended beyond Europe.[2]

For humans, the first use of RFID implants has been in elderly people suffering from degenerative diseases in the United States, and around 70 people have now been implanted to enable carers to locate them easily and prevent them from wandering and possibly endangering themselves.[3] Researchers and technological enthusiasts have also been implanting themselves with chips for several years now in order to be able to automatically perform small household tasks (turning lights on and off etc.)[4] At least one chain of Spanish nightclubs has offered patrons the chance to have cash and access privileges held on implanted chips (Graham-Rowe 2004).

A further step-change in RFID application occurred in February 2006 when a security company in Ohio, USA, implanted two of its workers with RFID chips to allow them to access company property (Waters 2006). Although such an invasive procedure was carried out voluntarily, it raises enormous questions about the integrity of the body and privacy in relation to employers. It is also not entirely surprising that the call for everyone to be implanted is now being seriously debated on some technology websites.

The influential ZTH Zurich group argue that RFID tags 'enable the implantation of a wide range of novel ubicomp applications by bridging the gap between the physical world (i.e., tagged real-world objects) and the virtual world (i.e.: application software or service infrastructure)' (Röhmer *et al* 2004, 689). However, the developers' argument ultimately boils down, like the more overtly economic example referenced earlier, to 'mass customizing' (p.689) and supply-chain management. Some of the examples of useful technological applications also betray a fundamental misunderstanding of the way in which humans exist within societies. For instance,

---

2 For details, see the Department of Environment, Food and Rural Affairs (DEFRA) PETS website: <http://www.defra.gov.uk/animalh/quarantine/pets/index.htm> (accessed 31 July 2006).

3 The company involved is Verichip Corporation.<http://www.verichipcorp.com/> (accessed 31 July 2006).

4 Amal Graafstra is one such high profile enthusiast and advocate of self-chipping. Explanations, pictures and videos can be downloaded from his website <http://amal.net/rfid.html> (accessed 31 July 2006).

the notion of a 'smart chef' system where vegetables are tagged with links to recipes etc. but 'to implement this functionality, the cook is identified by an RFID tag with the form factor of a credit card carried in his or her wallet' (p.690), is ultimately a vision of society where "the system" demands a change in human behaviour.

In contrast, when developers were still in the utopian period of pervasive computing it could be argued that 'Social interactions are the focus of our existence. We are social animals and for any technology to be useful, it must eventually support socialization; otherwise it will not survive' (Ark and Selker 1999, 506). However it seems clear that the mainstream of technological development is thoroughly biased towards a neo-liberal economic agenda, where "social concerns" are either highly delimited in the domain of other technicians or expert regulators, like privacy (e.g.: Beresford and Stajano 2003) and data protection, health or environmental management (e.g.: Köhler and Erdmann 2004), or else are simply externalities of economically rational or technologically-driven consumer behaviour[5] and when this begins to "demand" the compliant implantation of human beings one has to be highly concerned. Indeed, Arraya (1995) describes this as 'an emerging form of technological absolutism' in which 'the primacy of the unfolding of technology over the satisfaction of human needs, and the self-sufficiency of this unfolding are taken as *absolute givens*' (p.236).

Towards Spatial Protocol

Dana Cuff argues that in a society pervaded by computing, 'public life is spatially located but also displaced and dispersed, requiring new logics and new physical forms' (Cuff 2003, 48). Alex Galloway argues for the distributed network of the Internet as a more accurate diagram with 'protocol' as its system of rules. The Internet is the seemingly contradictory combination of enabling restrictions, utter vertical control but total horizontal liberty that increasingly define our societies and what we are. Protocol within this structure constitutes the negotiated and agreed standards that make up the vertical control: it shapes the boundaries.

For information, barriers and boundaries to mobility and flow are as important as flows. The geography of the pervasive surveillance society is a dynamic and contested process of the spatial control, or territoriality, of spaces that are at once both physical and virtual. For all those entities that have been made surveillable, and whose politics appears absent or vanished along with the machines that make them surveillable, there is still territoriality, but it relates not just to the "outside", the physical public space, but also to the "inside", the virtual space of databases and networks.

They are perhaps better understood as topologies rather than spaces in the usual topographic sense (Mol and Law 1994). For Mol and Law, their use of the term topology derived from the mathematical definition: topology is the study of properties of objects which are preserved through deformations; it is the study of

---

5 Indeed one finds articles written by those in field with title that imply a concern with 'social implications' which are almost entirely devoid of anything social (see, e.g.: Raisinghani *et al.*, 2004).

spatial objects (Weisstein nd.). However topology has another relevant meaning here: in computer technology, it refers to the physical pattern of connectivity within a computer network or between processors, memories, and peripherals.[6]

This patterning is determined by protocols which can make up both physical and informational barriers. However there are also more subtle forms of boundary-marking, which are vital to discovering the increasingly spatial nature of protocol and the location of the politics of the pervasive surveillance society. These boundaries are largely socio-cultural, produced by networks formed through learning and work-based activity, or 'epistemic communities'. Star (1995) shows that it makes sense to portray computer-based social groups as communities of practice, both in terms of the production of situated knowledge and the technologies themselves (the engineers); and the experience of the those systems (the users). In the case of the pervasive surveillance society, the direct users are other computer systems (though of course the ultimate although indirect users are human); our attention is drawn immediately to the producers of the systems. These producers are often software engineers, or research groups that include software, hardware and network engineers.

The politics of pervasive surveillance is really only regularly played out in these producer groups. For those who commission and buy digital surveillance systems, the necessity of such things remains unquestioned. However debates do take place within the knowledge-based communities of programmers, a new priesthood for the digital age whose access to the arcane world of code and algorithms provides them with the position from which to speak about the technologies they produce. These collectives are relatively opaque to outsiders. Their arguments do not spill out much beyond particular internet webzines or books only likely to be read by the already computer literate. In territorial terms, there are some physical barriers (lack of access to appropriate computer hardware), but the boundaries are created more by assumptions of knowledge, interests and other practice-based qualifications. Thus while the programming community may be committed to freedom of information, the methods by which they distribute information are intensely circumscribed.

In the case of surveillance, the arguments are also there because the digital surveillance world exists on the edge of several others: a curious mixture of 'geeks and guards' – computer programmers and security personnel often drawn from ex-police, intelligence or military backgrounds. It has been often noted that there is a prevailing 'default' libertarianism within the former grouping (Davis 1999) whereas a clear concern with secrecy and information control in the latter. Hence the debates about surveillance tend to be either about crime, risk and danger (for the latter), or about liberty and privacy (for the former).

But a world governed by spatial protocol does not necessarily mean a world of control for control's sake in the manner of the more excitable accounts of surveillance. It also does not necessarily mean the world of absolute institutional control suggested by Lianos, nor of automated spatial production as Thrift and French posit. However as Alex Galloway suggests, 'the limits of a protological system and the limits of *possibility* within that system are synonymous' (p.52). There is nothing

6 Harcourt Academic Press Dictionary of Science and Technology, <http://www.harcourt.com/dictionary/browse/19/>.

possible outside protocol. If one is to see distributed and pervasive computing as the diagram of power for a post-modern digital subjectivity, one cannot simply regard this as coterminous with security – in other words, as dystopian. Protocols are simply reflective of the agreements that go up to make them, with all their arguments, discussions, assumptions and values. In movements like the open-source software one finds suggestions for protocol that are far from restrictive. Open-source, collaborative socio-spatial design could likewise create spatial protocols that actually open up better forms of civility and interaction and new domains of possibility for a wider range of citizens.

References

Ark, W.S. and T. Selker (1999), 'A Look at Human Interaction with Pervasive Computers', *IBM Systems Journal*, 38:4, 504-507.

Arraya, A. (1995), 'Questioning Ubiquitous Computing', *Proceedings of the 1995 ACM, 23rd annual conference on Computer science,* 230-237.

Ball, K. (2002), 'Elements of surveillance: a new framework and future directions', *Information Communication and Society*, 5:4, 573-590.

Bannon, L.J. (2006), 'Forgetting as a Feature not a Bug: The Duality of Memory and Implications for Ubiquitous Computing', *CoDesign* 2:1, 3-15.

Barry, A. (2001), *Political Machines: Governing a Technological Society* (London: Athlone).

Beresford, A.R. and Stajano, F. (2003), 'Location privacy in pervasive computing', *IEEE Pervasive Computing* 2:1, 46- 55.

Bohn, J., Coroamă, V. Langheinrich, M., Mattern, F. and Rohs, M. (2004), 'Living in a World of Smart Objects: Social, Economic and Ethical Implications', *Human and Ecological Risk Assessment* 10, 763-785.

Bowker, G. and Star, S.L. (1999), *Sorting Things Out: Classification and its Consequences* (Cambridge MA: MIT Press).

Brown, J.S. and Duguid, P. (2000), *The Social Life of Information* (Boston MA: HBR Books).

Burrows, R. and Ellison, N. (2004), 'Sorting Places Out: Towards a social politics of neighbourhood informatisation', *Information, Communication and Society* 7:3, 321-336.

Burrows, R. and Gane, N. (2006), 'Geodemographics, Software and Class', *Sociology* 40:5, 773-791.

Callon. M. (1991), Techno-economic networks and irreversibility, in Law, J. (ed.)

Chabert, J. (ed.) (1999), *A History of Algorithms: From the Pebble to the Microchip* (Berlin: Springer-Verlag).

Crampton, J. and Elden, S. (eds.) *Space, Knowledge and Power: Foucault and Geography* (Aldershot: Ashgate).

Cuff, D. (2003), 'Immanent Domain: Pervasive Computing and the Public Realm', *Journal of Architectural Education*, 57:1, 43-49.

Davis, E. (1999), *TechGnosis: Myth, Magic and Mysticism in the Age of Information* (London: Serpent's Tail).

Deleuze, G. (1992), 'Postscript on the societies of control', *October* 59, 3-7.

Dodge, M. and Kitchen, R. (2005), 'Codes of Life: identification codes and the machine-readable world', *Environment and Planning D: Society and Space*, 23, 851-881.

Donaldson, A. and D. Wood (2004), 'Surveilling Strange Materialities: categorization in the evolving geographies of FMD biosecurity in the UK', *Environment and Planning D: Society and Space*, 22:3, 373-391.

Elmer, G. (2003), 'A diagram of panoptic surveillance', *New Media and Society* 5:2, 231-247.

Fleisch, E. (2004), 'Business Impact of Pervasive Technologies: Opportunities and Risks', *Human and Ecological Risk Assessment*, 10, 817-829.

Foucault, M. (1977), *Discipline and Punish: the Birth of the Prison* (Harmondsworth: Penguin).

Foucault, M. (1990), *The Care of the Self (The History of Sexuality Vol. 3)* (Harmondsworth: Penguin).

Galloway, Alexander (2004), *Protocol: How Control Exists after Decentralization* (Cambridge MA: MIT Press).

Galloway, Anne (2004), 'Imitations of Everyday Life: Ubiquitous Computing and the City', *Cultural Studies* 18:2/3, 384-408.

Gandy, O.H. Jr. (1993), *The Panoptic Sort: A Political Economy of Personal Information* (Boulder CO: Westview Press).

Graham, S. (1998), 'Spaces of Surveillant Stimulation: new technologies, digital representations, and material geographies', *Environment and Planning D: Society and Space* 16, 483-504.

Graham, S. and D. Wood (2003), 'Digitising surveillance: categorisation, space, inequality', *Critical Social Policy*, 23:2, 227-248.

Graham-Rowe, D. (2004), 'Clubbers chose chip implants to jump queues', *New Scientist* (published online 21 May 2004) <http://www.newscientist.com/article.ns?id=dn5022> accessed 31 July 2006.

Haggerty, K. and R. Ericson (2000), 'The surveillant assemblage', *British Journal of Sociology*, 51:4, 605-622.

Harel, D. (1992), *Algorithmics: the Spirit of Computing* (Reading MA: Addison-Wesley).

Hucklesby, A. and Mair, G. (2005), *Issues in Community and Criminal Justice – Monograph 5* (London: National Association of Probation Officers).

Ifrah, G. (2000), *The Computer and the Information Revolution (The Universal History of Numbers Vol.3)* (London: Harvill).

Islam, N. and Fayad, M. (2003), 'Towards Ubiquitous Acceptance of Ubiquitous Computing', *Communications of the ACM*, 46:2, 89-91.

Jones, R. (2000), 'Digital Rule: Punishment, Control and Technology' in *Punishment and Society*, 2:1, 23-39.

Knorr-Cetina, K. (1999), *Epistemic Cultures: how the Sciences make Knowledge* (Cambridge MA: Harvard University Press).

Köhler, A. and Erdmann, L. (2004), 'Expected Environmental Impacts of Pervasive Computing', *Human and Ecological Risk Assessment* 10, 831-852.

Krogh, P.G. and Gronback, K. (2001), 'Architecture and Pervasive Computing: when Buildings and Design Objects Become Computer Interfaces', *Nordiske Arkitekturforskning*, 3, 1-10.

Kurzweil, R. (2006), *The Singularity is Near: When Humans Transcend Biology* (London: Penguin).

Latour, B. (1991), 'Technology is society made durable', in Law, J. (ed.).

Latour, B. (1999), *Pandora's Hope: Essays in the Reality of Science Studies* (Cambridge MA: Harvard University Press).

Latour, B. (2005), *Reassembling the Social: An Introduction to Actor-Network-Theory* (Oxford: Oxford University Press).

Law, J. (ed.) (1991), *A Sociology of Monsters: Essays on Power, Technology and Domination* (London: Routledge).

Lessig, L. (1999), *Code – and Other Laws of Cyberspace* (New York: Basic Books).

Lianos, M. (2001), *Le Nouveau Contrôle Social: toile institutionnelle, normativité et lien social* (Paris : L'Harmattan-Logiques Sociales).

Lianos, M. (2003), 'After Foucault', *Surveillance & Society* 1:3, 412-430.

Lyon, D. (1994), *The Electronic Eye: the Rise of the Surveillance Society* (Cambridge: Polity Press / Blackwell).

Lyon, D. (2001), *Surveillance Society: Monitoring Everyday Life* (Buckingham: Open University Press).

Lyon, D. (2003), *Surveillance after September 11th* (Cambridge: Polity Press).

Lyon, D. (ed.) (2003), *Surveillance as Social Sorting: Privacy, Risk and Automated Discrimination* (London: Routledge).

Marx, G.T. (2002), 'What's new about the 'new surveillance'? Classifying for change and continuity', *Surveillance & Society* 1:1, 9-29.

Mol, A. and Law, J. (1994) 'Regions, Networks and Fluids: Anaemia and Social Topology', *Social Studies of Science*, 24: 641-71.

Murakami Wood, D. (2007), 'Beyond the Panopticon? Foucault and Surveillance Studies', in: Crampton, J. and Elden, S. (eds.).

Murakami Wood, D and Graham, S. (2006), 'Permeable Boundaries in the Software-sorted Society: Surveillance and the Differentiation of Mobility', in Sheller, M. and Urry, J. (eds).

Nellis, M. (2005), 'The electronic monitoring of offenders in England and Wales: a critical overview,' in Hucklesby, A. and Mair, G. (eds.).

Norris, C. (2003), 'From personal to digital: CCTV, the Panopticon and the technological mediation of suspicion and social control', in Lyon, D. (ed.).

Norris, C. and G. Armstrong (1999), *The Maximum Surveillance Society: The Rise of CCTV* (Oxford: Berg).

Norris, C., J. Moran, and G. Armstrong (eds.) (1998), *Surveillance, Closed Circuit Television and Social Control* (Aldershot: Ashgate).

Raisinghani, M.S., Benoit, A., Ding, J., Gomez, M., Gupta, K., Gusila, V., Power, D. and Schmedding, O. (2004), 'Ambient Intelligence: Changing Forms of Human-Computer Interaction and their Social Implications', *Journal of Digital Information* 5:4, Article No.271 (no pagination) (published online 24 August

2004) <http://jodi.tamu.edu/Articles/v05/i04/Raisinghani/> accessed 31 July 2006.

Röhmer, K., Schoch T., Friedemann M. and Dübendorfer T. (2004) "Smart Identification Frameworks for Ubiquitous Computing Applications", *Wireless Networks* 10(6): 689-700.

Rule, J.B. (1973), *Private Lives, Public Surveillance: Social Control in the Information Age* (London: Allen Lane).

Sheller, M. and Urry, J. (eds) (2006), *Mobile Technologies of the City* (London: Routledge).

Star, S.L. (1995), *The Cultures of Computing* (Oxford: Blackwell / The Sociological Review).

Stephenson, N. (1999), *In the Beginning was the Command Line* (New York: Avon Books).

Thompson, C. (2004), 'Everything is Alive', *IEEE Internet Computing*, Jan/Feb, 83-87.

Thrift, N. (2004), 'Electric Animals: New Models of Everyday Life, *Cultural Studies*, 18:3/4, 461-482.

Thrift, N. and S. French (2003), 'The Automatic Production of Space', *Transactions of the Institute of British Geography*,.

Thrift, N. and Graham, S. (2007), 'Out of Order: Understanding Repair and Maintenance', *Theory, Culture & Society*, 24:3, 1-25.

Torpey, J. (2000), *The Invention of the Passport: Surveillance. Citizenship and the State* (Cambridge: CUP).

Urry, J. (2000), *Sociology Beyond Societies: Mobilities for the Twenty-First Century* (London: Routledge).

Waldrop, M. (2003), 'Pervasive Computing. An overview of the concept and an exploration of the public policy implications', Paper for the Future of Computing Project, Foresight and Governance Project. (published online 2003) <http://www.thefutureofcomputing.org> accessed 31 July 2006.

Waters, R. (2006), 'US group implants electronic tags in workers', *Financial Times*, (published online 12 February 2006) <http://www.ft.com/cms/s/ec414700-9bf4-11da-8baa-0000779e2340.html> accessed 31 July 2006.

Weiser, M. (1991), 'The Computer for the 21st Century', *Scientific American*, 265:9, 66-75.

Weiser, M. (1993), 'Ubiquitous Computing', *IEEE Computer*, 26:10, 71-72.

Weisstein, E. (nd.), 'Topology', *Eric Weisstein's World of Mathematics*, (published online no date) <http://mathworld.wolfram.com/Topology.html> accessed 31 July 2006.

PART 2
Augmenting Communities

Fiorella De Cindio

The first section of the book has considered, from different perspectives, how digital technologies reshape urban *space* to augment it. But whom is augmented space for, and how is it used? The goal of the second section is – basically – to answer questions such as: what do people do with and within the augmented city? Does augmented space enrich the networks of local, social relationships which are the ultimate *raison d'etre* of the city, or does it annihilate them ?

The answer is not straightforward, and it is not even unique. What is clear is that the transformations induced by digital technologies challenge the very nature of the city. As Gumpert and Drucker recall from Goldberger (2001),[1] 'The urban impulse is an impulse toward community – an impulse toward being together, and toward accepting the idea that however different we may be, something unites us.' They note that 'cities provide for common interests and needs historically often associated with protection of inhabitants from threat and attack. Concentrating a population within walls was a defensive advantage providing security by keeping some in and keeping others out'. But digital technologies make these walls "permeable" and this creates the need to reconsider the notions of community and citizenship.

The impact over communities, rather than on the individuals, is therefore the key issue for understanding how people transform an *urbe* into a *civitas*. However, the term "community", in a digital context, evokes straight away the numerous virtual community experiences which have populated cyberspace. Nonetheless, although effective and quite popular as a consequence of Howard Reinghold's book (1993), it can be argued that the term "virtual community" can be misleading, as it suggests that communities which take place and develop online are not "real" compared with the communities developed through physical proximity. This is not at all true – see, for instance (Levy 1998) since the value and the strength of the binds which get created online quickly becomes itself part of everyday life for many people, hence something very "real". Moreover, very often we can observe an interplay between the online and the offline "worlds" that makes the two dimensions just aspects of a "whole" which manifests itself both physically and digitally. The idea of hybrid communities, as groups of 'people who interact together socially using both on-line and off-line methods of communication' – introduced by Gaved and Mulholland

1 Goldberger, P. (2001), 'Cities, Place and Cyberspace', *Paul Goldberger* [website], (published online 1 February 2001), <http://www.paulgoldberger.com/speeches.php?speech=berkeley#articlestart>, accessed 30 January 2008.

(2005),[2] and explicitly or implicitly used in several papers of this section – aims precisely to highlight this interplay which involves the same people meeting up in either of the dimensions according to circumstances.

These remarks that hold for any online community are even more pertinent and relevant when the virtual/online/web/network/digital (however one might prefer to call it) community develops and exists within a well defined territory, typically urban. These experiences, called "community networks" – and also known as free nets or civic networks – were particular popular during the 1990s. Dechief et al. describe community networks as pioneer 'grass-roots, community-based experiments in the use of new ICTs to empower individuals and urban communities and promote local civic participation and social inclusion'. Although many of them have failed in terms of their own financial sustainability and eventually declined, they have provided (time bounded) successful examples of how feasible it was creating, using the poor net technologies of those years, a "continuum" between the "online" and the "offline", between physical and digital territories. Community networks can be seen as the precursors, or the first prototypes, of augmented cities, built and shaped with the active participation of their digital inhabitants. And it is not accidental that community networks represent the explicit or implicit background of many of this section's chapters, which offer the added value of being strongly rooted into concrete experiences of augmented communities.

As an ideal preamble to the reflections drawn by these field experiences, the section opens with the inspirational paper by Gumpert and Drucker which offers the readers evocative views, and asks intriguing questions on the ways in which people conceive and live within the augmented city. Attention is given to space and places and their representations through maps; to people's relationships and mutual trust. Extremely stimulating is the notion of "permeable walled city", as the authors observe how traditionally city walls defined the social and economic life of the city, and how media technology now makes city "walls" more and more permeable. This is something that can challenge the very idea of city as a place for communities.

These concepts – identity, place, common good, mutual trust, and so on – as well as new concepts, such as "subcommunities" and "knowledge commons", become then the tools for reading the field experiences presented in the other chapters of the section.

The papers by Navarrete, Huerta and Horan and by De Cindio, Ripamonti and Di Loreto have several points of analogy and one relevant difference that is worth pointing out. Both look at a local hybrid community: Navarrete, Huerta and Horan study the '*Oaxaca* Web Community (Oaxaca is a state in southern México, and also the name of the state's capital city) [which was] established with the purpose of providing information about natural and cultural attractions in Oaxaca (México) and as a meeting point for Oaxacans to share their interests regarding their hometown.' De Cindio, Ripamonti and Di Loreto focus their attention on the Milan Community Network, a 'technological environment within which the local community could gain

2 Gaved, M. and Mulholland, P. (2005), 'Grassroots Initiated Network Communities: A Study of Hybrid Physical/Virtual Communities', in *Proc. 38th Hawaii International Conference on System Sciences* (IEEE Computer Society).

experience and learn hands-on how to exploit online interaction. [It] was conceived and designed [as] a means to give voice to those who had never had a chance to take part in civic and political arenas.'

Both papers focus on people's identities and their network of relationships – key issues widely studied in the virtual communities' literature, for which the interested reader will find a rich bibliography in the references. Both papers agree that people do not divide their worlds into physical and online worlds. However, the relation between physical and digital identities occurs in the two described cases in a very different way. In the Oaxaca Web Community (OWC) 'no registration is required and OWC makes no effort to verify the identity of the people posting information', while 'RCM members must register using their actual names.' In OWC 'some postings are clearly from people using fake identities. Nevertheless, participants do not seem to be bothered by this situation and ... when they were asked whether the OWC should require registration to control for identity deception, all interviewees rejected the idea.' In RCM 'People's names plainly appear when they send public messages to a forum.' Despite this radically different choice, also in RCM the 'policy of requiring members to use their actual identity ... has never been questioned, since it was accepted and perceived as good by the vast majority of RCM participants' and 'appropriate for a community of citizens focused mainly on city problems, where the ability to clearly recognize, say, a neighbour, a public official or a local politician has obvious significance'.

It seems that, despite the attention given to the choices concerning the registration policy, they end up having less impact than one (namely, the community designers) would expect. One could argue that the hybrid nature of these communities makes the choice somehow irrelevant: in both cases, the mutual trust among the community members comes from the interplay between the physical and the digital dimension, between the physical and the digital identity.

Both these papers explicitly or implicitly show the presence, within the overall – local hybrid – community, of subcommunities: for instance, 'Oaxacans [...] distinguish themselves as different social groups': native Oaxacans still living in Oaxaca, Oaxacans living abroad, foreigners living in Oaxaca. In this case the distinction is also strongly related to people's physical location.

Chapters 10 and 11 investigate further on the concept and examples of local subcommunities. Dechief, Longford, Powell and Werbin share the focus on community networks, namely on the opportunities they provide 'for increased civic participation among disadvantaged groups'. They present two cases: the first one describes how the Vancouver Community Network (VCN) has been a "gateway to community" for a group of VCN volunteers who had recently immigrated to Canada. The second one presents Île Sans Fil (ISF), a grassroots, community wireless networking group in downtown Montreal. Here we learn that 'wireless groups serve as a kind of gateway to community and civic participation for technologically-savvy but socially disengaged youth'. In both cases the positive results come from the interplay between the technical and the social dimensions, namely from the possibility of creating new social relationships starting from a technical skill. Again, the hybrid nature of these initiatives is what really matters for augmenting the community.

Community wireless networking is also considered by Gaved and Mulholland, who describe the Consume wireless network, created in the mid 1990s in London by a group of artists and electronic enthusiasts, as an example of one of the three kinds of subcommunities they identify. The authors provide a rich set of examples of 'grassroots activism with urban residents appropriating ICTs to augment their experiences of the physical city' and propose a taxonomy which distinguishes among "pioneers", "subcultures" and "cooperative". Beyond their differences, all the three kinds of subcommunities envisage ways of using 'networked technologies to enhance physical spaces and create hybrid "great good places" of the future, offering local residents the opportunity to define their own agendas and rules of engagement with new technologies and new ways of envisaging the urban environment.'.

Without forgetting the grassroots nature of community networking, the last two chapters explicitly draw the attention on the need for establishing partnerships among different local actors. If the city is the place for developing social relationships and the net is first of all a communication technology, the augmented city fails if it does not provide opportunities for experimenting new forms of cooperation among the various social actors.

Pang, Denison, Williamson, Johanson and Schauder, explore – through studying two libraries in Victoria (Australia) – the role public libraries can play for augmenting communities with knowledge resources in the contemporary media environment. These enormously increase individuals' as well as communities' ability to generate "knowledge commons", intellectual goods produced by communities of people 'that are made freely available for all in society to build relationships, culture, and democracy'. They recognize that 'a significant finding from these case studies is the emergence of partnership pursued by public libraries', although the strategy adopted in the two cases is different.

The issue of partnerships is even more crucial in the case of the OTIS (Opening the Information Society) project which is presented by Powell and Millward. The OTIS project took place from 1999 to 2001 in the city of Sheffield, UK, which had to undertake a radical transformation of its economy in order to regenerate it. Established under the umbrella of an institution not by chance named Sheffield First Partnership, the OTIS project aimed at shaping the information architecture of the city by 'encouraging the diverse stakeholders in a set of issues to work together to develop and deliver common solutions'. While carrying on the project, what became clear was the gap between 'the community and the voluntary sector [that] had pioneered aspects of an "information society"' and other components of the local community – such as the City Council or some larger voluntary sector organizations – with little history and competences in this kind of initiatives. This lagging back of official institutions and authorities is an issue that implicitly emerges within several chapters, and which threatens – in the medium term – the sustainability of the grassroots initiatives which have populated the augmented city. As Powell and Millward remark, 'many more people are [now] connected to the Internet but its use for collective economic, social or political engagement is limited. There are many fewer grass roots experiments now, compared to ten years ago.'

If a successful city needs the cooperation among most of its local actors – people, groups, voluntary sector organizations, public institutions and the private/business sector as well – this section's chapters, as a whole, show how a variety of experiences have proved effective for the local (sub)communities, but have also encountered difficulties in consolidating through time and in establishing strong relationships with the public institutions. These fail to try hard to join the 'open and transparent partnership committed to quality' that the Information Society demands.

The possibility to facilitate these partnerships is probably one of the real, big challenges for urban planners in the new millennium, something the third section of the book will also try and address.

Chapter 7

The City and the Two Sides of Reciprocity

Gary Gumpert and Susan Drucker

Over the past several years we have attempted to articulate the paradigm of "the city" and its evolving form as it has been reshaped and redefined by the accelerating impact of new communication technology. In 'Privacy, Predictability or Serendipity and Digital Cities' (Gumpert and Drucker 2005) we argued that the human being seeks surprise within that structured form organized and called the city. In that article we recognized the tension that exists between the need for regulation, control and safety and the opposing forces of play, surprise and individuality. In Digital Cities III we articulated the 'Perfection of Sustainability and Imperfection in the Digital Community: Paradoxes of Connection and Disconnection' and examined the nature of invention and their lack of neutrality. In that essay we argued that the "fixed features" of specific technologies shape who and what we are.

It is difficult to look at what one has written. Sometimes the arguments have grown stale and uninspiring. Sometimes, it appears as if the words are repetitive and undistinguished. Infrequently, the ideas are impressive and we do not recognize them as our own.

The Media Landscape Meets the Urban Landscape

In looking at our obsession with cities and their current state of being it is necessary to distance oneself for a moment from what has been said and to isolate and focus on fresh and newer ideas without depending upon former thoughts as a crutch. The search for these ideas is difficult because of our own entrapment in the process of technology. We are not alone in our attempt to re-examine the nature of the city at time it is undergoing radical change. Paul Goldberger, former New York Times architectural critic echoed this theme in an address given at the University of California, Berkeley in 2001 when he asked:

> What does have to do with our sense of place, and with the question of the meaning of the city right now? We are torn, it would seem, between believing that the city is irrelevant in the age of cyberspace, and believing that it has more urgency than every—or that place matters more than ever (Goldberger 2001).

We are so immersed in a technological world that it becomes virtually (in the old sense of the word – almost or nearly) impossible to examine ourselves and the transformation of self and city by increments of communication technology. Perhaps this is a variation of Heisenberg's "law of indeterminacy" – that we cannot readily understand the impact of communication technology upon us because we cannot detach ourselves from the very phenomenon we seek to understand. Explicit in our previous examination of the issues is the notion that the rise of the digital city does something (a rather unscientific choice of words) to the traditional conception of city. It is a simplistic binary approach to a very complex process that reduces or contrasts the new and the old, the digital with the pre-digital, the modern with the traditional. Our preoccupation has been with assessing the impact of new communication technology upon older conditions, structures, values and habits. It is a concern that may be unrealistic and simplistic.

The introduction of innovation is often linked to a sense of apprehension and uncertainty. In the 15th century uncertainty surrounded the introduction of moveable type upon the nature of knowledge and worship. The invention of television as a mass form of entertainment and news was seen by some as a threat upon the more traditional forms of performance and news dissemination – the stage, the motion picture, and the newspaper. And the impact of each new medium was radical and changed the nature of social interaction. Each of these inventions did alter the urban place and the intricate nature of information, interaction, and public space. We continue to ask similar questions regarding the impact of the digital revolution upon the traditional conceptualization of the city. We express concern about the impact of communication innovation upon the urban landscape. What impact will Wi-Fi have upon social spaces? How does the ubiquitous mobile telephone alter our sense of privacy and our relationship with others? Will the integration of the telephone and text messaging plus Voice over Internet Protocol devices alter the concept of the office and the workplace? To what extent will increased telecommunication mobility be reflected in the design and development of housing? How is public space augmented?

The accelerated invention of communication technology over the past few decades and the integration into the fabric of consciousness and use has been so gradual as to appear seamless, but has clearly been the product of careful economic planning. Social planning and design has been more or less accidental and coincidental.

Media and The Allegory of the Urban Cave

When considering social planning we return to The Republic in which Plato relates the allegory of "The Cave." Socrates (in a dialogue with Claucon) tells of men who live in a cave and are chained with their backs toward the light at the entrance so that they cannot change their position and must look at the world only through the shadows that are thrown by the reflection of a fire that exists between themselves and the wall of that cave. And then Socrates poses the following question:

> Consider now what would occur (...) if one freed them from their chains (...) Suppose that one of them has been released, and suddenly forced to stand erect, and turn his head, and walk with open eyes toward the light. Suppose that all these actions caused him pain,

> and that the dazzling brightness renders him incapable of looking at the object of which hitherto he saw the shadows?
>
> What answer, think you, would he give if anybody told him that all he saw before was empty mockery, but now, being somewhat closer to reality, and turned toward things more real, he sees more truly? (...) Don't you think that he would be perplexed, and would regard the things he saw before as truer than the ones they show him now? (Plato 1955, 334-335).

One interpretation of that allegory is a perception of reality that is illusory, while the other less illusionary, using the entire sensorium, would have been momentarily, at least, rejected in favor of a life of projected shadows, in a way, a virtual reality. For the purists here, those who know their Plato, this interpretation is slightly askew. There is a slight irony built into the argument. According to Plato, the phenomenal world (tables, chairs, people, etc.) is a "virtual world" - the world of shadows in the cave. The "real" world for Plato is the realm of form transcending reality - the world of ideas. Today, the reality of the street and the city has been transplanted by the illusions of virtual reality and mediated communication. The "see and hear contact" settings are being transcended through media contact. The cave is the augmented city, it is the digital city. It is a reality, whether virtual, digital, or an electronic construction. The cave may have been a communication innovation too early for its time.

In a similar vein, Paul Goldberger calls attention to a lecture given by Frank Lloyd Wright in 1901 at Hull House in Chicago entitled 'The Art and Craft of the Machine.' He argues that Wright saw that architecture would use the machine as an aesthetic inspiration and 'also denounced the hypocrisy of embracing technology as a model means to achieve a traditional end. Wright believed passionately that the machine somehow had to shape the aesthetic as well; it had to be a generator of architectural form. Architecture would not only have to be made differently in the age of the machine; it was essential that it look different too.' (Goldberger 2001)

Obsolescence, cities and media

Simultaneously and almost forgotten has been the increasing obsolescence of once innovative and revolutionary forms of communication technology. Technology spawns technology. The focus of critics has been upon the impact of technology and not upon the increasing replacement of one technology with another.

The focus has been upon the impact of the Gutenberg's moveable letter of type and yet little has been said about the end of moveable type and its replacement by lithography, offset printing, the mimeograph, xerography, and digital printing. Kodak has recently announced that it will cease manufacturing some versions of kodachrome film. In all probability photographic film and film cameras will disappear from the market place to be replaced by their digital progeny. Other examples of communication technology obsolescence come to mind. In terms of audio recording media the wire recording, the 78 rpm shellac recording, the 45 rpm vinyl recording, the long playing record, have entered the museum stage of development and in all probability the audio tape and cassette will soon follow. Other than for the purists, investors, and others with vested interest, the transition has been fairly smooth and seamless. Obsolescence is considered essential to progress. The question we pose is whether the traditional city has become obsolete in the face of the augmented city? How does this phenomenon of obsolescence relate to the issue of the traditional and digital city? While it is compelling to continuously compare traditional city values with digital city dreams, it may be more accurate to suggest that the older romantic notion of the city has, in part, become obsolete – rendered and modulated into a different form by newer technologies layered upon previous ones.

The Networks of Communication

Digital cities are a logical extension of networks of communication - a concept long associated with the traditional city.

> The city, as it develops, becomes the center of a network of communications: the gossip of the well or the town pump, the talk of the pub or the washboard, the proclamations of messenger and heralds, the confidences of friends, the rumors of the exchange and the markets, the guarded intercourse of scholars, the interchange of letters and reports, bills and accounts, the multiplication of books—all these are central activities of the city. (Mumford 1960)

The earliest concept of a network originates around 1560 and referred to any net-like combination of filaments, lines, veins, passages or the like.

Later the term referred to any network of arteries, for example the sewers under the city, or a system of interrelated buildings, office, stations found over a large area. And, of course, later, the term was forever linked to a group of radio and television stations connected by wire or microwave relay in order to broadcast the same program (Gumpert and Drucker 2003).

Before there were technological networks there were social networks. Today they confront one another. The nature of the relationship of these distinct types of networks, one interpersonally interactional, the other technologically integrational, has significant consequences on our attitudes toward cities and towards community. But the media generated transformation of institutions is not new (Gumpert and Drucker 2003).

The transformation of place and space

The impact and associated fears brought on by any new technology were eloquently expressed in the Hunchback of Notre-Dame through the voice of the Deacon who points out that until Gutenberg's time architecture was the principal universal form of writing. A medieval cathedral was a permanent, unalterable source of information but Gutenberg's invention liberated knowledge making it available to the masses and stimulated literacy. It weakened the power of institution and place. While 'printing will kill architecture' is a hyperbole, it inspires a twenty-first century addendum 'electronic space will kill the map.' (Gumpert and Drucker 2003). Once again we return to Wright who referred to the most important works of architecture as 'great granite books,' noting that Gutenberg's letters of lead superseded Orpheus's letters of stone (Goldberger 2001).

The traditional spatial dimensions of place are obliterated in Cyberspace. Geographical distance becomes irrelevant – is transmogrified from a location to a directory – a list of functions united through nodes and gateways. Ironically, global association redefines local orientation. Global becomes local because local is defined by ease of access to formerly distant points. As global is perceived as local, global is homogenized. We are coupled to sources of information with no location and we find ourselves in non-physical relationships constructed in the new geography of connections. The question has arisen whether the popularity of software like Google Earth, Yahoo! maps, Microsoft Local Live, and other map-hacking tools actually reintroduce the significance of geography. Intuitively we think not. For the traveler, such software is an extraordinary planning tool but for the day in day out online habitué, space remains a matter of connection and linkage. Geography is a curiosity, not a limitation. Space and distance remain irrelevant.

We are attempting to move from a binary mentality of then and now and the eloquent Victor Hugo thesis of 'this will kill that' expressed in 'Notre Dame de Paris' (Hugo 1978) to a much more nuanced developmental process in which communication technology becomes integrated, modifies and becomes part of the traditional. The built-in comparison of an ideal pre-technological communal village with a complex technologically interconnected multiple community becomes antiquated and non-productive. Thus, while it is patently clear that the existence of the telephone alters the role of public space and the nature of social interaction, the rise of the cell phone and mobile telephony is built upon the pre-existent notion of telephony. Short of catastrophe, one cannot un-invent a communication device (we wait for the next stage of development) and the mysterious infrastructure that supports the device is simply taken for granted – unless a mal-function occurs. This dynamic sense of technological communication innovation is implicit in the notion of 'media grammar and generation gaps.' It should never come to this drastic point, but ask a pre-teen to choose between the park and the mobile telephone, the choice is clear and the playground becomes irrelevant. The process of connection is more important than the place where one congregates.

Technology is constructed upon the accumulated layers of previous innovations. Each period of such technological change and innovation creates expectations and assumptions and we refer to those differences as media generations. Cities and their media infrastructure are built by layering one media era upon another. As archeologists unearth layers of history through the process of excavation, layers of technologies accumulate, reshaping and sometime obliterating the previous media generation. Each successive media layered upon the other leads to either one medium "killing" another, or more frequently, one medium changing or modifying the medium that came before. In turn, human activity changes, reflected in what we have previously introduced as the concept of "displacement" - the 'reciprocal and defining interdependence of place modified by communication technology.' (Gumpert and Drucker 2005).

"Displacement" is a temporal concept; defined by a difference in time or time usage. That is, the amount of time spent in media activity fundamentally alters the amount of time available for other events. It is thus self evident that the contemporary individual spends an increasing amount of time electronically connected with others in lieu of physical interaction on a face-to-face basis. The process of mediated connection brings with it a degree of "replacement" as well as "displacement." "Replacement" is a spatial concept, a difference of environment, the substitution and or alteration of one or more locations for another. Both displacement and replacement portend significant consequences on the nature and use of augmented cities. "A-location" is a consequence of media mobility and the encompassing experience of being "in the moment" in a "non physical place." The online and telecommunicative experiences available today (particularly those offered by mobile technologies like mobile phone, Bluetooth and Wi-Fi enabled connection) have the ability to reduce full sensory and psychological awareness of physical place. A-location refers to the "un-defining" of social space and psychological presence with its potential emancipation from physical place. A-location is enhanced and linked to mobile communication in which the connection of what is being communicated is disconnected from place and setting (Gumpert and Drucker 2007). Reciprocal interactivity between person and the environment increases the sense of presence and it is this interactivity or reactivity offered in the electronic online or telecommunicative environment which facilitates the shift of psychological presence. With ubiquitous, flexible, and connected mobility (both technology and person), awareness and interaction with the physical environment is modified and even becoming irrelevant at times. Psychological presence in physical space is altered. What is in the foreground is the illusion of being there, whether 'there' exists in the physical space or not (Biocca 1997).

Media Accessibility

The mobile phone and the development of Wi-Fi extend the parameters of location. The telephone has always extended geography, but the mobile phone redefines the individual in relationship to both extending oneself over space, but also redefines the self in terms of immediate location. It is not simply a matter of calling a distant place, but refining the place from which one calls and the relationship between oneself and others in that location. Large-scale municipal experiments with Wi-Fi broadband Internet access available to the entire community are taking place. The technological infrastructure required for a wireless environment is receiving the attention of planners and urban developers who view Wi-Fi as a device to attract commerce and individual users back to the city – hoping that such an innovation might serve as a tool for urban renewal in which cities offer free wireless access to downtown areas and provide a means of reconciling community and 21st century urban/technological developments. City authorities are extending old notions of utility regulation, highway metaphors, and the government's appropriate role as a player in protecting and promoting economic and social welfare to support city-subsidized municipal wireless.

Diverse models are emerging for the equitable deployment of broadband and Wi-Fi. The world's biggest municipal wireless rollout is Taipei, Taiwan. Japan and

Korea have been at the forefront of public policy supporting public broadband rollout. These countries along with Canada have successfully combined municipal systems with privately deployed networks to wire their countries. The European Union's stated policy encourages member states to make a public investment in broadband and Wi-Fi despite its record of encouraging market competition. Sweden and the Netherlands have emerged as leaders in the "muni" movement. Some credit Sweden with developing the most successful business model – a "wholesale" model where the municipality owns the network, but private sector companies provide the actual broadband services running over the network. British Telecom announced a hefty expansion of its hotspot network in London and across parts of the United Kingdom.

City authorities are extending old notions of utility regulation, highway metaphors, and the government's appropriate role as a player in protecting and promoting economic and social welfare to support city-subsidized municipal wireless. From a regulatory perspective, does the regulatory environment support or even permit public muni-rollout? In the United States there have been contradictory legislative actions. Fifteen states have passed legislation prohibiting public municipal broadband systems. There has even been a bill proposed by a Texas congressman to prevent any city in the country from providing their citizens with Internet access if a private company offers service nearby. On a Federal level there has been momentum coming out of 2006 strongly on the side of protecting municipal broadband activity. Three other bills died at year's end – one would have protected the right of local communities to offer high-speed broadband service; another would have harmfully empowered the private sector to block community Internet; and a third would have completely banned cities and towns from offering service. Both telecom bills from 2006, HR 5252 and S 2686, would have overturned bad state legislation and protected the right of municipalities in every state to offer broadband. In the Senate, Senators. Frank Lautenberg (D-NJ) and John McCain (R-AZ) introduced the 'Community Broadband Act of 2005.' (GovTrack.us 2005). This bill would specifically permit municipalities to offer low-cost broadband service. If this bill is reintroduced and passes, it would overturn all state legislation prohibiting municipal broadband systems.

Opponents to municipal provision of service argue that private providers can more efficiently provide services to competitive markets. It has been asserted that it is not wise or fair to allow government to provide service and therefore compete with private service providers. Some 'proponents point out that the social good coming from providing greater access than a monopoly or near monopoly private marketplace might provide is indeed a public good that government may appropriately foster.' (Jassem 2007). Public/private models have been promoted as well. Philadelphia, Pennsylvania is one of the largest American cities getting into muni-wireless. The city entertained proposals from diverse private vendors to build a Wi-Fi system and ultimately chose to partner with Earthlink as part of the agreement with the City of Philadelphia. Earthlink promises to provide other competing ISPs leased access to the network. Earthlink will pay the costs of building and operating the system in a revenue sharing partnership with the city. A non-profit group is to oversee the network (Jassem 2007). In California, San Francisco is planning to partner with Google to provide an advertiser supported system.

The permeable walled city

'The urban impulse is an impulse toward community—an impulse toward being together, and toward accepting the idea that however different we may be, something unites us.' (Goldberger 2001). Cities provide for common interests and needs historically often associated with protection of inhabitants from threat and attack. Concentrating a population within walls was a defensive advantage providing security by keeping some in and keeping others out. Cities are civic constructions that result from negotiating for wellbeing, safety, and community. All cities, whether of mortar and brick or creations of digital connection are built around the foundation of security, exclusion, community and control.

City walls defined the social and economic life of the city. The infrastructure of a city shapes and is shaped by the need for communication. The original infrastructure of cities was the physical environment; paths, roads, streets, market places, meeting places, and city walls. In The City in History, Lewis Mumford noted:

> The wall, then served as both a military device and an agent of effective command over the urban population. Esthetically it made a clean break between city and countryside; while socially it emphasized the difference between the insider and outsider ... (Mumford 1960, 66).

Land was used for the protection of people and the city served as fortress, or prison. Walls shaped the form and a city and set limits to expansion, served both symbolically and practically in the shaping of cities. The openings in the city wall were carefully controlled. Restricted ingress and egress marked life in the walled city from ancient times onward. Walls created a communication divide between those within and outside.

Digital cities are defined by controlled access, sometimes by limited resources and at other times by firewalls. Ironically the Trojan Horse was a type of virus constructed to breach virtual walls by introducing a harmful program embedded inside an apparently harmless one. Walls are symbols of authority, reinforcing power, defining identity. They were built to last.

Walls aren't what they used to be. From the thirteenth century on, the dread of plague prompted a periodic exodus from the city (Mumford 1960). Throughout history people have fled from cities to avoid exposure to illness, crime, enemy raids or bombing. Advances in transportation facilitated escape from cities and communities. Communication is always about flow, movement, and transportation. The vehicle for moving information, ideas, interaction, emotions is capable of scaling city walls and passing through (and around) city gates.

City walls, whether physical or psychological have always been measured by permeability, facilitated by the rise of communication technologies with the capacity to emancipate people from place, information from location and human relationships from surroundings. All walls are, to some degree, more or less permeable. They are measured by the amount and type of things passing through from one side to the other. And it is this quality of porous-ness that allows for growth and dissemination of community. Media technologies long ago rendered city walls permeable. Permeability alters community through extension and choice. The more porous the wall, the more choice of community or communities. Increased mobility, whether through human contact or mediated presence, facilitates the redirection of the community impulse transcending the traditional urban impulse. As community becomes less dependent upon physical place, choice of community becomes an option. The citizen of the 21st century is measured by his/her connection or affiliation with multiple communities. The digital divide has been replaced, in part, by community divide.

Trust and the Digital Cities

City living, from the walled city onward inferred an agreement made for protection. Individuals surrender some degree of freedom in exchange for security provided by collective living. Similarly, even today we enter into a series of contracts: a social contract between government and the governed and a media contract between a technology and the individual. Much has been written about social contract theory; the justification for politics (Baldwin 1999; Rawls 1999). So too, we suggest that a media contract exists between the user and communication technology that rests upon a conditional degree of authenticity. Digital communities and cities must be understood within the larger context of the relationship with a medium, with an eye on the significance of the media contract.

Each medium extends data over time and place. The transmission of information is either delayed or immediate. We attribute to a medium both the notion of replication and the issue of authenticity upon which it is judged. There is an illusion of transparency, a sense of permeability of function and medium, obliterating traces of production which construct mediated connection or message. A medium (often in combination with a group of other media) always has characteristics that shape and alter one's understanding and perception of that which is transmitted. It is not neutral. The differences between mediated and direct experiences have become less distinct. The less apparent or obtrusive the medium is, the more transparent the influence of the medium. The degree to which an image or message is perceived to be trustworthy is associated with the perception of the neutrality in a medium's transmission. Analog media made it difficult to mask alterations. With the shift to digital media it becomes increasingly difficult, if not impossible, to detect alterations.

Obsolescence has generated and stimulated a digital environment characterized by convergence. The convergent nature of a broadband environment in which we glide from medium to medium, in which we move with transparent ease from one function to another, signals not only the impact of technology upon technology but also of technology upon the city. The traditional city has evolved into the augmented city, the Tertium Quid City previously suggested by the authors (Gumpert and Drucker 2005). The augmented city includes a digital civic component manifested in e-government initiatives.

With e-government the social contract and media contract come together. The illusion of transparency between the governed and governing fostered by an ever-growing environment of press coverage of the governors and institutions of government has been coupled with more sophisticated use of media technologies to make the government appear more accessible and open to the governed. Each branch of government, from national to local is theoretically located a mouse click away. E-government and digital cities offer "one-stop" public access for local information. From news channels broadcasting 24/7 to the proliferation of online publications, the illusion is apparent increased coverage and information about government. The assumption has been that e-government and e-democracy has the ability to replicate government functions yet the critical question remains whether mediated replication provides an authentic qualitative experience. Perhaps mediated experiences are functionally equivalent but are they experientially identical, substitutable without consequence?

According to Stephen A. Garcia (2002) writing in the Journal of Business and Psychology, the 'illusion of transparency is the tendency for individuals to overestimate the extent to which their internal states and intentions are apparent to an outside observer. Thus, this illusion equals the difference between perceived and actual transparency.' The perception of access, interactivity and transparency of government can define the relationship between augmented city and the people.

Changes in communication technology, the shift from analog to digital, should lead to a greater awareness of manipulation and suspicion and therefore less trust in mediated interaction and information. An analogical medium brings with it an assumption of replication with traces of alteration. A digital medium brings with it an assumption of potential manipulation potentially undetectable. All mediated communication are constructions but the hand of construction has become less apparent, perhaps rendered invisible. Implicit in our processing expectations is the matter of trust. The paradox however is that by erasing traces of manipulation, digital media foster more trust and less suspicion.

Trust of communication technology and the press as a means of checking on government is flawed and based upon unwarranted assumptions of social and technological trust.

Conclusions

The city is not a static phenomenon nor is it a vague non-definable dream of social interaction. It is a place where people live, work, and play. It is a place reflecting and projecting technological change, particularly in terms of transportation and communication. The history of the city has always been entwined with the history of media developments.

We return to Paul Goldberger's remarks because he calls our attention to the larger issues when he states that he does not want to talk about: '….digital architecture…. since I don't really want to talk about the way technology affects individual buildings so much as the way the technology we have now affects our sense of urbanity, and our sense of place' (Goldberger 2001). It is this very sense of urbanity and community at heart of the future of cities, digital cities and augmented public space. The two sides of the city reflect a pattern of permanence and change, connection and disconnection. In an augmented city, the more we connect through communication technology, the more we disconnect from traditional community. For the modern city to exist it must be emancipated from place while at the same time connecting to some sense of the physical here and now. The modern city without connection cannot exist because to a great extent the telecommunication infrastructure defines the city. A city without connection would decay. Cities require two forms of connection; the emotional and the technological. Each builds the city but the technological is a double-edged sword in that it also dismantles the city by loosening the emotional and psychological hold of the city on its citizens.

As durable and eternal as the form and function of the city can appear, the rise of new technologies incrementally erodes the form, function and hold a city can exert. For many individuals un-augmented cities are outdated relics of the past while other look to augmented cities and digital communities as the means of salvation supporting the continued value of cities. Augmented or traditional, the fundamental function of a city is still the heart of the matter. The challenge is not to simply turn the city into the next generation of a media-filled landscape in which those severed from immediate community co-exist in an urban physical environment. The challenge is to find a way to harness the potential of the augmented public space as a place for community, for common interest, for social exchanges. If augmented public space serves as more than a backdrop for people living simultaneous existences in divided communities, the digital city will not "kill" the traditional city.

Acknowledgements

This chapter is an adapted version of a research article originally published as Gumpert, G. and Drucker, S. (2007), "The City and the Two Sides of Reciprocity", The Urban Communication Reader (Cresskill, NJ: Hampton Press), 35-45.

References

Baldwin, J. 'What Contract?'. ‹http://debate.uvm.edu/NFL/rostrumlib/ldcontractbaldwin 0395.pdf›, accessed 29 January 2005.

Biocca, F. (1997), 'The cyborgs dilemma: progressive embodiment in virtual environments.' *Journal of Computer-Mediated Communication* 3:2, <http://jcmc. indiana.edu/vol3/issue2/biocca2.html>, accessed 9 August 2006.

Garcia, S.A. (2002), 'Power and the Illusion of Transparency in Negotiations', Journal of Business and Psychology, 17:1, http://www.indiana.edu/~tisj/readers/ topics.html, accessed 23 October 2006. Goldberger, P. (2001), 'Cities, Place and Cyberspace'. *Paul Goldberger* [website], (published online 1 February 2001), <http://www.paulgoldberger.com/speeches.php?speech=berkeley#articlestart>, accessed 21 March 2005.

GovTrack.us. (2005), S. 1294 [109th]: Community Broadband Act of 2005, GovTrack.us (database of federal legislation) http://www.govtrack.us/congress/bill.xpd?bill=s109-1294, accessed 16 January 2008.

Gumpert, G. and Drucker, S. (2003), 'From Locomotion to Telecommunication, or Paths of Safety, Streets of Gore', in Strate et al. (eds.), *Communication and Cyberspace: Social Interaction in an Electronic Environment*, 2nd edition (Cresskill, NJ: Hampton Press).

Gumpert, G. and Drucker, S. (2005), 'The Perfections of Sustainability and Imperfections in the Digital Community: Paradoxes of Connection', in Tanabe, M., Van den Besselaar, P. and Ishida, T. (ed.), *Digital Cities III*, Lecture Notes in Computer Science 3081 (Berlin: Springer Verlag).

Gumpert, G. and Drucker, S. (2007), 'The Parable of the Mobile Rock: Displacing Place Mobile Communication in the 21st Century or "Everybody, Everywhere, At Any Time"', in Kleinman (ed.), *Displacing Place: Mobile Communication in the 21st Century* (Peter Lang Publishing).

Hugo, V. (1978), *Notre-Dame of Paris*, translated by Turrock, J. (New York: Penguin Books).

Jassem, H. (2007), 'Municipal Wi-Fi-ing of the United States', in Burd G. et al. (eds.), *Urban Communication Reader* (Cresskill, NJ: Hampton Press).

Mumford, L. (1960), *The Culture of Cities* (New York: Harcourt, Brace).

Plato (1955), *The Republic,* translated by Cooper, L. (Ithaca, NY: Cornell University Press).

Rawls, J. (1999), *A Theory of Justice* (Belknap Press).

Chapter 8

Social Place Identity in Hybrid Communities

Celene Navarrete, Esperanza Huerta and Thomas A. Horan

Introduction

Much has been said about the effects of Internet technologies on people's identities. One point of view is that cyberspace allows the manipulation of multiple pretended identities. On the other hand, cyberspace is seen as just another medium through which the identity of a person is expressed (O'Brien 1999; Wynn and Katz 1997). The relationship between technology and identity has been studied in a variety of social groups. For instance, Lamb and Davison (2002) studied the use of e-mail, teleconferencing and file transfer protocol applications (FTP) to manage identity in socio-technical networks of scientists. Wynn and Katz (1997) also studied professional settings but on the context of web pages. Identity has also been analyzed in the multicultural, political and economic exchanges that take place in discussion forums of diasporas (Adams 2004; Mitra 2000).

Having an identity is part of feeling a sense of virtual community, which in turn is essential to the sustainability of the virtual community. When members feel they have an identity, they are more conscious of their participation; therefore the quality of participation increases. Virtual community studies have approached the creation of identity as the processes of individuation from the group (Blanchard and Markus 2004; Donath 1999; Stryker and Burke 2000). Identity in these studies means identification and allows other members of the virtual community to identify who posted a message. However, identity is broader than identification, it also emerges from *self-categorization* or identification in terms of membership in social categories or groups (Stets and Burke 2000). The Internet is just the communication means used to present the multiple characteristics of the self. Group membership offline is anchored to the place where social interaction is hold. People online disclose their place-based identity to construct their online identities. On such identity relates to "place" and how that is facilitated or communicated over the Internet (Hampton 2002; Horan 2001).

Despite the interest that studying identity in online interactions generates, studies that employ a theoretical framework of identity to situate empirical results are still limited (Koh et al. 2007; Ridings and Gefen 2004). This chapter analyzes identity in hybrid communities from the perspective of social identity theory (Stets and Burke 2000). In particular, we study how members of the community of interest (Oaxaca Web Community, or OWC) invoke their off-line/online identities in virtual spaces with specific interest in how they communicate their place-based identity over the Internet. OWC is considered a hybrid community because their members rely upon both electronic and face-to-face means for their communication (Gaved and Mulholland 2005).

Exploring the ways in which people are linked to hybrid communities through social identities provides a more fully integrated view of the self. Social identity theory is a well-established theoretical framework and have a long tradition in Social Psychology research, from which identity research in virtual community studies can benefit (Burke 1980; Hogg et al. 1995; Stets and Burke 2000; Stryker 1968; Stryker and Burke 2000). By exploring how identities are enacted in online settings we assess first, the enduring role of physical place association (both present and past) in creating and maintaining identities and the extent to which hybrid communities can facilitate new forms of physical/virtual identities. A related exploration relates to the ways in which the physical communities (including social networks) influence online identity, thereby creating new forms of digital communities (Horan 2000b). Second, we explore whether identity salience leads to greater commitment to the online/physical community translated into increased interest, participation and cooperation. Previous research has shown that having a social identity in open source communities can extend the capabilities of the community –e.g. establishes a friendly environment for novices to ask questions, enhances understanding and solution of complex problems, facilitates the formation and duration of cordial interactions, among others (Jin et al. 2006).

This chapter is organized as follows. First, we describe foundations of social identity. Second, we present an overview of research on hybrid communities. Third, we describe the research methodology employed for the collection of data in this study. Fourth, we examine the hybrid community OWC to determine how social identities are negotiated in off-line/online settings. The chapter ends with a discussion of the results and implications for further research.

Identity

The term identity is used with 'considerable variability in both its conceptual meaning and its theoretical role' (Stryker and Burke 2000, 284). Internet studies also vary on their use of the term identity. Some studies equate identity to being identified by others (Blanchard and Markus 2004; Donath 1999; Stryker and Burke 2000). Identity is different, therefore, in the physical world and on the Internet. 'In the physical world there is an inherent unity to the self, for the body provides a compelling and convenient definition of identity. The norm is: one body, one identity' (Donath 1999, 29).

However, the Internet, lacking a physical verification of the self, allows people to create multiple identities if they wish to. Even if a person does not want to create multiple identities but a single one, the online identity does not necessarily match the physical identity due to the use of pseudonyms (Donath 1999). From this perspective, identity is constructed through participation in terms of frequency and quality of the postings (Donath 1999), and digital signatures (Blanchard and Markus 2004).

Other studies point out that identity is important for community building. For instance, identity is necessary to motivate participation (Kollock 1999). Participants who willingly help community members expect that their needs will be satisfied sometime in the future. However, for reciprocation to happen, members need to be identified. Identity is also needed to develop trust and increase accountability (Donath 1999). Yet again, in these studies identity equates identification.

We differ from this perspective for several reasons. First, people do not divide their worlds in physical and online worlds (Turkle 1995; Wellman and Gulia 1999). The relationship is what is important. The Internet is but another communication means (Wellman and Gulia 1999). Second, we believe identity is more than identification. Understanding what identity is must be based on sound theory. Third, research has shown that the self is composed by multiple organized identities (Stets and Burke 2000; Stryker and Burke 2000).

We based our analysis on social identity theory. According to social identity theory an identity (more precisely a social identity) is a 'person's knowledge that he or she belongs to a social category or group' (Stets and Burke 2000, 225). Persons can have as many identities as groups they belong to. However, membership to a group can be based on a specific place (e.g. Canadians, Chileans). People who are defined by a location usually develop an emotional attachment to a particular place (Castells 2004; Curry and Eagles 1999; Proshansky et al. 1983). Thus, social identity is not only defined by socialization but also refers to 'perceptions in the form of images, memories, facts, ideas, beliefs, values, and behavior tendencies relevant to the individual's existence in the physical world' (Proshansky 1982, 1).

This study uses social identity theory to understand how participants in hybrid virtual communities negotiate their group (or place) identities. The following section describes previous research on hybrid communities and identifies the characteristics of this type of communities that are relevant to establish identities.

Hybrid Communities

Hybrid communities are groups of 'people who interact together socially using both on-line and off-line methods of communication' (Gaved and Mulholland 2005, 2). In hybrid communities, members hold physical and virtual relationships simultaneously, and their interaction takes place through multiple forms of communication: face-to-face, traditional means (e.g. telephone and fax) and Web based tools (e.g. e-mail, chat rooms, bulletin boards). The network of relationships among members can be established before the creation of the online community – i.e. members of the physical community extend their existing levels of face-to-face contact to include online spaces. For example, an e-mailing list of a local community group or association extends face-to-face interaction to online interaction (Gaved and Mulholland 2005; Putman and Feldstein 2003). However, Internet-based tools can be used for the formation of the community itself. In the latter communities, personal relationships are first established in online settings and later extended to the physical world (Boase and Wellman 2004). For instance, a newsgroup of people interested in sports establishes online communication and extends it to the physical world (Blanchard and Markus 2004). Other examples of hybrid communities include e-mailing lists of scientific associations (Churchill et al. 2004; Lamb and Davison 2002), civic institutions (Horan 2000a), open source developers (Jin et al. 2006), and bulletin boards of diasporas (Navarrete A. and Huerta 2006).

In hybrid communities, voluntary disclosure of identity depends on the social context in which the online interaction unfolds, and contributes to increased participation, sense of community and social capital. Previous research shows that keeping public the personal profile of participants in online forums of scientists encourages online and face-to-face community participation (Chan et al. 2004; Churchill et al. 2004). Similarly, Blanchard and Markus (2004) found that the overlapping of online and offline social networks in a community contributed to the creation of a sense of community. Churchill and colleagues (2004) report that a 'persistent and consistent identity lets people identify and find out more about one another and provides a foundation for social accountability in the future' (p.41). Even though identity plays an important role in hybrid communities, anonymity might sometimes be desirable. For instance, Cardenas Torres (2005) shows that members of an online community of immigrants chose anonymity when they denounced government corruption. In this case, anonymity is used with the purpose of disassociating the person's identity with the comment because of the potential negative outcomes for the person posting the message. Anonymity will increase participation in this setting because it will be a means to express an opinion without the fear for further consequences. Similarly, anonymity might be useful for communities directed to members who, for some reason, are reluctant to participate (Andrews 2002). Here, anonymity will increase participation because for some people the fear of being identified prevents them to participate, regardless of the sensitivity of the topic discussed. However, under normal circumstances, it can be expected that members of hybrid communities want to be identified. People want to be identified because they want to be contacted by their acquaintances and they want to bring to the online community their reputation from the physical settings.

In hybrid communities, participants' voluntary decision to reveal their identities (including personal and private information) in spite of the public nature of the Internet might be related to the already existent base of trust established in the real world relationships among the members of the community. Deception or false identities are perhaps less likely because information about members flows between the virtual community and face-to-face community and vice-versa (Blanchard and Horan 1998).

Moreover, when there is a dense network of social interaction the potential costs for negative behavior increase (Putman 2000). Thus, inconvenient posting might not only affect participants' relationship in the online community but also their long-term relationship with members of the physical community.

This chapter will address these issues by examining the activation of social and identities in a hybrid community. The following section describes the research approach employed in the collection of data.

Method

The hybrid community studied in this chapter is Oaxaca Web Community (OWC) (Comunidad Web de Oaxaca http://www.oaxaca.com). Oaxaca is a state in southern México with a population of almost 3.6 million inhabitants (Instituto Nacional de Estadística Geografía e Informática 2005). Oaxaca is also the name of the state's capital city. OWC was created in 1998 for the purpose of providing information about natural and cultural attractions in Oaxaca (México) and as a meeting point for Oaxacans to share their interests regarding their hometown. OWC is appropriate for the examination of identity since it is a well-established community where constant and long-term exchanges between members take place. At the time of this study the community had 4,452 registered users and most of its participants were born in Oaxaca (4,004). Archives dating back eight years show the history of individual members contributions in discussion forums. Records of participants' activities and behaviors can be helpful to understand under which circumstances a particular identity was invoked and whether such identity was stable across time (Churchill et al. 2004; Donath 1999).

OWC was examined using an ethnographic approach (Hunter 2004; Myers 1999). We observed the interactions of community members in OWC for 18 months. Messages posted in forums were downloaded and reviewed. To ensure that interpretation of the postings was done through the eyes of community members themselves rather than the researchers, we conducted seventeen (n=17) semistructured interviews (Sharf 1999). With the authorization of the community web master we invited community members to participate in our study. The webmaster posted an invitation to participate in the study in the main page and the discussion forums, and sent an email to members in the community's newsletter (about 2,500). Interested participants contacted the researchers through e-mail and scheduled phone interviews. Members interviewed were Oaxacans living in different regions of México and the United States. Interviews were conducted over the telephone for about half an hour each. A more detailed description of the community and its web site is presented in the next section.

Findings

Overview of Oaxaca web community

OWC is a virtual venue where people learn and share information related to Oaxaca. OWC provides various services for members: access to a chat room, nine discussion forums, a directory with search capabilities, a link to a collection of Web pages with tourist information, and a newspaper (The Grasshopper). OWC is open to everyone who wishes to join as participant without the need to register. We identified three types of members: Oaxacans living in Oaxaca, Oaxacans who have migrated to other regions in México and other countries, and foreigners. A hit counter in the directory section shows that Oaxacans constitute the majority of the web community (4,004 of a total of 4, 452 members) geographically dispersed in different regions of México and fifty-three countries. Messages posted in forums suggest foreigners visit forums mostly looking for information related to tourist destinations and attractions in Oaxaca. In OWC, communication among members takes place primarily in Spanish. However, it is also common to find postings in English.

Observation of the community and interviews indicated that OWC met the criteria proposed by Gaved and Mulholland (2005), for a "hybrid physical/virtual community". The site was based on a physical place. The chat room and bulletin boards support the online interaction among members. Discussion in bulletin boards takes place asynchronously – members post and read messages at their time of convenience. Messages are organized in chronological order. The type of messages exchanged include: discussions concerning current and past social, cultural, and political events in Oaxaca (e.g. the 2006 teacher strike); asking and providing information about tourism (e.g. accommodation, transportation and attractions), advertising services for the community (e.g. looking for people in the U.S.), announcing of events in the physical community (e.g. Guelaguetza festival), and sharing of anecdotal information (e.g. immigrants' experiences in the U.S.).

Interviews and messages posted in the forums suggest that interactions also take place in off-line settings. For example, a member posted a message looking for people from his hometown who lived in Oaxaca. He received a reply from a friend whom he had not seen in a long time. The community has been also used to make new contacts or develop new off-line personal relationships with others in the group. For instance, seven of the participants interviewed living in the United States reported that they have extended their network of relationships with other Oaxacans in the U.S. through OWC. Previous research has addressed the significant role of the Internet in increasing, supporting and reinforcing social networks of communities of immigrants (Navarrete and Huerta 2006).

The OWC's site does not offer any mechanisms to protect the privacy of its members (personal data and contributions are publicly accessible). Privacy is protected by that fact that personal data are voluntarily disclosed. People do not share their information if they do not want to. However, OWC seems to provide a trusting and supportive atmosphere for members to disclose great deal of personal information. Then again, since no registration is required and OWC makes no effort to verify the identity of the people posting information, some postings are clearly from people using fake identities. Nevertheless, participants do not seem to be bothered by this situation. Interviewees expressed that postings that are clearly faked are usually ignored. Moreover, when they were asked whether the OWC should require registration to control for identity deception, all interviewees rejected the idea.

Our observation of discussion forums suggests that participants determine the boundaries for acceptable identity deception based on the set of rules and social conventions already established in the physical community. For example, the use of pseudonyms (or anonymity) is accepted in the virtual community when contributions might somehow endanger participants' offline identity. The following message thread shows how a participant chose anonymity when she denounced corruption among high school teachers:

Participant 1:

> Is it true that there are bullies in the high school under the orders of the principal who uses them to control everything in the school and getting in reward good grades and certificates without attending classes or taking any exam? [translation from Spanish by the authors].

Participant 2:

> Unfortunately it is true, it really happened long time ago, but it is important to say that there were some really good teachers and you have to be in the shoes of young people (at that age everything is fun, parties, etc) and of course they try to break the rules. There are some groups who yes, they take control of the institution and they sell and buy grades, you can get a bachelors degree without even assisting to school ... I was never involved in that kind of business but it is still a common practice.
>
> WARNING: Do you know that telling you all this is risky for me and they can track me down, that's why if you want to TRACK me down, don't try it please, I will send you a VIRUS to your computer if you try to send me a direct email [translation from Spanish by the authors].

We also found that identity verification is particularly relevant for those participants interested in extending their network of social relationships to offline settings. For instance, some participants posted their contact information to show their interest in a message announcing a language exchange program in Oaxaca. In a different thread the coordinator of the language program replied to one of her students from the U.S. who participated in the program: 'I'm happy because I can read that you have passed a good time in Oaxaca City and you could improve your Spanish. I hope that you'll be able to continue practicing the language and I'll be ready to correct the few mistakes every time.' Thus, the verification of identities not only contributes to the creation of an atmosphere of respect and trust where participants feel comfortable in revealing more details about their offline identities, but also can increase interest and participation in the hybrid community. Participants who receive respect and status from other participants are more likely to activate the verified identity in other situations and invest more time in their network of relationships.

Social identity and Oaxaca web community

As mentioned above people participate in the forums for several reasons, but all of them have in common the physical place of Oaxaca. Even though the common link is Oaxaca, participants invoke different identities when participating in the community.

Community members have different social identities: Native Oaxacans still living in Oaxaca, Native Oaxacans living abroad or in México in states other than Oaxaca, and foreigners living in Oaxaca. It is important to note that researchers did not artificially create these social identities. Rather, community members specify the group they belong to in their postings. Members use statements such as: 'I am a Mixtec [native from the Mixtec region in Oaxaca] who left the place I was born searching for better opportunities' [translation from Spanish by the authors].

In short, community members distinguish themselves as belonging to a specific group. Moreover, when members do not post that information, they are prompted to state where they were born and where they live. We do not consider sporadic visitors to be part of the community. Tourists searching for information do not conform a social group. Their contributions are limited to search for what they need and their permanence in the community is short.

Native Oaxacans still living in Oaxaca consider themselves as more knowledgeable about the local traditions, food, people, government, and other issues, than other groups. When they express their opinions they highlight their Oaxacan identity to strengthen their arguments. For instance, a member expressed: 'For most of us born and raised in the city of Oaxaca and nearby, the zocalo [central plaza] really was a special place to spend a peaceful time talking with friend, taking a coffee with friends while the children played, etc.'

Oaxacans thus distinguish themselves as different social groups based on their physical location express mixed feelings about Oaxacans living abroad or in other cities within México. Some of them praise migrants and others reject them as traitors. In 2002 a discussion arouse about McDonalds wanting to establish a restaurant in Oaxaca's zocalo (Central Plaza). The issue was publicized nation-wide and American newspapers discussed the topic. Discussion centered not only on the issues, but also on what group had a right to express its opinions. Native Oaxacans living in Oaxaca regarded Oaxacans living abroad as a social group with lower status, or considered migrants' arguments weak because they were not living in Oaxaca. A participant posted the following message to a Oaxacan living abroad: 'If you are so proud of Oaxaca, why did you leave? Come back home and be the great patriot you pretend to be.'

In the same discussion, Oaxacans living abroad expressed that they left their hometown searching for better opportunities. They expressed their right to be concerned for their local communities and they asked for understanding on their motives for migrating. Some Oaxacans residents agreed with the migrants and praised them as Oaxacans too. The following posting illustrates this argument:

> We owe you and other people like you, not only from Oaxaca but from all México, for your courage on migrating to an unknown country. The money they [migrants] send to their families, earned with their hard work, is most times determinant for developing our most marginalized communities [translation from Spanish by the authors].

Oaxacans thus distinguish themselves as different social groups based on their physical location. Their identity as Oaxacans is associated to a physical place. Their identity as Oaxacans is associated to the state of Oaxaca because they were born there. At the same time, it matters whether they continue living in Oaxaca or not.

Foreigners living in Oaxaca identify themselves as another social group. Foreigners assume their opinions should be the least valuable of all regarding Oaxacan matters. They assume a role of passive spectators rather than active members. In the discussion about allowing a McDonalds restaurant in the zocalo, a foreigner living in Oaxaca expressed:

> I would like Oaxaca to remain the picturesque colonial city that I have come to love. But the decision is not ours; it belongs to the citizens of Oaxaca. No matter how long I live here I am always going to be a guest and the first rule of being a guest is to avoid criticisms of my host. The Americanization of the world appalls me, but I have no right to impose my beliefs on the citizens of my adopted country.

When foreigners try to express their opinions on Oaxacan matters, members of the community point out their foreign identity and instruct them about the proper behavior. Remarking the out-group status to foreigners is also common when foreigners do not behave according to Oaxacan customs.

According to social theory, social identity is reinforced through a process called self-enhancement (Hogg et al. 1995; Stets and Burke 2000). In this process, people make positive evaluations of the group and minimize its negative characteristics. We found evidence that the community engages in self-enhancement processes. Postings with negative evaluations of the group are strongly rejected. However, reactions differ when negative evaluations come from an in-group member (Oaxacans) and form an out-group member (foreigner). When foreigners criticize Oaxaca, Oaxacans quickly remind them of their foreigner status. When Oaxacans criticize Oaxaca they are labeled as traitors to the group. When a Oaxacan complained about how dirty the city was, another Oaxacan reacted negatively and posted:

> Our city is going downward because of the opinions of people like you, that scare tourists away, ok, nobody denies that Oaxaca is a dirty city, but get real! Most states in México are the same [translation from Spanish by the authors].

A foreigner who agreed that Oaxaca was a dirty city got this for response, all in capital letters:

> If you don't like our country go back … nobody forces you to be here … in your posting you didn't say anything valuable, you just … complained that there are dirty people on the streets, they are dirty because they are poor … ! Who do you think you are? [translation from Spanish by the authors].

Members are able to recognize themselves from the hometown they were born. Even if members have not met before, they are able to identify themselves as part of the hometown social network. Knowledge of the social network legitimizes participants as members of a group and reinforces their social identity. A community member posted the following message:

> 12 years ago a I left my beloved hometown, I miss its people, its parties, and the *clayudas* [native Oaxacan food] in the market. I hope I will return someday to my hometown. I live in Los Angeles California. I would like to contact people from the Istmo [Oaxacan region] to find friends [translation from Spanish by the authors].

Someone in the community from the same region recognized the name of the person who posted the message. Moreover, he identified the family he belong to and provided him further information from people on the same social network. Another community member boasted about being in contact with the hometown social network and offered to provide information to others. Similarly, five of the interviewees commented that thanks to OWC they have extended their network of friends in their local communities in the U.S. Again, there is no difference between the social network in the offline and online community. Even members of the physical community who might not even participate in the OWC are involved through others who participate in the virtual community. As one of the interviewees commented, 'Well, if you want to know everything about everybody just ask me because I know everybody, just let me know the names of who you want information from and I will pass it to you [translation from Spanish by the authors]'. In response, a participant asked him about a friend. 'Hey … I want news about a friend, his name is [name omitted], … I hope you can tell me only good things about him [translation from Spanish by the authors]'.

In sum, the Internet has not created new social identities; rather it is only a communication means where community members bring along their off-line social identities. The boundaries between the online and offline world blur because social identities are linked to a physical place. The social identity invoked is linked to the place people were born, raised, and live.

Discussion and Concluding Remarks

The purpose of this chapter was to analyze the gradations in place-based social identity as realized via internet communications in hybrid communities. Specifically, we applied social identity theory to the interaction of individuals in Oaxaca Web Community (OWC). We argue that the dual form of interaction in OWC (physical and virtual) facilitates the representation of a various place-based identities in online settings. This chapter shows that individuals in OWC shape their social identities based on the set of meanings and expectations they hold as community members i.e. members perceive their identity in terms of self-affiliation to the physical community of Oaxaca.

In addition, findings suggest that for OWC members the self is composed of multiple identities and the salience of these identities is unique for each member. In some cases the context of the interaction determines the invoked identity. However, the nature of the relationship to members within the physical community also determines identity choice. Offline identities are invoked in online settings providing the identification required for generating the trust and commitment necessary for the sustainability of the virtual community. In OWC, members' identity is a natural consequence of the affiliation with the physical community. Thus, the Internet acts only as an instrument to present the various characteristics of the self. That is, while the Internet allows someone to still "belong" to the Oaxacan community via the Internet, this virtual community distinguishes between different types of Oaxacans those living in Oaxaca, those visiting Oaxaca, and those who previously lived in Oaxaca. Moreover, the physical place determines the status of each group.

This study has the following implications. First, understanding the impact of identity disclosure in online interaction can guide the convenience of using technological tools to enable people to communicate and verify identities. Shared perceptions of attitudes, beliefs, norms and tasks among community members are important conditions for social identification and group cohesion (Stets and Burke 2000). Previous studies provide evidence of successful cases where Internet-based technologies are used to trace identity information (e.g. person's behaviors) (Churchill et al. 2004).

Second, this study shows that the reinforcement of social identity enhances participation in hybrid communities (e.g. reduces inconvenient posting). Knowledge of the social network legitimizes participants as members of a group and reinforces identity. Community leaders could explore ways to strength social identity by motivating the interaction of participants in face-to-face community events. In addition, community designers can include advanced search capabilities for the community's directory. This can help participants to be more aware of the depth and strength of their social network of relationships. Participants can also benefit from this since social networks are rich in information and other resources (e.g. social capital, emotional support) (Putman 2000).

Finally, previous literature have approached identity as the ability to recognize other community members (identification) (Blanchard and Markus 2004; Chan et al. 2004). This study extends that view by applying a socio-psychological theory in the examination of identity in hybrid communities. The Internet allows one to maintain a virtual form of place-based identity through such networks, although it is not possible to completely erase the different roles that are taken by those who physically live in versus outside a physical community.

Acknowledgements

The authors would like to thank Juan Antonio Ruiz Zwollo, webmaster of *Comunidad Web de Oaxaca*, for his generous support in this research. We also extend our gratitude to the people who volunteered for the interviews. Esperanza Huerta thanks the *Instituto Tecnológico Autónomo de México* for its support during the early stages of this research.

References

Adams, A. (2004), 'Diaspora, Community and Communication: Internet Use in Transnational Haiti', *Global Networks,* 4:2, 199-217.

Andrews, D. C. (2002), 'Audience-Specific Online Community Design', *Communication of the ACM,* 45:4, 64-68.

Blanchard, A. and Horan, T. (1998), 'Virtual Communities and Social Capital', *Social Science Computer Review,* 16:3, 293-307.

Blanchard, A. and Markus, M.L. (2004), 'The Experienced "Sense" of a Virtual Community: Characteristics and Processes'. *The Data Base for Advances in Information Systems,* 35:1, 65-78.

Boase, J. and Wellman, B. (2004), 'Personal Relationships: On and Off the Internet', in Perlman, d. and Vangelist, A.L. (eds), *Handbook of Personal Relations* (Oxford: Blackwell).

Burke, P.J. (1980), 'The Self: Measurement Requirements from an Interactionist Perspective', *Social Psychology Quarterly,* 43:1, 18-21.

Cardenas Torres, M. (2005), 'Las Comunidades Virtuales de Migrantes en los Estados Unidos, su Impacto y su Vinculación con el Lugar de Origen. El Caso de San Martín de Bolanos, México', in Massé, C. (ed.), *La Complejidad de las Ciencias Sociales en la Sociedad de la Información y la Economía del Conocimiento. Trastocamiento Objetual y Desarrollo Informacional en Iberoamérica.* (México, El Colegio Mexiquense).

Castells, M. (2004), *The Power of Identity* (Malden, MA, Blackwell Publishing).

Chan, C.M.L., Bhandar, M., Oh, L.-B. and Chan, H.-C. (2004), 'Recognition and Participation in a Virtual Community', in *Proc. 37th Hawaii International Conference on System Sciences* (IEEE Computer Society).

Churchill, E., Girgensohn, A., Nelson, L. and Lee, A. (2004), 'Blending Digital and Physical Spaces for Ubiquitous Community Participation', *Communication of the ACM,* 47:2, 39-44.

Curry, M. and Eagles, M. (1999), 'Geographies of the Information Society: Place and Identity in an Age of Technologically Regulated Movement' (Santa Barbara, California: National Center for Geographic Information and Analysis).

Donath, J.S. (1999), 'Identity and Deception in the Virtual Community', in Smith, M. A. and Kollock, P. (eds), *Communities in Cyberspace* (London/New York: Routledge).

Gaved, M. and Mulholland, P. (2005), 'Grassroots Initiated Network Communities: A Study of Hybrid Physical/Virtual Communities', in *Proc. 38th Hawaii International Conference on System Sciences* (IEEE Computer Society).

Hampton, K. (2002, 'Place Based and IT Mediated Community', *Planning Theory and practice,* 3:2, 228-231.

Hogg, M.A., Terry, D.J. and White, K.M. (1995), 'A Tale of Two Theories: A Critical Comparison of Identity Theory with Social Identity Theory', *Social Psychology Quarterly,* 58:4, 255-269.

Horan, T. (2000a), *Digital Places: Building Our City of Bits* (Washington: Urban Land Institute).

Horan, T. (2000b), 'A New Civic Architecture: Bring Electronic Space to Public Place'. *Journal of Urban Technology,* 7:2, 59-84.

Horan, T. (2001), 'The Paradox of Place', *Communications of the ACM,* 44:3, 55-59.

Hunter, G.M. (2004), 'Qualitative Research in Information Systems: An Exploration of Methods', in Whitman, M. E. and Woszczynski, A. B. (eds), *The Handbook of Information Systems Research* (Hershey, PA: Idea Group Publishing).

Instituto nacional de estadística geografía e informática (2005), 2005 Census, México.http://www.inegi.gob.mx/est/contenidos/espanol/rutinas/ept.asp?t=mpob02&c=3179.

Jin, L., Daniel, R. and Boudreau, M.-C. (2006), 'Exploring the Hybrid Community: Intertwining Virtual and Physical Representations of Linux User Communities', *Administrative Sciences Association of Canada (ASAC)* (Banff, Canada).

Koh, J., Kim, Y.-G., Butler, B. and Bock, G.-W. (2007), 'Encouraging Participation in Virtual Communities', *Communication of the ACM,* 50:2, 68-73.

Kollock, P. (1999), The Economies of Online Cooperation. In Smith, M. A. and Kollock, P. (eds), *Communties in Cyberspace* (London/New York: Routledge).

Lamb, R. and Davison, E. (2002), 'Social Scientists: Managing Identity in Social-Technical Networks', *Proc. 35th Hawaii International Conference on System Sciences* (IEEE Computer Society).

Mitra, A. (2000), 'Virtual Commonality: Looking for India on the Internet', in Bell, D. and Kennedy, B. M. (eds), *The Cyberculture Reader* (London/New York: Routledge).

Myers, M.D. (1999), 'Investigating Information Systems with Ethnographic Research', *Communications of the Association for Information Systems,* 2:23, 1-20.

Navarrete A., C. and Huerta, E. (2006), 'Building Virtual Bridges to Home: The Use of the Internet by Transnational Communities of Immigrants', *International Journal of Communications, Law and Policy,* 11, published online in November 2006 <http://ssrn.com/abstract=949626>.

O'Brien, J. (1999), 'Writing in the Body: Gender (re)Production in Online Interaction', in Smith, M. and Kollock, P. (eds), *Communities in Cyberspace* (London/New York: Routledge).

Proshansky, H.M. (1982), 'Place-Identity in Urban Settings', <http://www.eric.ed.gov/ERICWebPortal/custom/portlets/recordDetails/detailmini.jsp?_nfpb=true&_&ERICExtSearch_SearchValue_0=ED223746&ERICExtSearch_SearchType_0=no&accno=ED223746>, accessed 8 September 2007.

Proshansky, H.M., Fabian, A.K. and Kaminoff, R. (1983), 'Place-identity: Physical world socialization of the self, *Journal of Environmental Psychology,* 357-83.

Putman, R. (2000), *Bowling Alone* (New York, Simon and Schuster).

Putman, R. and Feldstein, L. (2003), *Better Together. Restoring the American Community* (New York: Simon and Schuster).

Ridings, C. M. and Gefen, D. (2004), 'Virtual Community Attraction: Why People Hang Out Online, *Journal of Computer Mediated Communication,* 10:1.

Sharf, B.F. (1999), 'Beyond Netiquette: The Ethics of Doing Naturalistic Discourse Research on the Internet', in Jones, S. (ed.), *Doing Internet Research: Critical Issues and Methods for Examining the Net* (Thousand Oaks: Sage Publications).

Stets, J.E. and Burke, P.J. (2000), 'Identity Theory and Social Identity Theory', *Social Psychology Quarterly,* 63:3, 224-237.

Stryker, S. (1968), 'Identity Salience and Role Performance: The Relevance of Symbolic Interaction Theory for Family Research, *Journal of Marriage and the Family,* 30:4, 558-64.

Stryker, S. and Burke, P.J. (2000), 'The Past, Present and Future of an Identity Theory, *Social Psychology Quarterly,* 63:4, 284-297.

Turkle, S. (1995), *Life on the Screen* (New York: Simon and Schuster Paperbacks).

Wellman, B. and Gulia, M. (1999), 'Virtual Communities as Communities', in Smith, M.A. and Kollock, P. (eds), *Communities in Cyberspace* (London/New York: Routledge).

Wynn, E. and Katz, J.T. (1997), 'Hyperbole over Cyberspace: Self-Presentation and Social Boundaries in Internet Home pages and Discourse', *The Information Society,* 13:4, 297-327.

Chapter 9

Interplay Between the Actual and the Virtual in the Milan Community Network Experience

Fiorella De Cindio, Laura Anna Ripamonti and Ines Di Loreto

Introduction

Virtual communities were one of the first experiments in net-based social interaction. From the outset, most virtual communities have shown tight interplay between what happens online and offline. In *Qu'est ce que le Virtuel* (1995), Pierre Levy points out that the *virtual* (world) stands in opposition not to the *real* but to the *actual* (or *physical*) world. The *virtual* is just as real as the *physical* is. It is precisely to avoid implying that virtual communities are somehow less *real* that Preece (2000) and other authors prefer to term them *online communities*.

Community networks can be seen as virtual (or online) communities whose focus of interest is a local area, often a city or an urban area (Schuler 1996). The members of a community network have different professional interests, political views, and social environments: what they share is the fact that their lives inhabit the same urban space. Here, the interplay between the online and the offline dimensions becomes strikingly clear.

This paper explores the interplay between the actual and the virtual, taking as a test bed one of the first – and still active – European community networks, namely the Milan Community Network (hereafter RCM for *Rete Civica di Milano*), operating since 1994. This paper analyses RCM by adopting a series of metaphors to discuss how ICT (Information and Communication Technology) applications that support private as opposed to public dialog differently augment the chance for citizens to participate in public affairs.

The interplay between actual (or "offline") and virtual (or "online") worlds, although a phenomenon rooted in the essence of the internet – which originally aimed to support interaction among the members of the worldwide research community – is now assuming importance at a skyrocketing pace. This growing importance is partly due to several concurrent factors (such as the worldwide spread of environments like Second Life[1]) that straddle the strong relationship that binds the actual and virtual dimensions together.

---

1 www.secondlife.com.

While studying post-modern society, sociologists (e.g. Maffesoli 1996) have explained the role of (online or offline) "tribes" as a means for reconstructing a fragmented society through new forms of community that link together *identity*, *relationships*, and *space*.

Because these three concepts are the basis on which social relations appear to be built in (online) communities, this paper first briefly discusses identity, relationships and space and the interrelations of the three concepts. These concepts are then used as a key to interpret and describe the evolution of the Milan community network. We aim to show how the interplay between the online/virtual dimension and the offline/actual dimension has extended, enriched, and augmented the city of Milan, as well as the private and professional lives of those who live or stay there.

The Interplay Between Virtual and Actual: Identity, Relationship and Space

The relationship between online and offline life has been widely studied in recent years, and the key concepts of identity, relationship, and space have emerged as central to this field. We briefly summarize the main research streams for each of these terms and then sketch their mutual relationships.

Identity

Personal identity is socially constructed through communication and interaction: we perceive everything in relation to its relevance and context, never as the thing itself, due to the 'interconnectedness of all things' (Bateson 1999). To be more specific, identity is socio-culturally constructed through interaction with social and cultural environments, while the body acts as the *mediator* through which identity is developed and stored.

In the Cyberculture framework (see for instance Markham 1998; McKenna and Bargh 1998), it was assumed that technology allows people to detach from the actual world, inventing a completely different online identity, unconnected to the actual one: the physical/actual world is left behind when entering the cyberspace. However, it has quite recently emerged (see for instance Graham, 2002) that personal identity is based on the interaction between physical and virtual elements even when identity is considered in terms of the online world (McLuhan 1964), thus leading to a completely different conclusion from that proposed by the "traditional" cyberculture perspective.

Indeed, in the actual world, our body is the *mediator* in creating our personal identity, but when the body is put aside – precisely as it happens in online social interaction – technology replaces it. Paraphrasing Marshall McLuhan, we can see 'technology as an extension of man' (Lister et al. 2003).

In this vein, Manuel Castells says that people with online identities are still nevertheless 'bound by the desires, pain and mortality of their physical life' (Castells 2002, 118) while several case studies support the assertion that online identity extends offline identity: see, for instance the analysis of RumCom.local newsgroups in Rutter and Smith 1999. Hence we can say that identity is socio-culturally constructed for both environments: the virtual and the actual.

Moreover, identity in the actual world is continuously evolving, due to interaction with the multiple socio-cultural contexts we come across during our lives (Maffesoli 1996). Online, this phenomenon is enforced by the fact that the internet is intrinsically "global", thus supporting and multiplying worldwide cross-cultural social interactions.

Strictly related to personal identity is the manner each of us adopts while introducing themselves to other people ("self-presentation" or "impression management"). In the actual world people – both consciously and unconsciously – offer verbal and non-verbal information about their personality, attitudes, and abilities. This information is usually adjusted so as to show their best and to gain acceptance in a specific social environment (Goffman 1959; DePaulo 1992; Leary 1993). Applied to online environments, these theories show that online and offline impression management work in very similar ways. The "cyberselves" are built through *presentation*, *negotiation*, and *signification* (Waskul and Douglass 1997), and evolve over time due to the ongoing interactions with others, exactly has happens to our "actual selves".

Studies in this area seem to indicate that, although people like to indulge in some experimentation with their self-projection, identity play decreases with time. In other words, the longer people use MOOs (MUD *Object Oriented*)or chats, the more likely they are to produce self-presentations that are more "authentic". Thus, people's virtual personality tends to stay increasingly the same, or at least to change over time at the same pace as actual personality[2] (Schiano and White 1998, Becker and Mark 1999, Cheng et al. 2002). Hence, even if one is not "authentic" in her first online self-presentation, over time 'her true self will seep through' (Leary 1993; Turkle 1995; Curtis 1997; Roberts et al. 1996).

Relationship

The online world also has relevant effects on *relationship*, on the natural tendency of people to gather in associations, and – more in general – on community life.

These effects can be seen through dystopian or utopian lenses. On the one hand, the internet is seen as a means to increase social alienation and the erosion of community life (see for instance Dreyfus 2001, Putnam 1995b), even though it acknowledgedly helps build social relations, because such relationships cannot be compared to those of actual life from which they subtract time. On the other hand, the internet is seen as a social glue binding collective intelligence, the matrix on which the global village germinates and develops (de Kerckhove 1997).

2 The Bechar-Israeli (1996) study confirms this hypothesis: they queried regular participants in several discussion forums on the nature and meaning of their online nicknames, finding strong indications not only that people used nicknames as impression management devices, but also that a relevant relation between the nickname and the actual personality exists for nearly half the sample (45 percent).

Both positions appear too deterministic. As is often the case, the truth may lay in the middle: the internet should be looked upon as a "space of possibilities" supported by technologies that are unable – on their own – to built or disrupt social networks (Wellman 2005). People need to maintain their social relations in the actual world; the net's impact depends on the use people – and organizations – make of it.

Space

The internet has tickled the interest of a large number of different disciplines (geography, architecture, urban planning, informatics, etc.), from which alluring suggestions can be drawn about the role of actual vs. virtual *space*. In the virtual world, people generate a "sense of place" – exactly as it happens in the actual world (Mitchell 1995) – and tend to interact with virtual space using the same metaphors adopted for the actual world. Cyberspace, like its actual counterpart, can be zoned, trespassed upon, interfered with, and split up into a small landholdings similar to actual property holdings (this is fairly evident in 3D environments like Second Life, where people arrange virtual spaces according to the same principles they are accustomed to in their actual lives).

These effects are sometimes emphasized when they involve online communities: just as actual communities need an appropriate mix of private and public spaces to prosper, their online versions need analogous spaces carefully designed to effectively support the social interactions that underlie community life. It is through the balance of these two types of space that we encourage spontaneous conversation and social-network building among "neighbours." Such interaction is the terrain upon which strong relationships, sense of community, and identification germinate (Wenger et al. 2002).

We observe that the three concepts – identity, relationship, and space – are strongly linked and enforce each other in both environments (the actual and the virtual). The use of effectively designed spaces actually enforces (and is enforced by) the building of strong social networks – that is to say a net of relationships – but strong social networks constitute an ideal environment for expressing and evolving personal identities.

This implies that an appropriate "use" and mix of these three elements may serve as a fulcrum to achieve noteworthy results when dealing with online communities that aim to promote and support citizens participation at the local level. In the following sections, by looking at the case of RCM through this lens, we derive some hints that help understand and explain its success and evolution, providing empirical support for our claim.

Milan Community Network – RCM

The Milan Community Network (*Rete Civica di Milano,* RCM for short), is based in Milan, the capital of Lombardy, which is the most developed of Italy's 20 regions. It began operating in September 1994 as the first initiative of the Civic Informatics Laboratory – LIC at the Department of Computer Science and Communication of the University of Milan. Its aim was, and still is, to promote the enlightened use of (ICT) in the local community through specific projects and initiatives that address the various social actors (private citizens, local public bodies, nonprofit associations, schools, and private companies, particularly professionals and small businesses) who play a role in the City of Milan and surrounding areas, especially within the Province of Milan (De Cindio 2004).

Salient features of RCM's history (interwoven with Italy's)

The first community networks in Italy were set up in 1994-95, a unique period in the nation's history. In 1992-93, a group of public prosecutors at the Milan Bench laid bare the corruption that had characterized Italian democracy since the end of the Second World War. Italian politics had been immobilized for some 45 years, due to the dominance of a small set of political parties, two of which dominated the scene: the Christian Democrats, who headed the national government without interruption from 1948 onwards, and the Communist Party, the main opposition throughout the period. These two parties were surrounded by a number of smaller satellite parties, including the Socialist Party, which acquired a major role in the 1980s. During the period following the corruption scandals and the related investigations, most leading statesmen were indicted and, in several cases, imprisoned. The political class underwent radical change, while civil society – then known as the "fax people" because citizens sent faxes to newspapers in order to shout out their point of view – demanded a more significant role. These events brought about the transition from Italy's so-called *First Republic* to the *Second Republic*.

RCM was conceived and designed in this socio-political context. Building a community network was deemed a means to give voice to those who had never had a chance to take part in civic and political arenas. The net was seen as a tool to counter Italy's closed society, dominated by a "membership" culture, where only those who belonged to a certain "sect", basically a party or a faction within it, were to be trusted. The net represented a new way to open uncharted domains of possibility for everybody: it meant not only giving access to everyone, but also taking a step forward in the direction of reinventing citizenship and democracy (De Cindio 2000).

Against this backdrop, RCM was opened as a "declaration of possibility": a group of people within the University of Milan, namely the Department of Computer Science and Communication, fully aware of the profound impact that ICT was to have on society as a whole, decided to offer their skills to the local community by adopting an approach, inspired by North American community networks, that was quite unusual in Italian academia. They set up a technological environment within which the local community could gain experience and learn hands-on how to exploit online interaction. The community could seek support while developing its own ICT-based projects.

In 1997 after three years of operation, having been recognized as a standard for developing the information society in Milan and beyond, RCM asked local institutions to provide it with support and stability. A year later, RCM became a participatory foundation, the *Fondazione Rete Civica di Milano* (FRCM). Local government bodies (the Region of Lombardy, the Province of Milan, the Milan Chamber of Commerce) and the University of Milan, where RCM maintained its operating headquarters, became charter members of FRCM. Each of these organizations has a representative on the FRCM board. Other constituents of the local community (businesses, citizens, schools, nonprofit organizations) were also included as members of the Foundation. Both the community network and the Foundation are still active today.

In its early stages, RCM was managed by a small – but highly committed – staff: a university professor, a computer professional who had then just graduated in computer-supported cooperative work (CSCW), and a computer science student with ten years of direct experience in the early bulletin-board systems (BBS). Over the years, RCM kept its staff as small as possible, while developing a network of professionals available to cooperate with RCM on specific projects. Many of these professionals are former students who got their degrees at the Civic Informatics Laboratory, supervised by RCM staff. Graduating students have always been, and still are, a major resource for RCM because they are, so to say, engines of innovation.

RCM's long real-world experience made it perfectly clear that the most important element for sustainability is not merely the "right" staff size but primarily having the "right" mix of skills distributed among members of its staff: legal, technical, managerial, and social (Ripamonti et al., 2005).

About RCM's technical infrastructure

In 1994, the web was in its very early stages and web-based applications that supported social interaction were not user-friendly enough to be accessible to ordinary citizens. On the other hand, the software used by North American freenets, FreePort, with its text-only interface, was hardly going to be appealing to people, as windowed interfaces had already become standard.

In this context – between the obsolete past and the immature internet – RCM had little choice but to use BBS software, namely FirstClass (by SoftArc, now OpenText), already adopted in Europe with great success by the Open University (UK). The essential requirements met by this platform were the following:

- highly interactive, supporting active citizens in producing content of local interest;
- easy-to-use windowed interface;
- support for a wide range of communication protocols, enabling exchange with other systems of interest (either linked to the internet and directly);
- client-side platform independence and, of utmost importance, a client application capable of reasonable performance even with cheap personal computers and modems, so as to be affordable to all potential users.

After adopting FirstClass, RCM strove mightily to integrate it into the rising worldwide web. However, this was increasingly problematic because of the barriers, typical of proprietary software, to accessing data formats and internal protocols. This forced RCM into an endless chase after the web (for example, FirstClass currently still does not support generating RSS feeds).

Table 9.1 RCM's Main Projects Grouped by Type of Target User

Macro-category	Project	Main purpose	Year(s)
Local government	*Collaboration with the City of Milan*	Teaching the basics of computer-mediated public communication to municipal employees and executives	1994 – 2001
	"Linee Dirette" with the City of Milan	Public forums, 'hotlines,' where citizens can dialogue with public administrators	1996 – 2001
	"Linee dirette" with the Province of Milan	Public forums, 'hotlines,' where citizens can dialogue with public administrators	1995 – present
	Appuntamenti metropolitani	Online service for the Province of Milan where events of local interest are collected (events can also be posted by citizens)	2000 – present
	TruE-vote (www.true-vote.net)	EU-funded project aimed at testing experimental e-voting software used in certified consultation with citizens	2001 – 2003
Citizens	*ComunaliMilano (www.comunalimilano2006.it)*	On occasion of municipal elections in 2006, platform for public discussions and giving candidates visibility during the municipal elections that took place in Milan during spring 2006	Nov. 15, 2005 - Dec. 2006
	partecipaMi (www.partecipami.it)	Successor to "ComunaliMilano", online interactive environment where citizens can dialogue with members of the Milan City Council and neighborhood councils	January 2007 - present
Schools	*Scopri il Tesoro (www.retecivica.milano.it/tesoro)*	Providing children and teachers with an environment for learning ICT by playing online treasure hunts	2000 – present
Nonprofit associations	*AssociazioniMilano (www.associazioni.milano.it)*	Helping associations manage their online presence independently, a project launched on behalf of the Province of Milan bringing together about 300 nonprofit organizations	2000 – present
Small and medium enterprises	*Partecipate: Just do it! (www.partecipate.it)*	To overcome low rate of Internet use among SMEs, provides "JLI!" e-learning environment and a community of practice for exchanging experience (40 small businesses registered in project's first year)	2003 – 2005

Projects developed by RCM

RCM has promoted, hosted, and enabled several projects. All of them have always been carefully vetted in order to guarantee consistency and synergy with the community network's primary goals. As a result, unlike what happened to other community networks during the net-economy boom (e.g. the Amsterdam Digital City, see Lovink and Riemens 2002), RCM has always avoided turning into a website design farm, even when this could have been boosted its bottom line.

RCM's main projects can be grouped into macro-categories according to the specific segment of local society they address, as summarized in Table 9.1.

It should be noted that ComunaliMilano2006 and partecipaMi, although listed in Table 9.1 to underline the reluctance of the Municipality to open a true dialogue with the citizenry, are not standalone projects, but represent an extension of RCM.

Profile of RCM population

Citizens log on to RCM without charge, each using a personal account, through either a standard web browser or the client application. To become members of the RCM community, citizens must register, via a procedure whose essential features are:

- providing true first and last names;
- providing a short public self-description (e.g. with personal interests);
- signing RCM's *Galateo*, a statement of ethics that – along with common netiquette and Italian law – governs and structures relations among RCM members (De Cindio et al. 2003).

More than 20,000 citizens have joined RCM since it opened. About 2,000 (10%) are now active members of the online community, and this percentage is slightly less than that established by theories about degrees of community participation (Wenger et al. 2002; Lave and Wenger 1991). As Wenger et al. pointedly remark: 'We used to think that we should encourage all community members to participate equally. But because people have different levels of interest in the community, this expectation is unrealistic.' (Wenger et al. 2002, 55). They also distinguish three main levels of participation in communities of practice. Although communities of practice are but one type of community and can count on mechanisms that promote involvement, the different levels of participation remain valid in general, and the percentage indicated in Wenger et al. (2002) for each level fits quite well with RCM data (Figure 9.1):

- At the first level lies a tiny "core group": people who actively participate in discussions and identify with the community. This group consists of the community manager and the staff managing the community, but also extends to other members who have over time become "auxiliary" to the staff, acting as moderators in specific discussion areas.
- At a second level is the "active group", generally comprising 15-to-20 per cent of community members (in RCM's case) This group of people actively participates in community life though less regularly than the core group;

- At the final level stands the "peripheral group", which is the largest. People at this level rarely participate actively in community life. Nevertheless they are not outsiders, because they regularly observe – "lurk" in virtual community jargon – what goes on among other members. In many cases, peripheral participation should not be deemed a negative attitude, since it anyway fosters learning and a sense of community (Lave and Wenger 1991). Moreover, we have noticed that lurkers do occasionally become active, whenever they come across topics they are highly interested in.

We have also identified a fourth group, which we have called "returnees", the former RCM members who left for a while – they usually do not log on for a few years – and then return to their active membership. This group adds up to about one percent of total RCM population and some 20 percent of it then becomes part of the active group.

Last but not least, it is interesting to note that – at least until 2003 – the RCM population's distribution across age and literacy classes was generally more similar to that of the Italian population than was the comparable distribution of all Italian internet users (for details, see De Cindio et al., 2007b).

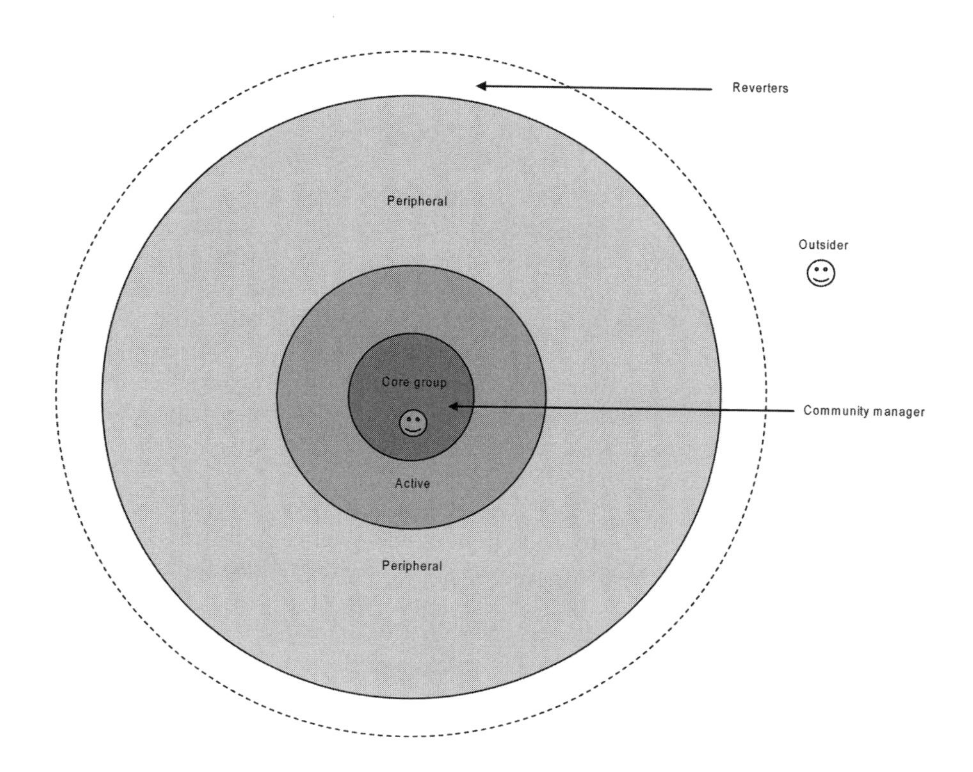

Figure 9.1 Degrees of Community Participation (adapted from Wenger et al., 2002)

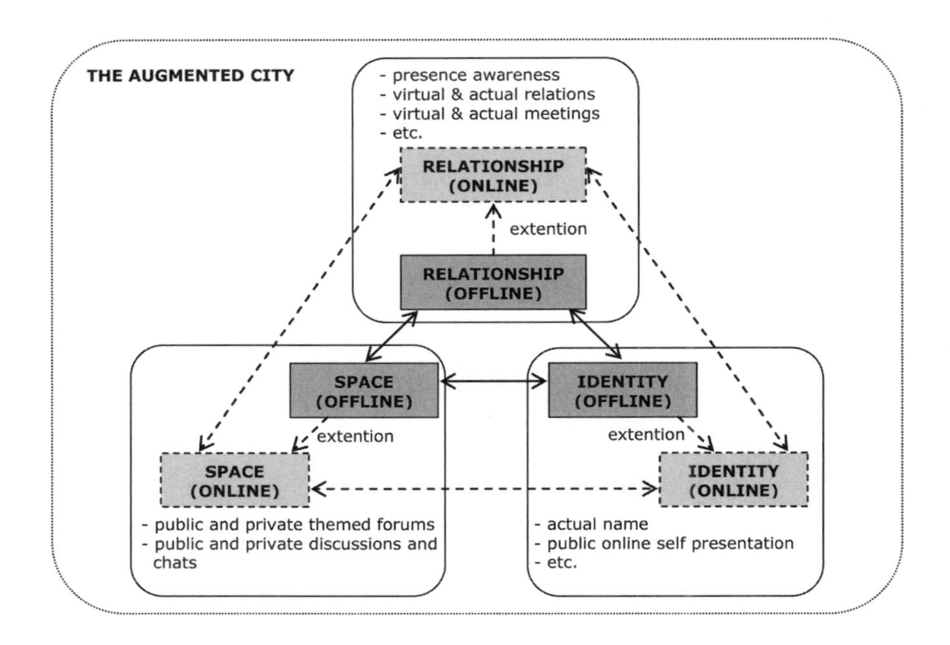

Figure 9.2 The Augmented City: Relationships Between Actual and Online Dimension

RCM Supporting the Augmented City

RCM members are – in the vast majority of cases – also citizens of the *actual* Milan. In view of this, it is interesting to analyze how actual citizenship can be mapped onto online membership. The framework defined above, implicitly describes two superimposed circles of interaction: the first in terms of actual "citizenship" and the second in terms of its online counterpart, where – as we saw – actual identity, relationship, and space are extended by their online manifestations. Applying this model to RCM yields a rather complete mapping with the structure of the community network, since all three components are appropriately mirrored in RCM's socio-technical infrastructure (see Figure 9.2). We are firmly convinced that this superimposition of the virtual dimension onto the actual dimension is crucial to increasing commitment and participation in online civic environments, creating a sort of "augmented city", which has been one of the critical success factors in several RCM projects.

To demonstrate these assumptions, let us briefly dig into RCM's past and present by analyzing it in terms of identity, relationship, and space.

Actual and virtual personal identity in RCM

As we saw above, online personal identity and self-presentation are extensions of their offline counterparts, except perhaps for the initial phase of online "life", which, for most people, is devoted to "experimentation" and learning by trial and error. In RCM, the interplay between actual and online identity is supported by a mixture of technological features and the community network's behavioral rules (De Cindio et al. 2003). RCM members must register using their actual names. People's names plainly appear when they send public messages to a forum. In addition, they must provide a self-presentation visible to other community members (but not to outsiders), which can be thought of as an impression-management device. The self-presentation generally emphasizes their interests, qualities, etc., providing therefore an online extension of their actual identity.

In this sense, the mapping between online and actual in RCM is even stricter than average for online communities, where actual names are often replaced by nicknames. This is quite appropriate for a community of citizens focused mainly on city problems, where the ability to clearly recognize, say, a neighbour, a public official or a local politician has obvious significance.

The policy of requiring members to use their actual identity dates back to the founding of RCM and has never been questioned, since it was accepted and perceived as good by the vast majority of RCM participants. The only changes that have taken place over time involve the procedure for providing the information. Initially new members had to fax their ID card, but the procedure has now been simplified by replacing faxing with a fairly detailed online form. The information collected through the form is then manually vetted by the community manager in order to guarantee, as far as possible, that it is genuine, though an awareness that abuse based on artful false pretenses is still possible.

Several remarks are in order about the rationale for – and effects of – requiring people's actual identification as a prerequisite for becoming RCM members:

Names: When RCM began, Italian laws about managing an online community and about the internet generally were fuzzy; it was 1994 after all. To protect themselves and release the University, which hosted RCM, from any responsibility, RCM's founders came upon the option of requiring people to appear in the community with their actual identities (each post in every RCM forum is signed with the name of its creator) and to provide their actual identification when registering. Thanks in part to such precautions no significant problem has ever arisen. Moreover, the interplay between actual and virtual citizenship allowed several online disputes to be handled more easily, so they could be settled in actual life instead of just affecting virtual life.

Profiles: In addition to appearing online with their actual names, RCM members are expected to provide a short, publicly accessible profile (self-description). This combination makes each RCM member highly recognizable, which has both positive and negative effects. On the one hand people may know exactly who their interlocutor is. On the other hand, someone well-known in the actual world can risk being attacked in the virtual environment every time she expresses strong positions, often not because of her opinion but for what she represents. This effect has led some people to leave the RCM community;

Reputation: Extending actual identity with virtual identity has given several people in RCM a chance to rebuild their "reputation". One outstanding example is a fellow who, in spite of previous trouble with the law – not mentioned in his profile, hence unknown to most of other RCM members – has managed to create and run a nonprofit association that assists the homeless: RCM has been a significant factor in rebuilding his life.

It is not always easy to discuss specific problems using someone's actual name: there are topics for which anonymity is a prerequisite to any discussion truly free from conditioning. For this reason, in few very special cases, such as a forum about the relationship between homosexuality and society, RCM has guaranteed anonymity to its participants. In the few instances when users managed to registered under false pretenses, they were very soon detected by the community manager. In accordance with the literature about self-presentation and identity, detection in each case was due to the fact that it is not easy to assume a different identity in a community where actual selves are so important. Given the importance of identity, 'your true self seeps through' in very short order and you are easily discovered by a keen eye, used to managing the community, or by your co-citizens.

Actual and virtual relationships in RCM

The net is a space full of possibilities whose eventual outcomes largely reflect the use that people and organizations make of it. In this sense, the fact that RCM members are easily recognizable has some implications that mainly support the optimistic point of view about the impact of online social relations on offline community life. It is undeniable that the social network of a Milanese who belongs to RCM is reinforced by their shared online community, at least insofar as in-town relationships are concerned. Old schoolmates, former neighbours, etc. meet again on RCM, immediately recognize each other, and get in touch – as often has happened. Groups of long-time members have organized into a welcome committee that helps newbies, and so forth.

But what really makes the difference is that people – thanks to relationships begun on RCM – have often been able to overcome otherwise insurmountable barriers. This is true not only for private experience (several members have found new houses through acquaintances made online or have easily contacted public officials), but especially when projects begun within RCM are considered. For example, the *Scopri il Tesoro* (Find the Treasure) project (see Table 9.1) was launched relying entirely on the social network of a group of teachers who belonged to RCM. Furthermore, the RCM community has also made the difference in many "formal" projects: for example, in the European-funded TruE-vote project (see Table 9.1) the existence of a well-established social network among members provided an effective, proactive testing environment for software developers, unlike what happened with other partners in the consortium, whose efforts to actively involve citizens through their municipalities basically failed (De Cindio et al. 2007b). Further additional examples of this positive impact of the virtual on actual relationships could be mentioned, far surpassing the scope of this paper.

Actual and virtual space in RCM

As we remarked above, an online community develops a certain sense of space that has to be appropriately supported through the careful design of public and private areas. When the first civic telematics initiatives appeared in Italy, two major approaches were adopted – quite unconsciously – to represent the spaces of the city: one strove to reproduce the spaces found in the actual city (for instance, Iperbole's homepage[3] offered citizens of Bologna a reproduction of the city's most important buildings), while the other adopted metaphors to map concepts meaningful to citizens onto the virtual city.

The second approach is well represented by RCM, which – due in part to technical constraints – avoided any metaphor inspired by the actual urban landscape, creating instead collections of forums grouped into macro-areas according to their semantic meaning to the inhabitants. This led, for example, to the creation of: "School Town," "Kidsville," "Sports Arena," "Fountain of Knowledge," and other macro-areas. It also gave rise to areas named after the actual buildings that house government offices: "Palazzo Marino," named after the ancient downtown building occupied by city government, "Palazzo Isimbardi," housing the Province of Milan, and "Il Pirellone," housing the Region of Lombardy. People commonly use such names as metonymic sobriquets for specific government offices; hence the names of these buildings are both significant and *evocative* to citizens.

Both approaches support the extension of the actual town into the online version, but they use different perspectives. In the former approach the extension is somewhat more topographic, while in the latter it reflects the explicit and tacit knowledge citizens have of their town. This different focus seems to have profound implications: a less formal and more evocative approach to local public institutions fosters citizens' creativity, turning them into a precious wellspring of suggestions and ideas. The evocative approach manages to harness social capital, which would not be possible through a formal, institutionalized website, a medium generally perceived as only a top-down source of information.

A final remark: as time went by and people grew more accustomed to the RCM metaphors, they gradually disappeared – except for few cases – making room for more standardized names.

Conclusions

In his work *Du Contrat Social* (1762) Jean-Jacques Rousseau writes: '*Les maison font la ville, mais les citoyens font la cité'*. His words recall the distinctions ancient Romans traced between the *urbe* and the *civitas*. Both refer to the city, while adopting different perspectives: the *urbe* is a territory equipped with a set of buildings, streets and related infrastructures, while the word *civitas* refers to the community of people living on that territory.

3 www.comune.bologna.it

Although this distinction remains valid nowadays, both dimensions have been supported and augmented by the technological evolution, and especially by the ICT.

Digital technologies can be employed to augment the *urbe*, the *civitas* or (hopefully) both. In the first case the focus is on the development and empowerment of electronic services (i.e., the augmented version of "traditional" infrastructures: wi-fi, broadband connections, online services, etc.). In the second one, the *civitas* is augmented by exploiting the power of internet-based communication channels to consolidate existent social aggregations as well as to create new ones (i.e., public forums in the public bodies' websites for dialoguing with citizens, non-institutional discussions in online social networks of citizens, etc.).

The different degree into which these two possibilities mix online maps into a continuum. This continuum starts from very institutional portals, directly managed by the municipalities, that provide official information and may offer e-government services (often denoted as "digital cities") and reaches a variety of websites that provide aggregation places for the community, often enriched with dedicated information and services. Community networks have been significant pioneers of this kind of grassroots sites.

Keeping in mind the dichotomy between the *urbe* and the *civitas* is also helpful for analytical purposes, as we have shown through the paper. In fact, when the goal is augmenting the *urbe* – i.e. the infrastructural services – it will be necessary to adopt more "traditional" and technical approaches (e.g. to network design, performance measures, load balance and the like). When, instead, the focus is on augmenting the *civitas*, we should look at the individuals and their social networks centred in the local community.

In the paper we have considered identity, relationship, and space and their interplay in the actual and virtual dimensions. Analyzing the history of RCM through these categories has enabled us to better understand some hidden dynamics underpinning its development. On this basis, we believe that they could be fruitfully put at work for designing purposes, namely for understanding how the possibilities offered by the new generation of web applications, often denoted as web 2.0, can foster the overcoming of the dichotomy between the *urbe* and the *civitas*.

Acknowledgments

We wish to thank Oliverio Gentile, RCM's community manager. Without his relentless work, RCM would have not been possible, thus demonstrating once again that the human component is essential to the functioning such socio-technical systems. Heartfelt thanks to Philip Grew, the patient and precise reviewer of our English.

References

Bateson, G. (1999, originally published 1972), *Steps to an Ecology of Mind: Collected Essays in Anthropology, Psychiatry, Evolution, and Epistemology* (Chicago: University of Chicago Press).

Bechar-Israeli, H. (1996), 'From bonehead to clonehead: Nicknames, Play, and Identity on Internet Relay Chat', *Journal of Computer-Mediated Communication*, 1:2.

Becker, B. and Mark, G. (1999), 'Constructing Social Systems through Computer-Mediated Communication', *Virtual Reality* 4, 60-73.

Castells, M. (2002), *The Internet galaxy: Reflections on the Internet, Business, and Society* (Oxford: Oxford University Press).

Cheng, L., et al. (2002), 'Lessons Learned: Building and Deploying Shared Virtual Environments', in Schroeder , R. (ed.), *The Social Life of Avatars: Presence and Interaction in Shared Virtual Environments* (Berlin: Springer) 90-111.

Curtis, P. (1997), 'Mudding: Social phenomena in text -based virtual realities', in S. Kiesler (ed.), *Culture of the Internet* 121-142 (Hillsdale, NJ: Erlbaum).

de Kerckhove, D. (1997), *Connected Intelligence: The arrival of the web society* (Toronto: Somerville House).

De Cindio, F. (2000), 'Community Networks for Improving Citizenship and Democracy', in Gurstein, M. (ed.) *Community Informatics* (London: Idea Group Publ.).

De Cindio, F. (2004), 'The Role of Community Networks in Shaping the Network Society: Enabling People to Develop their Own Projects', in Schuler, D. and Day, P. (eds.) *Shaping the Network Society: The New Role of Civil Society in Cyberspace* (Cambridge, MA: MIT Press).

De Cindio, F. et al. (2007a), 'Deliberative community networks for local governance', *International Journal of Technology, Policy and Management*, 7:2, 108-121.

De Cindio, F. et al. (2007b), 'Community Networks as lead users in online public services design', *The Journal of Community Informatics*, 3:1.

DePaulo, B. M. (1992), 'Nonverbal behavior and self-presentation', Psycholo*gical Bulletin* 111:2,203-243.

Dreyfus, H. L. (2001), *On the Internet* (New York: Routledge).

Goffman, E. (1959), *The Presentation of Self in Everyday Life* (New York: Anchor Books).

Graham, M. (2002), *Future Active* (New York: Routledge).

Lave, J. and Wenger, E. (1991), *Situated Learning: legitimate peripheral participation* (New York: Cambridge University Press).

Leary, M. R. (1993), 'The interplay of private self-processes and interpersonal factors in self-presentation', in J. Suls (ed.), *Psychological Perspectives on the Self* 127-55 (Nahwah, NJ: Erlbaum).

Levy, P. (1995), *Quest-ce que le virtuel* (Paris: La Dècouverte).

Lister, M., Dovey, J., Giddings, S., Grant, I. and Kelly, K. (2003). *New Media: A Critical Introduction* (London: Routledge).

Lovink, G. and Riemens, P. (2002), 'A Polder Model in Cyberspace: The Contemporary Amsterdam Public Digital Culture', in *Shaping the Network Society* (Cambridge, MA: MIT Press).

Maffesoli, M. (1996), *The Time of the Tribes* (London: Sage).

Markham, A. N. (1998), *Life Online: Researching Real Experience in Virtual Space* (Walnut Creek, CA: AltaMira).

McKenna, K. Y. A. and Bargh, J. A. (1998), 'Coming out in the age of the Internet: Identity "Demarginalization" through virtual group participation', *Journal of Personality and Social Psychology* 75:3, 681-694.

McLuhan, M. (1964), *Understanding Media* (New York: McGraw-Hill).

Mitchell, W. (1995). *City of Bits: Space, Time and the Infobahn* (Cambridge, MA: MIT Press).

Poster, M. (1995) The Second Media Age, Cambridge: Polity Press.

Preece, J. (2000), *Online Communities: Designing Usability, Supporting Sociability* (Chichester, UK: John Wiley).

Putnam, R. D. (1995), 'Tuning in, tuning out: The strange disappearances of social capital in America'. *PS: Political Science and Politics* 28, 664-683.

Ripamonti, L. A. et al. (2005), 'Online communities sustainability: some economic issue', *The Journal of Community Informatics*, 1:2, 63-78.

Roberts, L. D., Smith, L. M. and Pollock, C. (1996), 'Exploring virtuality: Telepresence in text-based virtual environments', paper presented at the Cybermind Conference, Perth, Western Australia: Curtin University of Technology.

Rousseau, J. J., (1762), 'Du Contrat Social', in: *Oeuvres Complètes* (Paris: Gallimard).

Rutter, J. and Smith, G. (1999), 'Presenting the Off-line Self in Everyday, Online Environment', *Identities in Action Conference* (Gregynog, UK).

Schiano, D. and White, S. (1998), 'The first noble truth of CyberSpace: people are people (even when they MOO)', in *Proceedings of CHI`98 Conference on Human Factors in Computing Systems* 352–359 (New York: ACM Press).

Schuler, D. (1996), *New Community Networks: Wired for Change* (Reading, MA: Addison-Wesley).

Turkle, S. (1995), *Life on the Screen: Identity in the Age of the Internet.* (New York: Simon & Schuster).

Waskul, D. and Douglass, M. (1997), 'Cyberself: The emergence of self in on-line chat'. *The Information Society* 13:4, 375-396.

Wenger, E., McDermott, R. and Snyder, W. M. (2002), *Cultivating Communities of Practice - A Guide to Managing Knowledge* (Boston: Harvard Business School Press).

Wellman, B. (2005), 'Connecting community: On- and offline', <http://www.chass.utoronto.ca/~wellman/publications/index.html>, accessed 10 July 2008.

Chapter 10

Enabling Communities in the Networked City: ICTs and Civic Participation Among Immigrants and Youth in Urban Canada

Diane Dechief, Graham Longford, Alison Powell, and Kenneth C. Werbin

ICTs and Urban Community

Social scientists and policymakers have been grappling for over a decade with an apparent decline in civic participation and community life in many western liberal democracies, particularly in the United States. *Civic participation* refers to individuals' active engagement with and involvement in their communities, and is a key determinant of individual and community development and well-being (Putnam 2000). There has also been lively scholarly debate about the role played by new ICTs, the Internet in particular, in contributing to downward trends in civic participation. Optimists argue that the Internet has transformative potential *vis a vis* civic participation and the augmentation of democracy (Lévy 2001; Mitchell 1999; Poster 1995; Rheingold 2000). Others argue that new ICTs erode and diminish social capital, civic participation and community (Nie et al. 2002; Kraut et al, 1998). Putnam and Sunstein, for example, worry that the Internet's tendency to foster the development of virtual "communities of interest" will lead to the balkanization of society (Putnam 2000; Sunstein 2001). A third approach views ICTs as supplementing existing social relationships and activities by facilitating coordination and communication, and extending those relationships and activities into cyberspace (Wellman and Hampton 1999).

The qualitative impacts of new ICTs on democratic life in high tech "networked" cities have also been debated. Castells warns, for example, that the networked city is an increasingly "dual city," one in which social inequality and exclusion are increasing (Castells, 1999). Others have documented the networked city's association with growing concentrations of wealth and economic power (Sassen 2001), gentrification and social exclusion (Solnit and Shwartenberg 2000; Zukin 1995), and privatization and the decline of public space (Sorkin 1992). Common to these critical perspectives is a treatment of the networked city and its development as the project of economic, political and planning elites, and of tectonic shifts and transnational forces in the global political economy.

The following essay reflects on the quality of democratic and communal life in the networked city through a different lens, however, emphasizing the countervailing role being played by citizens and civil society groups in shaping it for their own ends, particularly in ways that promote local civic participation, social inclusion and community. Through "community informatics" initiatives ranging from neighbourhood technology centres and public Internet access sites to community web portals and community wireless initiatives, civil society groups are engaged in diverse technology projects designed to augment local communities and public spaces (Schuler and Day 2004). Drawing on case studies of community informatics initiatives in urban Canada, conducted under the auspices of the CRACIN project, we examine grass-roots, community-based experiments in the use of new ICTs to empower individuals and urban communities and promote local civic participation and social inclusion.

Community Networking and Civic Participation

Much of the concern about the detrimental impact of ICT use on civic engagement and community stems from concerns about the displacement of face-to-face relationships by interaction within deterritorialized "virtual communities" (see for instance: Putnam 2000). Given that community networks use ICTs to increase civic participation within local, proximate communities, such concerns do not readily apply. As Carroll and Rosson argue, community networks '[entrain] quite different social protocols and consequences' than virtual communities, and may well help to ameliorate some of the technologically-induced shortcomings of contemporary civic life (Carroll and Rosson 2003). Community networks strive to promote civic participation and engagement by providing local information infrastructure, services and supports to communities, such as free Internet service provision, email/listserv/web hosting, ICT training, community information and directories, and electronic discussion forums.

A growing body of international research documents the contributions to local civic engagement and participation made by community networks (Gurstein 2000; Keeble and Loader 2001; Marshall et al. 2004; Schuler and Day, 2004; Tanabe et al. 2002). Case study research conducted at community networks in Amsterdam (Lovink and Riemens 2004), Toronto (Wellman and Hampton 1999), Blacksburg (Kavanaugh et al. 2005), Seattle (Silver 2004), and Milan (De Cindio 2004), among others, illustrates the role that community networks can play in fostering local civic participation and engagement in urban settings.

Of particular interest to community informatics researchers and practitioners are the ways in which community networks provide opportunities for increased civic participation among disadvantaged groups, including low-income earners, racial minorities, women, and youth, which are often negatively impacted by the pressures of dualization referred to by Castells. Community networking initiatives in U.S. inner-city African American neighbourhoods, for example, have revealed positive impacts in terms of increasing awareness of community issues and resources, increasing communication and information-sharing among residents, and expanding social networks (Alkalimat and Williams 2001; Hill 2001; Pinkett 2003). Green and Keeble (2001) and Vehviläinen (2001) have documented effective community networking initiatives targeting disadvantaged women, while projects offering Internet access and ICT training to children and youth have also been studied (Clark 2003; Sandvig 2003).

All told, existing research on community networks and civic participation suggests that the former help to stimulate the latter by: providing access to ICT equipment and training; fostering the development of local social networks and social capital; promoting local information sharing and community engagement; and affording access to and participation in the local public sphere, both on and offline.

A number of recent CRACIN studies conducted at community networking sites in Canada shed further light on the civic dividends of community informatics initiatives. While Canada appears not to have suffered the precipitous decline in civic participation and social capital that Putnam has found in the U.S., there are a number of worrying trends (Schellenberg 2004). While many Canadians belong to and participate in community groups, vote in elections, and maintain relatively extensive networks of friends and family members, only a small minority – less than 10 per cent – are actively engaged in their communities in a sustained and meaningful way. This group has been referred to as Canada's "civic core" (Reed and Selbee 2000). Furthermore, this "civic core" is a relatively unrepresentative group of Canadians who tend, on average, to be middle-aged, well-educated, and affluent (Reed and Selbee 2000). Among those groups that are significantly underrepresented in Canada's "civic core" are new immigrants and youth, groups that report lower than average rates of civic engagement, political participation and social capital. One CRACIN research project focused on the involvement of new immigrants at an urban community network in Vancouver, with a particular focus on how volunteering at community access sites helps ease the integration of new immigrants into Canadian society (Dechief 2005). A second project involved interviews with community network administrators from across Canada, with the aim of exploring how staff and users understand "community" and the conditions and processes of its creation (Werbin, 2006). A third research project examined youth participation in a community wireless networking project as an emergent form of civic engagement (Powell 2006). The remainder of this paper reviews the major findings of these studies.

The Vancouver Community Network: Building Community in the "Third Spaces" of public internet access

Recent CRACIN research by Dechief (2005) explored the role that urban community networks in Canada play in promoting civic participation by new immigrants and other marginalized groups. Through surveys, interviews, and participant observation at the Vancouver Community Network (VCN), Dechief explored 'how human and social capital is built at VCN, and how it contributes to the social inclusion and integration for immigrant volunteers' (2005).

The Vancouver Community Network (VCN) is an urban community network that offers a variety of free services to individuals and non-profit groups in Vancouver, including dial-up Internet service, public access computing space, computer training, and e-mail, listserv and web site hosting. VCN has 11,000 individual members and over 1,200 non-profit groups that take advantage of its services. VCN works closely with community groups and community centers to equip and train staff and volunteers with computing resources, and to develop interactive websites to make their programs better known and more accessible to the local community. Many of VCN's public computing initiatives focus on using new ICTs to organize and empower marginalized individuals and groups. VCN coordinates hundreds of public Internet access sites throughout the city, many of which are situated and designed to serve the poor, new immigrants, youth, and the homeless, including the residents of the city's Downtown Eastside neighbourhood, one of Canada's poorest.

Dechief conducted qualitative and quantitative research at VCN during the spring and summer of 2005. Qualitative research was comprised of immersion in the centre's activities and input gathered from community members and volunteers. A survey of VCN volunteers and analyses of organization-generated data and documents accounted for the quantitative components. VCN volunteers who had immigrated to Canada in the past five years shared their experiences through interviews, and many others contributed through surveys. The interviews focused on newcomers' reasons for volunteering at VCN, what they saw as benefits of their experiences, and how they accessed information.

Dechief's analysis of VCN's volunteer database revealed that recent immigrants constitute more than 60 per cent of VCN's volunteers. A key finding of her research at VCN is that while its new immigrant volunteers tend to be among the "digitally included," they are often economically excluded and socially isolated (Dechief 2005). Virtually all new immigrant volunteers possessed university degrees or technical diplomas and many had previous work experience in a computer-related field in their country of origin. The vast majority also reported a high level of skill and comfort using new ICTs, including computers and the Internet. Having said that, however, new immigrants are often among the economically excluded and/or socially isolated in Canada. While no income data on study participants was collected, other data collected suggest the precariousness of their financial situation. 50 per cent reported that they were unemployed and seeking work, while a further 27 per cent were attending school. In addition, more than 70 per cent reported having no Canadian work experience. In this respect, the precarious economic situation of new immigrant volunteers at VCN reflects the wider plight of new immigrants in Canada. Numerous studies point to deteriorating labour market outcomes and growing poverty among new immigrants in Canada (Schellenberg and Hou 2005).

The economically marginal situation of new immigrants helps explain their attraction to volunteering at community networking projects like VCN. Interviews with VCN's new immigrant volunteers revealed that most initially approached the community network for instrumental reasons related to finding employment. Most saw volunteering as an opportunity to gain Canadian work-related experience, to practice speaking and working in English, and to access information on employment opportunities. In other words, the initial attraction to VCN was the potential to establish a network of more or less instrumental relationships – or "weak ties" in Granovetter's sense (Granovetter 1973). Most of VCN's new immigrant volunteers have technical backgrounds in the field of computers, IT, and software development, and see their involvement in VCN as a way to maintain their skills and acquire new ones, including greater facility in speaking English (Dechief 2005).

Having said that, most volunteers reported that, over time, they valued the social benefits of volunteering at VCN more than the instrumental ones, particularly in terms of expanding their social networks, fostering a sense of community and, ultimately, easing their integration into Canadian society. Because of the diversity of the staff, volunteers and clients that pass through VCN's various locations every day, volunteers gained an appreciation for the diverse nature of Canadian society, as well as exposure to a community based on something other than shared first-language or home-country culture. Two volunteers described their interactions in VCN's heterogeneous setting in the following way:

> Every week I meet people from many different origins. It's the most interesting.

and:

> It is already a year since I started and I have found many friends here. I have friends from Yugoslavia, Germany, China, Austria, from France, from everywhere. Most of them have found jobs, but I keep in touch and sometimes we email.

Volunteers also describe VCN as a place where they feel socially supported. In the absence of full-time work, volunteering is one way of being engaged and feeling useful. Interacting with others in the shared circumstances of job-seeking and being a newcomer contributes to feelings of comfort and solidarity.

> You have to help each other. Because everyone is a foreigner here, it is easier if you help each other and get to know each other. That way you don't feel as depressed that you have left all of your friends behind.

A number of scholars have turned to the concepts of civic participation, social capital, and social inclusion to shed light on the settlement experience of new immigrants (Breton 1997; Caidi and Allard 2005a,b; Kunz 2003). Breton suggests, for example, that 'social participation can … sensitize group members to the fact that they are subject to the same economic, political, cultural or social conditions–such as immigrant status' (Breton 1997, 6). He suggests that through 'social involvement, people may realize that they share the same lot, are "in the same boat" as others in certain respects.' Newcomers can then identify with a "community of fate" (p.6). Putnam's distinction between "bridging" and "bonding" forms of social capital is also salient to the settlement process for immigrants. According to Kunz, 'bridging capital is … essential for immigrants to expand their networks beyond their own ethnic community and to acculturate into the receiving society' (Kunz 2003, 34). Thus, forms of participation, such as volunteering, that lead to increased and expanded social networks across a diverse range of groups and experiences can help facilitate and ease the process of settlement and integration.

Dechief also found that much of the social capital development and community-building among new immigrants took place during face-to-face interactions within the *physical* environs of VCN – administration offices, public access computing sites, training courses etc. While all of the volunteers interviewed have the digital skills required to keep in touch with friends and family and to find information online, they were all looking to connect with people in person. During interviews, volunteers said that meeting and getting to know people from diverse backgrounds, and making small talk or "chit-chat", are key rewards of volunteering. Although the volunteers are technically enabled and aware of opportunities for online interaction, they choose to make in-person contact with other volunteers and network members on a regular basis.

Oldenburg's (1989) concept of "third place" provides a way of understanding the way that VCN functions as a place of face-to-face interaction, and the role this plays in the settlement and integration process for new immigrants. "Third places" are locations outside the home or the workplace, such as coffee shops, parks or pubs which 'exist on natural ground and serve to level their guests to a condition of social equality'. It is in such places that people meet, chat and exchange information, and develop friendships. Third places nourish community not only by maintaining existing relationships, but also by enabling community members to make *new* acquaintances and connections with a wide variety of people. Oldenburg describes such third places as 'the core settings of informal public life.'

According to Dechief's research, while volunteers are drawn for instrumental reasons, they benefit enormously from the social affordances of VCN as a kind of third place that nourishes relationships and a diversity of human contact, creates a sense of place and community, encourages sociability, and that is accessible and welcoming to all members of the community. Third places like VCN also serve as community information hubs for staff, volunteers and clients. Participating in VCN activities connects new immigrants to formal and informal networks of information on a variety of subjects, ranging from immigration law and health services, to employment opportunities and tips on inexpensive places to shop. Possession of such knowledge empowers immigrants to take more control of their lives and insert themselves into and participate in their local communities (Dechief 2005).

Technology as a gateway to community

Dechief's findings are corroborated by research conducted by Werbin at a number of urban community networks in eastern and western Canada (Werbin 2006). In the course of exploring the meaning of "community" to various stakeholders in community networking initiatives, Werbin found that the affordances of physical "third spaces", where site users have the chance to meet others and develop social networks, plays a far greater role in fostering a sense of community than mere access to technology and opportunities for online interaction, particularly for new immigrants.

In his study Werbin conducted interviews with community networking practitioners about their experiences with community informatics initiatives. A key theme that emerged is that the sense of community fostered at community networking sites has less to do with the online activity they enable than one might expect. Contrary to widely held beliefs that civic participation and engagement are stimulated by providing public access to ICTs, study participants stressed the importance of the ways in which public internet access sites afford access to physical, in-person, "third spaces" organized *around* ICTs, where site users have the chance to meet others and develop social networks. Werbin"s research suggests that foreign-trained professional new immigrants place greatest value on community networks as physical third-spaces in which to meet others and develop social networks. As one community centre administrator puts it:

> They don't want to be in the corner programming, working on the computer, or surfing, they want to be talking, interacting, and learning to be comfortable in Canadian society, learning to speak better English ... It's like food is an excuse to get together with people, technology can also be an excuse to get together with people.

Practitioners report that the foreign-trained professional new immigrants do not see themselves as in need of ICT access and training; rather, they see themselves as people with specialized skills and experience who want to give something to their new communities and who hope to receive something in return - an opportunity to get connected with other people, to gain Canadian work experience, and to maintain their skills and expertise while trying to find jobs in their specialized fields. As another practitioner put it:

> this is not about access to the internet; it is about feeling like a part of a community (…) [Our site] is like a big family; you need help you come here; and if you can help you come here too! This is why [our site] is different from internet cafes or commercial computer schools, *it is a gateway to community.*

The results of both Dechief's and Werbin's research demonstrate that much the sense of community generated through community-networking initiatives in Canada is the result of the face-to-face interactions and social networking that take place within the "third-spaces" created around community informatics projects.

Considerable attention in the literature on community networking and civic participation is focused on community networks as platforms for or enablers of online civic participation. Less attention has been paid to the ways in which community networks serve as physical sites where staff, volunteers and users meet, mingle, develop social networks, and build community. Among the exceptions are studies by Clark (2003) and Sandvig (2003), who found that while youth users of public access computing sites seldom engaged in "civic" activities such as reading or discussing political news online, they nonetheless used the site as a place to meet, network, and share information and experience, and build community with peers and other members of the community. Dechief's and Werbin's research sheds much-needed light on the ways in which community networks facilitate the participation and inclusion of new immigrants by affording access not only to ICTs but, more importantly, to the communities and social ties that are built up around them and their use.

Building community while building something "cool": Île Sans Sil, Community Wireless Networking and Youth Engagement in Montreal

Powell conducted a participatory, ethnographic research project on a grassroots, community wireless networking group in downtown Montreal – Île Sans Fil (ISF) – between 2003 and 2007. Her work focused in part on the ways in which ISF helps to promote local civic participation and engagement among urban youth.

Community wireless networking: an overview The development, standardization and commercialization of wireless technology for amateur and home internet use in the last few years have enabled the ad hoc creation and sharing of wireless internet infrastructure in neighbourhoods and communities throughout North America, Europe and the UK. Wireless internet, also know as Wi-Fi for "wireless fidelity", is based on the Institute of Electrical and Electronics Engineers' (IEEE) 802.11 group of protocols. Since the introduction and standardization of this set of protocols there has been rapid growth and development in the wireless market. Wi-Fi enables the creation of wireless local area networks (WLANs), commonly refereed to as "hotspots". Much wider interconnected and overlapping "mesh" networks of hotspots can also be created to form a wireless "cloud" covering an entire urban area. While wireless internet technology has been taken up for a variety of commercial applications and services, including commercial internet access hotspot services at venues like Starbucks, it has also been adopted by amateur enthusiasts as well, who in a few short years have succeeded in creating an alternative, community-based infrastructure for internet access and use in dozens of major cities and smaller communities.

Community wireless network (CWN) began to emerge in North America, the UK and Europe around 2000, with many having roots in the wireless hobbyist and hacker groups of the late 1990s, such as the early "war drivers" interested in mapping open wireless networks (Sandvig, 2004). Early CWN groups (e.g. NYC Wireless) were initially preoccupied with the technical challenges of amateur wireless internet provision and signal sharing in local communities, and with experimenting with and pushing the limits of wireless equipment and unlicensed spectrum. A second wave of CWNs has emerged more recently that is concerned with using CWNs to support local content creation and community information infrastructures in order to foster local community engagement. These projects share an affinity with the community networking projects of previous generations that use computers and networks as a means to reinforce local communities.

Community wireless networking in Canada Research by Powell and Shade (2006) and Powell (2006) suggests that there are perhaps a dozen or so community-based wireless networks currently operating in Canada. CWNs in Canada take different institutional forms and have embraced a variety of organizational objectives and business models. Powell (2006) identifies and discusses four examples of CWNs in Canada. Île Sans Fil (ISF) focuses on free wireless internet access provision as well as location portal development and multimedia applications for its network of venue-sponsored Wi-Fi hotspots in downtown Montreal. ISF is volunteer-run and subsists on a limited number of grants from arts funding agencies, individual donations, and membership fees paid by hotspot venue owners. The Vancouver-based British Columbia Wireless Networking Society (BCWNS) focuses on social networking and CWN training for community members, with an emphasis on maintaining a network of aid and expertise to support wireless networking initiatives and capacity building in rural, remote and aboriginal communities. Another urban CWN, *Wireless Toronto* (WT), maintains a network of Wi-Fi hotspots deployed in public spaces (parks, public squares, etc.) in Toronto and develops content and multimedia projects to increase local community engagement (Cho 2006). *Ottawa-Gatineau Wi-Fi* (OG Wi-Fi), one of Canada's newest CWNs, has projects attached to social agencies, housing coops, and shelters that serve low income residents.

A number of researchers have begun to study various aspects of community wireless networking in Canada. Powell's work focuses on understanding technical development processes at Île Sans Fil as a form of civic and political engagement for its members (Powell 2006). Cho offers a profile of Wireless Toronto and a detailed ethnography of its members, whom she characterizes, after Florida's "creative class", as part of the city's "creative civic core" (Cho 2006). Wong and Clement have explored the feasibility of a neighbourhood wireless network in downtown Toronto (Wong and Clement 2006).

Powell's and Cho's ethnographic studies of the participants in CWN initiatives reveal the prevalence of technically skilled youth among them. While many volunteers initially become involved for instrumental reasons, such as acquiring new skills or experience for their work resumes, many describe the sense of community and belonging they experienced while pursuing the group's shared socio-technical goals as the most important aspect of their involvement. In fact, as the following case study demonstrates, CWN projects can serve as gateways to community and civic engagement for urban youth in much the same way that community networking sites discussed above serve as sites of community building for new immigrants.

Île Sans Fil: technical and organizational profile ISF is located in Montreal, Quebec, a bilingual (English and French) and multicultural city of two million people located on the St-Lawrence seaway in central Canada. The city hosts a vibrant "café culture" where cafés, restaurants and bars act as important "third places" that attract freelance workers, creative professionals, and artists to the city. ISF has two stated missions: the first, to create free wireless internet access points in public places; and the second, to use emerging technologies to build communities. ISF strives to achieve these goals via two main areas of activity: the development and maintenance of a network of free Wi-Fi internet hotspots; and the development of open source captive portal software with which it creates unique "location portals" at each of its hotspots featuring "hyper-local" information and cultural content to encourage local civic participation and engagement.

The early focus of ISF's activities has been the development and maintenance of a network of free wireless internet hotspots in public places throughout downtown Montreal. ISF has developed a *venue-sponsored* model, in which ISF volunteers recruit locations such a bars, cafés, and community centres that have an existing broadband connection to become members of ISF. Venue owners pay a modest annual fee, while ISF volunteers install and maintain networking equipment to share the venue's broadband service with visitors via wireless technology. Venue-owners and ISF enter into a kind of social contract, in which ISF maintains the network in exchange for the owners' commitment to share their signal with end users free of charge. ISF has been remarkably successful in developing and pursuing this model. Since being launched in 2003, ISF has expanded to a network of 140 Wi-Fi hotspots - primarily located in cafés and bars, but also in parks, restaurants, some public libraries and the public areas of universities and hospitals - with over 46,000 registered users as of October 2007. ISF has succeeded in creating an alternative infrastructure for internet access in downtown Montreal with a remarkably loyal cadre of volunteers and users. The ISF model is currently being replicated by volunteer groups in cities across Canada, and has had a significant impact on the plans and business models of commercial hotspot operators in Montreal.

From free public wiwi to wiwi-augmented urban community Free internet access provision it not ISF's only objective. As a second generation CWN, ISF members seek ways to contextualize Wi-Fi hotspots within their surroundings by developing applications, services and content designed to augment public space and build local community. Discussion of the mandate of the group has shifted from 'providing free public wireless internet access in Montreal' to 'bring[ing] people together and foster[ing] a sense of community … promot[ing] interaction between users, show[ing] new media art, and provid[ing] geographically- and community-relevant information' (http://www.ilesansfil.org).

The technical backbone for meeting this objective is a piece of open source authentication and captive portal software called Wi-Fi-Dog. Wi-Fi Dog acts as an authentication server and also provides a captive portal page – a unique "splash page" – for each hotspot that can be customized with local content, including news, local events notices and artwork by neighbourhood artists. This function permits the development and delivery of specialized, "hyper-local" content through community portal pages to the visitors at each hotspot.

In addition to providing venue and neighbourhood-specific information, such as local news and event announcements, ISF has collaborated with local artists and other community groups to develop community-based media content and services. For example, the *Hub des Artistes Locaux* (or HAL) project featured the work of local multi-media artists, whose works were made available to ISF users via the network's community portal pages. *City Wide* is a chatroom function enabled by Wi-FiDog that allows ISF hotspot users to chat online with other users and to see what users are talking about at other ISF locations. More recently, ISF collaborated with a number of other groups on its *Civic Sense* project, which provided RSS feeds to ISF portal pages featuring news about the spring 2007 provincial election in Quebec. Finally, in an effort to harness Wi-Fi hotspots as invitations to democratic participation, the developers of Wi-FiDog are creating a service that will enable users to access the voting records of local representatives, along the lines of similar projects such as *theyworkforyou.com* in the UK. In the course of developing and implementing these and other projects, ISF members build networks and bridging social capital between young "geeks" and other community members and organizations, including small business owners, artists, and community activists.

Having said that, some of the content and functionality that ISF wishes to support on the community portal pages remains to be developed. In this sense, the Wi-Fi-augmented, vibrant, and engaged local community that ISF aspires to create remains an imagined one. Nonetheless, its innovative Wi-Fi-Dog software and mission of building local community through Wi-Fi networks as an alternative communications infrastructure and form of community media has inspired numerous other CWN groups in Canada and abroad.

The community portal pages are the focus of ISF's current and expected development. Focusing on this type of software development is rationalized by the idea that Wi-Fi hotspots in local physical places can become a sort of community media: a means of reinforcing local community and culture, and of connecting the internet user to his or her location of access. By viewing images, artwork, and other information unique to a location, a user is reminded of where she is, of the culture of the place, and of her role as a citizen.

If ISF's goal of building local community and creating local electronic public space through its community portal network has yet to be fully realized, the group and its activities have had a significant impact on the way in which its members relate to and engage with their local communities. ISF has about 45 active volunteers, as well as hundreds of member subscribers to its mailing list. ISF's volunteer core is made up of technology workers, freelance consultants, students, community activists, and artists. The most active of these are software developers and hardware technicians, who are primarily male, between the ages of 20 and 40, and who volunteer in few if any other contexts. Many ISF volunteers initially become involved for instrumental reasons, such as acquiring job-related work experience, learning new technical skills, and working with cutting edge technologies (Powell 2006). However, when asked to describe what they felt was the greatest impact of their involvement with ISF, most volunteers stressed the way in which it has made them feel part of a community and more engaged with local issues, events and other members of the community.

ISF impacts on its members' lives in this way by creating opportunities for them to participate in their communities by using skills that have not traditionally been considered useful in volunteer or other civic contexts. These skills include software programming, hardware installation and network management. Interviews with volunteers suggest that they feel empowered by the way that their technical skills are valued in a community project, since otherwise they would not have felt they had much to contribute. Volunteering at ISF provides a way for technically skilled but otherwise disengaged youth to leverage their relationship with technology, expressed through a certain oppositional "hacker" ethic, into a more meaningful engagement with their local communities.

Do community wireless groups serve as a kind of gateway to community and civic participation for technologically-savvy but socially disengaged youth? Powell's interviews and informal discussions with ISF members suggest that this is the case. One young member describes his social engagement as a programmer and network manager: 'I want to build a community, but I don't have any other skills ... maybe later I will contribute to something else, like a hotline or something.' This man, who has just returned to university to study computer science after five years spent working, has never before volunteered, and explains that he spent his adolescence in his bedroom trying to program video games. Before talking to CRACIN researchers in the context of social research, he said he had never really described his activity in terms of building community. Instead he explained that he wanted to work on the group's software 'because it was cool'. In another instance, one core group member with an interest in high-tech but no other hobbies or activities, wrote on the ISF mailing list:

> I'm very happy at how Wireless internet has taken me away from my indoor computer to the outside world. Today I meet many people, discuss how this technology can help communities, develop new potentials for people.

Community wireless networks use wireless internet technology to create alternative communications infrastructure. In Montreal, Île Sans Fil demonstrates how building this infrastructure also acts as a way to engage groups of people who might otherwise not participate in the civic life of their community.

Some limitations of CWNs as civic participation As promising as CWN movements may be as sites of civic participation, both Cho and Powell question the degree to which the kind of social capital built by the members of CWN groups extends beyond the groups themselves (Cho 2006; Powell 2006). In addition, Powell observes an uneven fit between the technical and social goals of ISF, manifesting itself in tensions between the inward-looking, meritocratic culture of software hacking and the egalitarian and participatory ethos of community networking and engagement. Powell also observed a highly skewed gender profile and division of labour within ISF, where the group's three female members engaged in "soft" activities such as marketing and development, whereas more valorized, "hard" activities like coding were male-dominated (Powell 2006).

Nonetheless, as they evolve beyond inwardly-focused technical groups, Canadian CWNs are becoming unique sites of civic participation shaped by the culture and geography of their locations. At ISF, young technophiles are able to use technical development as a means to express their social engagement, and, in the process, are coming into contact with actors and situations that extend their social world. Like other voluntary projects, CWNs are ways for members to build their own social capital. The project-oriented organizational structure of Canadian CWNs incites volunteers to contribute to building something tangible, something that works, something "cool".

Conclusion

The foregoing survey of a number of CRACIN research projects has focused on the ways in which community informatics initiatives promote the civic participation and engagement of potentially marginalized groups – new immigrants and youth – within the networked city. By providing volunteer opportunities, job-related experience and a sense of community for new immigrants, who are often technically skilled but under- or unemployed, VCN promotes their integration into the local community. ISF, meanwhile, has succeeded in mobilizing technically skilled youth in Montreal around community Wi-Fi projects, encouraging civic participation within a demographic that is typically disengaged. Most noteworthy of the findings from this research, and seldom explored or appreciated in existing research on community networking and civic participation, is the extent to which much of the civic participation and community building that CRACIN researches found was taking place in physical places "around" ICTs and their use, rather than through ICTs and the online communities to which they afford access. Community informatics research on the social affordances of ICTs would do well to pay closer attention to this phenomenon, which has been obscured by recent fascination with the nature and impact of virtual community.

Acknowledgments

Research for this paper was supported by the Canadian Research Alliance for Community Innovation and Networking (CRACIN). CRACIN is based in the Faculty of Information Studies at the University of Toronto (Principal Investigator: Dr. Andrew Clement) and is funded through a grant from the Social Sciences and Humanities Research Council (SSHRC) of Canada.

References

Alkalimat, A., and Williams K. (2001), 'Social capital and cyberpower in the African-American community: A case study of a community technology centre in the dual city', in Keeble, L. and Loader, B. (eds.), *Community Informatics: Shaping Computer-Mediated Social Relations* (New York: Routledge) 177-204.

Breton, R. (1997), 'Social Participation and Social Capital', in *Immigrants and Civic Participation: Contemporary Policy and Research Issues* (Montreal: Department of Canadian Heritage) 4-11.

Caidi, N. and Allard, D. (2005a), 'Social Inclusion of Newcomers to Canada: An Information Problem?', *Library & Information Science Research*, 27:3, 302-324.

Caidi, N. and Allard, D. (2005b), 'The Information Needs and Uses of New Immigrants and Their Implications for Social Inclusion.' *International Settlement Canada: Research Resource Division for Refugees*, 18:4, 6-8.

Carroll, J. and Rosson, M. (2003), 'A Trajectory for Community Networks,' *The Information Society*, 19:5, 381-393.

Castells, M. (1999), 'The Informational City is a Dual City: Can it Be Reversed?,' in Schon, D., Sanyal, B. and Mitchell, W. (eds.), *High Technology and Low-Income Communities: Prospects for the Positive Use of Advanced Information Technology* (Cambridge MA: The MIT Press).

Cho, H. (2006), 'Explorations in Community and Civic Bandwidth: A Case Study in Community Wireless Networking', M.A. Thesis, Joint Graduate Program in Communication and Culture, York University.

Clark, L. (2003), 'Challenges of Social Good in the World of *Grand Theft Auto* and *Barbie*: a case study of a community computer center for youth', *New Media and Society*, 5:1, 95-116.

Dechief, D. (2005), 'Recent Immigrants as an 'Alternate Civic Core': How VCN Provides Internet Services and Canadian Experiences', CRACIN Working Paper No. 8, 2005.

De Cindio, F. (2004), 'The Role of Community Networks in Shaping the Network Society: Enabling People to Develop their Own Projects,' in Schuler, D. and Day, P. (eds.), *Shaping the Network Society: The New Role of Civil Society in Cyberspace* (Cambridge, MA: The MIT Press) 199-225.

Granovetter, M. (1973), 'The Strength of Weak Ties,' *American Journal of Sociology*, 78:6, 1360-1380.

Green, E. and Keeble, L. (2001), 'The Technological story of a women's centre: a feminist model of user-centred design', in Keeble, L. and Loader, B. Community Informatics: *Shaping Computer-mediated Social Relations*, London: Routledge.

Gurstein, M. (2000), 'Community Informatics: Enabling Community Uses of Information and Communications Technology,' in Gurstein, M. (ed.), *Community Informatics: Enabling Communities with Information and Communication Technologies* (Hershey,PA: Idea Group Publishing) 1-32.

Hill, L. (2001), 'Beyond Access: Race, Technology, Community,' in Nelson, A. et al. (eds.), *Technicolor: Race, Technology and Everyday Life* (New York: New York University Press) 13-33.

Kavanaugh, A., Carroll, J. M., Rosson, M. B., Reese, D. D. and Zin, T. Z. (2005), 'Participating in civil society: the case of networked communities', *Interacting with Computers*, 17, 9-33.

Keeble, L. and Loader, B. (eds.) (2001), *Community Informatics: Shaping Computer-Mediated Social Relations* (New York: Routledge).

Kraut, R.E., Lundmark, V., Patterson, M., Kiesler, S., Mukhopadhyay, T., and Sherlis, M., (1998), 'Internet paradox: A social technology that reduces social involvement and psychological well-being?', *American Psychologist*, 53:9, 1017-1031.

Kunz, J.L. (2003), 'Social Capital: A Key Dimension of Immigrant Integration', *Canadian Issues/Thèmes Canadiens: Immigration and the Intersections of Diversity* (Montreal: Association for Canadian Studies) 33-34.

Lévy, P. (2001), *Cyberculture,* translated by Bononno R. (Minneapolis: University of Minnesota Press).

Lovink, G. (2004), 'Polder Model in Cyberspace: Amsterdam Public Digital Culture', in Schuler, D. and Day, P. (eds.), *Shaping the Network Society: The New Role of Civil Society in Cyberspace* (Cambridge, MA: The MIT Press) 111-135.

Marshall, S., Taylor, W., and Yu, X. (eds.) (2004), *Using Community Informatics to Transform Regions* (Hershey PA: Idea Group Publishing).

Mitchell, W. (1999), *E-topia: Urban Life Jim, But Not as We Know It* (Cambridge MA: The MIT Press).

Nie, N., Hillygus, D.S., and Ebring, L. (2002),'Internet Use, Interpersonal Relations, and Sociability', in Wellman, B. and Haythornthwaite, C. (eds.), *The Internet in Everyday Life* (London: Blackwell Publishing) 215-243.

Oldenburg, R. (1989), *The Great Good Place: Cafés, Coffee Shops, Community Centers, Beauty Parlors, General Stores, Bars, Hangouts, and How They Get You Through the Day* (New York: Paragon House).

Pinkett, R (2003), 'Community Technology and Community Building: Early Results from the Creating Community Connections Project', *The Information Society* 19:5: 365-379.

Poster, M. (1995), 'CyberDemocracy: Internet and the Public Sphere', http://www.hnet.uci.edu/mposter/writings/democ.html, accessed 20 July 2008.

Powell, A. (2006), 'Last Mile' or Local innovation?: Canadian Perspectives on Community Wireless Networking as Civic Participation', paper for presentation the 34th Telecommunications Policy and Research Conference, George Mason School of Law, September 28 – October 1 (Arlington, Virginia).

Powell, A. and Shade, L. (2006), 'Going Wi-Fi in Canada: Municipal and Community Initiatives', *Government Information Quarterly*,23:3-4, 381-403.

Putnam, R. (2000), *Bowling Alone: The Collapse and Revival of American Community* (New York: Simon and Shuster).

Reed, P.B. and Selbee, K. (2000), 'Patterns of Civic Participation and the Civic Core in Canada,' presented at the 29th ARNOVA Annual Conference, November 16-18 (New Orleans, Louisiana).

Rheingold, H. (2000), *The Virtual Community: Homesteading on the Electronic Frontier,Revised Edition* (Cambridge MA: The MIT Press).

Sandvig, C. (2003), 'Public Internet Access for Young Children in the Inner City: Evidence to Inform Access Subsidy and Content Regulation', *The Information Society*, 19:2, 171-183.

Sandvig, C. (2004), 'An Initial Assessment of Cooperative Action in Wi-Fi Networking', *Telecommunications Policy* 28:7/8, 579-602.

Sassen, S. (2001), *The Global City: New York, London, Tokyo* (Princeton: Princeton University Press).

Schellenberg, G. (2004), *2003 General Social Survey on Social Engagement, cycle 17: an overview of findings*, Minister of Industry, Statistics Canada, <http://www.statcan.ca/english/freepub/89-598-XIE/89-598-XIE2003001.pdf>

Schellenberg, G. and Hou, F. (2005), 'The economic well-being of recent immigrants to Canada', *Canadian Issues/Thèmes Canadiens: Immigration and the Intersections of Diversity*, 49-52.

Schuler, D. and Day, P. (eds.) (2004), *Shaping the Network Society: The New Role of Civil Society in Cyberspace* (Cambridge, MA: The MIT Press).

Silver, D. (2004), 'The Soil of Cyberspace: Historical Archaeologies of the Blacksburg Electronic Village and the Seattle Community Network', in Schuler, D. and Day, P. (eds.), *Shaping the Network Society: The New Role of Civil Society in Cyberspace*. (Cambridge, MA: The MIT Press) 301-324.

Solnit, R., and S. Shwartenberg (2000), *Hollow City: Gentrification and the Eviction of Urban Culture* (London: Verso).

Sorkin, M. (1992), *Variations on a Theme Park: The New American City and the End of Public Space* (New York: Hill and Wang).

Sunstein, C. (2001), *Republic.com* (Princeton NJ: Princeton University Press).

Tanabe, M., van den Besselaar, P., and Ishida, T. (eds.) (2002), *Digital Cities II, Computational and Sociological Approaches*, Lecture Notes in Computer Science 2362 (Berlin: Springer Verlag).

Vehviläinen, M. (2001), 'Gender and citizenship in information society: Women's information technology groups in North Karelia', in Green, E. and Adam, A. (eds.), *Virtual Gender: Technology, Consumption and Identity* (London: Routledge) 225-240.

Wellman, B. and Hampton, K. (1999), 'Living Networked in a Wired World,' *Contemporary Sociology* 28:6.

Werbin, K.C. (2006), 'Where is the "Community" in "Community-Networking Initiatives"? Stories from the "Third-Spaces" of "Connecting Canadians"' CRACIN Working Paper No. 11.

Wong, M. and Clement, A. (2007), 'Sharing Wireless Internet in Urban Neighbourhoods', CRACIN Working Paper No.19.

Zukin, S. (1995) *The Culture of Cities* (Oxford: Blackwell).

Chapter 11

Pioneers, Subcultures and Cooperatives: the Grassroots Augmentation of Urban Places

Mark Gaved and Paul Mulholland

Introduction

Rather than bringing about the death of neighbourhood and driving the 'great good place' (Oldenburg 1989) from local café to online noticeboard, information communication technologies (ICTs) may augment and strengthen the sense of place felt by city dwellers. While some authors have relished the concept of a "city of bits" (Mitchell 1996) and community without propinquity, ICTs may reduce the "friction of space" but not the importance of place (Hampton 2004). ICTs may offer the opportunity for urban residents to reclaim rather than relinquish their neighbourhoods.

Since the 1990s, city planners, politicians and technological evangelists have envisaged and developed a wide range of digital city projects, seeking to revolutionise urban living through the deployment of ICTs. Proposals have frequently been couched in terms of technological progress and economic benefit, aiming to provide citizens with access to the new "information superhighways" (National Information Infrastructure 1992) drawing people into the new, post-industrial "knowledge economy". Often high budget and supported by national or international funding such as European Commission Framework Programmes, these initiatives have focussed on developing urban telecommunications infrastructures, building web based information systems to engage citizens and undertaking computing initiatives to overcome the "digital divide" (Ishida 2002; Tanabe, et al. 2002; Aurigi 2005). Many projects have been influenced by the telecoms industry "conduit and device model" (Lievrouw 2000), concentrating on the physical hardware needed to connect people, or the media industry influenced "broadcast model", emphasising the development of systems capable of broadcasting content to consumers and informing citizens of civic information. While the internet is increasingly becoming accepted as a non-extraordinary part of our everyday lives, such projects have often struggled to engage citizens who in many cases feel alienated by external interventions that disrupt rather than enhance their lives (Warschauer 2002).

At the same time, grassroots activists have been appropriating ICTs, adapting them to enhance urban living and undertaking local initiatives that are often minimally funded and little reported. These explorations may be driven by a variety of motivations, such as perceived lack of access to resources, a wish to improve one's own community or simply a desire to experiment. In this chapter we shall explore several such examples of grassroots appropriation of ICTs, exploring types of groups, their activities, and modus operandi.

Grassroots Appropriations of ICTs

Sandvig (2003) argues that as a new technology emerges there is an opportunity for amateur action: with no established profession or legislation to control the unstable and as yet unframed innovation, any interested party has the opportunity to participate in its development. This brief, chaotic period allows for bold innovators and entrepreneurs to shape new or disruptive technologies before larger, more organised financial and business interests step in (Hughes 1983). Such activity has been identified by academics across a wide range of scientific and technical domains, from electricity generation (Hughes 1983) through to telephone services (Fischer 1987) and radio (Sandvig 2003). Parallel "pro-am" innovation may also occur alongside more commercial and large scale explorations of technologies (Leadbetter and Miller 2004). Working alongside a number of grassroots community based ICT initiatives, undertaking fieldwork drawing from ethnographical and participatory action research methodologies (Friere 1970; Hammersley and Atkinson 1983; Ritas 2003), we have identified this kind of innovative activity being undertaken from small groups of hackers rigging antennae for wireless transmissions on tower blocks, through to skateboarders recording and uploading their exploits in urban environments and community activists running data cables between houses over municipal telegraph poles. As town planners and politicians envisage the possibilities of technological interventions to enhance cities, individuals and groups within local communities are also appropriating communication technologies in their own vision of how these tools can to augment city living.

A Simple Taxonomy

A broad range of grassroots ICT activism is taking place in urban environments, sometimes parallel to or preceding large scale policy lead "digital city" projects. We arrange these broadly into three groupings: pioneers, subcultures, and cooperatives.

Pioneers: technology experts exploring new possibilities

> Death to the communications monopolies! May ten thousand autonomous systems bloom! [1]

1 Consume General FAQ; retrieved from http://web.archive.org/web/20030207140645/consume.net/twiki/bin/view/Main/GeneralFAQ.

We refer to the first group of grassroots initiatives as *Pioneers* – innovators and early adopters appropriating new networking technologies. Historically, new communications technologies have been appropriated by pioneering enthusiasts and innovators. Telephone provision began with thousands of small companies offering their own services (Fischer 1992), and radio was similarly dominated in its early days by amateur operators (Sandvig 2003). In the last ten years, the standardisation of wireless networking protocols and inexpensive hardware has led to pioneers creating city-wide alternative communications infrastructures, providing low cost or free internet access and connecting likeminded innovators.

Originally developed to provide connectivity to home and office environments, low powered wireless networking equipment operating on unlicenced radio frequencies grew rapidly in popularity when a standard set of operating protocols was agreed in 1997. Often referred to as "Wi-Fi", the relatively inexpensive hardware and software tools offered businesses and individuals the ability to connect computing equipment in more flexible configurations and allow a limited amount of "roaming" within the workplace. Data could be transferred at much higher speeds within the workplace network than across the internet, and at no cost to the organisation. This made Wi-Fi attractive to innovators who took it out of the office and reconfigured it to provide alternative telecommunications networks in densely populated areas.

Across the world, groups such as Seattle Wireless, Île Sans Fil (Montreal), and Berlin's Freifunk network have sprung up. Forming a loose global movement, they are nevertheless firmly grounded in their own cities; local instantiations of a global ideal. A common rhetoric describes the desire to set up decentralised, often free access networks that encourage the participation of local residents and they are often described as "community networks" in contrast to the more commercial organisations that provide wireless access for profit (Verma and Beckman 2002). However they require a high level of technical expertise from participants; individuals must commit to maintaining their own connection and provide their own equipment, which requires time, learning and financial commitments.

Pioneer groups can tend towards 'social club(s) for technical elites' (Sandvig 2004), attracting members who wish to explore the limits of the technologies (Bina and Giaglis 2006b). Frequently there is a sense of playfulness or ludic investigation, engaging with the technologies as an autotelic activity (Csikszentmihalyi 1978; Steel 2004) and a willingness to explore interesting rather than commercially attractive avenues. Activists have run audio feeds from local bus stops live onto the internet, mused whether collaborating with graffiti artists might result in spray-on radio antennae that could extend the reach of their networks, and equipped roller skaters with Wi-Fi connected video lunchboxes for interviewing local residents. Wireless networking is firmly grounded within locality and engages directly with the physical environment. Connections need to be line of sight, and the legally permitted power levels of transmission limit the effective range of signals to a few hundred metres unless supported by expensive equipment. Hard surfaces may reflect radio signals, offering unexpected reception, tall buildings may block signals between apparently close neighbours, and the physical siting of radio antenna requires direct interaction with the substance of the city. Neighbours in high apartments or strategically placed properties overlooking further points of connection are particularly treasured and physical rigging skills are as valued as software or hardware expertise.

Case Study: Consume In the mid 1990s, a group of artists and electronics enthusiasts in London set up an internet connection into and around their workspace as a community resource. This 512kb connection, rented from the national telecommunications provider (BT) cost the group £40,000 per year. To connect another building 5 metres across the street, BT insisted on a second similarly priced connection. To save money, two of the members of the group, Julian Priest and James Stevens, purchased wireless networking equipment and set up a wireless link, forming the Consume wireless network. Neighbouring individuals and groups were invited to buy their own equipment and join the network, and gradually the network blossomed to over a hundred users, able to collaborate and share data at no transmission cost, and connecting to the wider internet. Consume began to offer workshops on antenna building and software configuration to help more people to participate and at a lower cost. Wireless access points run by members offered connections to their neighbours' wireless equipment forming a mesh of transmitters across the local area. At its height, the group was running nine mailing lists, holding regular meetings, acting as a wholesaler of equipment (negotiating favourable equipment prices for members through bulk purchasing), and had set up a web based portal (NodeDB) enabling individuals to register themselves as a "node" on this alternative networking infrastructure, and increasing its coverage. Consume inspired local offshoots in different localities across the UK, and the key members participated in international collaborations with other well known groups, promoting the ideals of grassroots networkers taking over from the major telecommunications providers through their self-provisioning: Consuming the Net. After several years, the group fragmented and individuals moved on to different activities, however the objective as noted by one of the founders 'to play and set the agenda' had been successfully achieved.

Subcultures

> Beneath the pavement, the beach! (May 1968 Paris slogan[2])

The second group of community activists we identify as *Subcultures* - early adopters focussed on a specific interest, based within a particular locality and utilising ICTs to enhance their activities. For these groups, the availability of ICTs has either enabled a practice not previously achievable, or enhanced the practice by improving communication, knowledge management, or community memory. The key actors in such groups may be domain experts who have developed ICT expertise, or technology experts who have taken an interest in the domain and work in collaboration with domain experts. Subculture participants appear to be motivated to use ICTs to better engage with their interests 'as a means to attain some separable outcome' (Bina and Giaglis 2006a) such as a platform to enable a community to share local information, or the creation of a high speed network to exchange work in an artistic quarter of

2 "Sous les pavés, la plage!" http://www.infoshop.org/wiki/index.php/May_1968

a city. We have found a variety of subcultures using ICTs to enhance their practice and share resources within urban areas, and we consider three particular groups: electronic artists, cartographers, and skateboarders.

Case study: electronic artists - East End Net One of the longest running wireless network in London is East End Net. However while it may have many similarities to the Pioneers, East End Net is distinctive because of the high number of electronic artists using the network to enhance their practice. The initiative covers a limited area in East London, including two former industrial buildings that are now multi-tenancy live-work apartments occupied by artists. Wireless connections provide the backbone of the network, with Ethernet cables run between individual apartments in the larger buildings, connecting neighbours to each other and out to the wider internet. This model of local wired "islands" connected by a wider wireless network is somewhat analogous to large scale commercial metropolitan wide area networks.

Whereas the participants within the Pioneer networks are using the networks to explore networking technologies (as well as self-provisioning their access to the internet and each other), here we find artists using the network to enable and enhance their professional practice. Through access to a high speed, local network, the artists are able to share work, exchange resources, and communicate upcoming local events to each other. One active group within the network, ambienttv.net, occupies an apartment in one of the former industrial blocks, and creates electronic art including gallery video installations, online streaming videocasts, and sound and music pieces. In constructing the works, individual members of the collective will develop, edit, and re-draft work, moving it between computers within the network and out to other art studios. Finished works may be put online, and in some cases presented at parties at the studio apartment, sometimes performed and re-mixed live while the party itself may be streamed live over the network.

These artists have created a hybrid virtual/physical quartier through their use of ICTs; their community is in part defined and bound together as much by the cables snaking over their shared warehouses and wireless signals bounced between studios as their meeting in local cafes and shared performances in industrial properties. In a similar manner to their physical performances and spheres of engagement within the city, so the virtual layer is limited by its capacity to negotiate the physical environment. The implementation of ICT by this group has created a network within a private space, grounded in the locality, occasionally leaking out to passers by and fellow residents within the neighbourhood.

Case study: community cartographers – OpenStreetMaps Our second "subculture" is the OpenStreetMaps community, a mixture of technology innovators and cartographic experts mapping their own localities. Drawing from the philosophy of the open source movement, a group of cartographers and technologists are creating their own maps, using ICTs: 'The project was started because most maps you think of as free actually have legal or technical restrictions on their use, holding back people from using them in creative, productive or unexpected ways'.[3] In the UK, published maps are subject to strict copyright laws and the national standard maps produced by the Ordnance Survey are strictly protected with re-use and repurposing very restricted.

With the advent of relatively cheap handheld GPS (Geographical Positioning System) devices and open source mapping software being developed to interpret data collected by these tools it is now possible for community activists to create their own highly accurate maps of their local areas, adding in elements that they are interested in, and uploading to the web for free public access. The OpenStreetMaps participants travel around their locality, walking, cycling, or driving the area street by street, gathering raw geographical data via handheld GPS units. The data is uploaded to computers and run through open source software packages such as JOSM[4] to produce simple traces of where the participants have travelled, adding to existing traces drawn down from a central database. These traces can then be manipulated in an image processing package like OSMRender[5] to produce maps suitable for display on the web or for printing. The maps are made available as they are created on the community website,[6] which acts as a portal for mapping groups internationally. Directly interacting with their physical locality, these map-makers seek to reclaim the geographical knowledge of their local environment for the residents themselves, allowing local people to map and define their own space, rather than being dependent on external professional bodies that reserve rights when manipulating the data and controlling its reuse. Whereas the electronic artists of East End Net could be described as creating a private hybrid physical/virtual space, the cartographers of OpenStreetMaps might be considered to be generating a public, alternative articulation of their neighbourhood space.

Case study: skateboarders, knowhere and YouTube Our final example of a "subculture" is the community of skateboarders. It could be said that the skateboarders interact more directly with their urban environment than any of the other groups so far described, often appropriating physical spaces in direct conflict with authorities (the slogan 'Skateboarding is not a crime!' is a popular sticker found attached to skateboards). As Borden notes:

> Skaters produce an overtly political space, a pleasure ground carved out of the city as a kind of continuous reaffirmation of the notion that beneath the pavement, lies the beach. (Borden 1996)

Skaters are also a subculture in the sociological sense of the word – a specific demographic (young people) with their own particular culture, rituals, and dress. While there is only a loose sense of organisation, a community formed around a shared interest, there is nevertheless the desire to share knowledge and build community memory. The development of the internet as a public content creation

---

3 *http://wiki.openstreetmap.org*

4 *http://josm.eigenheimstrasse.de/*

5 *http://svn.openstreetmap.org/utils/osmarender/*

6 *http://wiki.openstreetmap.org/*

and distribution tool has allowed this community to share resources more effectively: filming feats of expertise for posterity, to advise other skaters of good or bad places to skate, and to promote local events or political issues (e.g. authorities' responses to skating in certain areas, or the development or destruction of favourite skating locations). The skate community like other youth subcultures has a strong tradition of underground zines (small circulation, low budget fan magazines) and a number of high quality commercial magazines, and the internet has allowed individuals to publish their own thoughts and communicate with others in their community in a more decentralised fashion. Skateboarding shops and favoured skating locations allow for informal exchanges of knowledge at a face to face level, but the internet, combined with hand held video cameras, has allowed for more permanent storage of community memories and exchange of local knowledge and feats of skill at a global level.

In the UK, one of the first websites to consciously grow from the skating community was Knowhere, which 'started out as a list of places to skateboard in the UK. It is a compilation of (unedited) information and views supplied by users like yourself.'[7] The guide is divided by geographical area, and broad categories of interest, originally just for skating but now a broader range of street cultural activities, and requests user input through web forms, which allow individuals to upload their own recommendations and discuss or argue their opinions. The spread of easy to access discussion boards across the web means that now there are a large number of similar resources, though like many other territories occupied by youth cultures they are often used and discarded as short term temporary spaces. As social spaces for young people, they offer what Bey describes as Temporary Autonomous Zones or "TAZ" (Bey 1991); temporary spaces that elude formal structures of control. This echoes Borden's declaration that 'skateboarders are part of a long process in the history of cities: a fight by the unempowered and disenfranchised for a distinctive social space of their own' (Borden 1996). The internet offers opportunities for skaters to create a hybrid physical/virtual TAZ, beyond the control of others, yet able to be discarded when no longer useful. Like the wireless networking pioneers, the skateboarders do not need to consider long term sustainability of their community, but operate in a more fluid temporal space. The internet does however offer a means of capturing their activities and giving permanence to their actions through online publishing sites such as YouTube and MySpace. Combined with the increasing availability of video cameras and video enabled mobile phones, skateboarders are recording their activities, specifically setting up shots of their best moves to record for posterity and distribute to a global audience. A search on YouTube reveals over 101,000 entries tagged "skateboard" and evidence that mobile multimedia increasingly allows youth subcultures to 'document, edit, and upload their lives' with the hand held self produced images offering credibility and authenticity, and 're-territorialising' experiences and communication (Hjorth 2006). Skateboarders can be seen to be utilising ICTs in urban spaces as a means of creating an alternative articulation of a public space.

---

7 *http://knowhere.co.uk/skindex.html*

Cooperatives

> the very fact that the project is not dependent on external money means there is nothing to run out of! (Davies 2004)

We call our third group within our taxonomy *Cooperatives*: local residents and technology enthusiasts working alongside residential groups such as housing or neighbourhood associations to provide shared low cost internet access and intranet services within clearly defined neighbourhoods, using ICTs to support community activities and residents' own objectives .They engage a broad spectrum of participants and seek to reach the whole population of their defined locality, often achieving near-ubiquitous coverage within their defined neighbourhood, using ICTs to support local social interactions. While some authors despair that neighbourhoods are no longer so important in peoples' lives, with less communal activities undertaken (Putnam 2000) and theorise a move from local socialisation in urban areas towards 'glocalization' or networked individualism (Wellman 2002), Cooperatives are an example of grassroots activism proving the contrary, that place is still important for people (Hampton 2004), and that ICTs can enhance rather than reduce the opportunities for residents to communicate with one another (Foth 2004). The motivation for such groups is to support a broad demographic in a defined area, enabling people to better undertake their activities when given access to an ICT infrastructure. In many cases these Cooperatives have been driven by lack of appropriate commercial provision: following earlier technological developments such as the telephone, previously marginalised groups are usually the last to be offered access to new innovations (Fischer 1992). Lead members of Cooperatives seek to engage as many residents within a defined area as possible, and are willing to support those who are not technically able to support their own connection. While the Pioneer groups are highly decentralised, Cooperatives tend to be highly centralised, with a small number of experts supporting a wider demographic.

Cooperative groups are very much defined by their locality, and their ICT usage reflects the needs of the community and the affordances of the environment. Cooperative groups will be more likely to use whatever technology is available to them to achieve their goal of networked neighbourhoods, rather than seeking cutting edge solutions, and work with the environment to achieve this goal, whether it is running computer cabling across rooftops, using the highest properties to mount antennae, or appropriating existing telegraph poles to run their own cables between streets. The Cooperatives we have identified have highly centralised structures, often a small number of technology experts from within the community working alongside community activists to set up and maintain the network infrastructure. Users of the networks are more like subscribers of a commercial internet service provider, though they have often made a conscious decision to support their neighbourhood initiative, seeing it as more responsive to local needs, more understanding of the local situation and more accountable. Generally running at very low budgets with little or no outside support, the Cooperatives operate from the subscriptions raised and in some cases "sweat equity" offered by members. They offer shared internet access at low cost, but place great emphasis on the provision of local intranet services; mailing lists, discussion boards, file sharing servers and the like, seeking to provide tools that can enable local social interactions, build local knowledge repositories and support community memories. The groups do not seek to replace social interaction between neighbours, but offer means of enhancing these through the affordances provided by ICTs. We have found members to be pragmatic in their usage of these services, utilising the networks as means of improving local communications (e.g. for sharing photographs with a large number of neighbours) but maintaining their existing communication channels, such as walking round to their neighbours for casual conversations.

Case study: Digcoop, London Digcoop is a community network initiative based in a housing association in London. At the time of formation (2001), broadband internet connections still required a significant expenditure and degree of technical expertise to set up. The Cooperative grew from a highly active neighbourhood that had previously set up their own housing association and purchased and renovated 29 Victorian properties in two adjacent streets. A group of the residents decided to create a shared network to provide low cost internet to members of the housing association, and to build a server that would act as a management tool to help run the neighbourhood and as a community asset enabling the sharing of resources and documents amongst the residents. The team offered their work as "sweat equity"; regeneration of the locality though voluntary work rather than external loans or grants.

Each house in the housing association has since been connected to a central server, originally linked by cables run across roofs and gardens but now with a hybrid of both wired and wireless links, using equipment loaned from a nearby networking group. A content management system was set up on a recycled PC in order to store minutes from committee meetings, host a noticeboard informing residents of local activities, and run a neighbourhood discussion board. Residents were asked to help with the basic manual tasks such as running data cables to their own properties, and charged a low monthly subscription. Most of the work is carried out by a small group of the residents who are technological enthusiasts, with the help of one resident managing the accounts. The team provides informal training, troubleshooting when problems arise. Support is offered on an ad hoc basis when requested, for example connecting webcams or setting up anti-virus programs. Rather than phoning a call centre in the USA or India for technical support, residents pop round to their neighbours, or resolve problems over a pint in the pub at the end of the road, the unofficial "support centre" for the Cooperative.

The Cooperative has become an integral part of the housing association infrastructure, with work, learning, socialisation and communication taking place over its network. Its presence has affected the internal layout of properties, with the domestic PC, formerly partitioned in spare bedrooms as the home office device, migrating into the living room, and becoming the home entertainment centre, providing radio, music and video. Take up of intranet services within this Cooperative, however, has been slow, echoing Wright's findings in an Australian neighbourhood intranet (Wright 2005), and maybe for similar reasons. While used well for specific tasks, such as uploading minutes of housing association meetings, and providing a central knowledge repository for the technical team themselves to hold information, it has yet to be adopted by the broader range of subscribers. It may be that such services have not yet been "domesticated" (Ward 2003) and accepted as an aspect of the residents' everyday life, or may be seen as otherwise alien or irrelevant (Arnold, Gibbs et al. 2003). The one instance where it became heavily used was when the local council proposed the construction of a large tower block opposite to the street. At this point, residents began to post copies of the letters that they had sent to the council on the discussion board, using the board as a place to keep others updated of the latest council proposals, and a passionate debate flourished. Interviews carried out with subscribers to the network (Gaved and Mulholland 2005) suggest that users are pragmatic in their choice of media tools when communicating with their neighbours, using them as part of a broader 'ecology of communication' (Altheide 1994). Only when a crisis occurred within the community did the affordances of intranet make it more useful than those offered by existing communication media.

Discussion

In conclusion, we see a wide range of grassroots activism, with urban residents appropriating ICTs to augment their experiences of the physical city. Whether hackers exploring the possibilities of new technologies, subcultural groups enhancing their own practices through the use of ICTs, or local neighbourhoods supporting social interactions and the sense of community, networked technologies are being used to augment city living. Developed in response to residents' own goals and needs, they may be more enduring than external projects imposed upon the communities: as one community activist noted, the very factor that they have had no outside funding means there is none to run out of (Davies 2004). Alongside policy led and exogenous interventions into communities that offer grand visions of digital cities, local residents are undertaking parallel innovations, augmenting the city with their own visions and actions. Their goals may be different from those of the larger scale schemes: cooperative groups may aim to permanently enhance the communications infrastructure of their neighbourhood, members of pioneer groups may engage in playful exploration and seek to set a broader agenda than offered by the dominant commercial model, and members of subcultures might offer alternative articulation of public and private space through their engagement with their practice through the use of ICTs.

We have identified a range of activists and see three broad groupings: Pioneers, Subcultures, and Cooperatives.

- Pioneers are explorers, investigating the cutting edge of new technologies. Often motivated by a desire to play and to set a broader agenda, they represent a technological elite who may move onto the next innovation as it emerges.
- Subcultures represent a specific demographic within a locality drawn together by a shared interest defined by geographical area and moderated through technology. Membership, like with the pioneer groups, can be highly fluid.
- Cooperatives tend to be highly centralised and emphasise service to a broader community. They have a high commitment to long term sustainability, and geographically narrow but demographically wide membership. They strongly

identify with a specific locality and are highly embedded within it.

These groups can be further understood by examining their key characteristics (see Table 11.1).

Table 11.1: Key Characteristics of the Different Types of Grassroots Networking Groups

	Pioneers	Subcultures	Cooperatives
Organisational structure	Decentralised	Decentralised	Centralised
Purpose	Network within neighbourhood	Domain plus neighbourhood	Neighbourhood
Motivation	Experimentation, Play	Furthering practice	Community enhancement
Commitment to sustainability	Low	Low	High
Membership obligation	Self-provision	Participation	Subscription
Membership demographic	Narrow (based on technical knowledge)	Narrow (based on technical and /or domain knowledge)	Broad (open to all in the geographical area)

Drawn from local communities these activists may be more likely to develop a more meaningful, fully engaging augmentation of the city than exogenous projects due to their grounding within the local population, responding to specific needs and desires. These grassroots activists may be able to utilise networked technologies to enhance physical spaces and create hybrid "great good places" of the future, offering local residents the opportunity to define their own agendas and rules of engagement with new technologies and new ways of envisaging the urban environment. We would expect this to increase in the future as technologies become more commonplace, affordable and mobile; while the case studies we have presented have been mainly focussed on personal computer based artefacts, increasingly citizens' engagement with technologies is more pervasive and ambient: through mobile phones, entertainment devices and other artefacts increasingly embedded to the point at which they are not perceived as technology per se.

As predominantly volunteer run collectives, the grassroots digital activists in these pioneer, subculture and cooperative groups may not have the broader impact of high budget city or even national level projects, however through their rapid appropriation and innovation of new technologies they may provide exciting and unexpected outcomes and are likely to contribute to and influence the wider discourse.

References

Altheide, D.L. (1994), 'An Ecology of Communication' *The Sociological Quarterly* 35:4, 665-683.

Arnold, M., et al. (2003), 'Intranets and Local Community: "Yes, an Intranet Is All Very Well, but Do We Still Get Free Beer and a Barbeque?", in Huysman, M., Wenger, E. and Wulf, W. (eds.), *Communities and Technologies 2003* (Amsterdam: Kluwer Academic Publishers).

Aurigi, A. (2005), 'Tensions in the Digital City', *Town and Country Planning*, April 2005, 143-145.

Bey, H. (1991), *T.A.Z.: The Temporary Autonomous Zone, Ontological Anarchy, Poetic Terrorism* (Brooklyn: Autonomedia).

Bina, M. and Giaglis, G.M. (2006a), 'A Motivation and Effort Model for Members of Wireless Communities', in Ljungberg, J. and Andersson, M. (eds.), *Proc. 14th European Conference on Information Systems (Ecis) 12 -14 June* (Göteborg, Sweden).

Bina, M. (2006b),'Unwired Collective Action: Motivations of Wireless Community Participants', *Proc. 5th International Conference on Mobile Business (Icmb'06), 26-27 June* (Copenhagen, Denmark: IEEE Computer Society).

Borden, I. (1996), 'Beneath the Pavement, the Beach: Skateboarding, Architecture and the Urban Realm', in Borden, I., Kerr, J., Pivaro, J., and Rendell, J. (eds.), *Strangely Familiar: Narratives of Architecture in the City* (London: Routledge) 82-86.

Csikszentmihalyi, M. (1978), *Beyond Boredom and Anxiety: Experiencing Flow in Work and Play* (Cambridge: Cambridge University Press).

Davies, W. (2004), *Proxicommunication: ICT and the Local Public Realm* (London: The Work Foundation). http://www.theworkfoundation.com/Assets/PDFs/proxicommunication.pdf

Fischer, C.S. (1987), 'The Revolution in Rural Telephony, 1900–1920', *Journal of Social History* 21:1, 5-26.

Fischer, C.S. (1992), *America Calling: A Social History of the Telephone to 1940* (Berkeley, CA: University of California Press).

Foth, M. (2004), 'Designing Networks for Sustainable Neighbourhoods: A Case Study of a Student Apartment Complex', in Stillman, L. (ed.), *Proc. of CIRN 2004 Conference and Colloquium: Sustainability and Community Technology, 29 Sep – 1 Oct* (Prato, Italy).

Friere, P. (1970), *Pedagogy of the Oppressed* (New York: Continuum).

Gaved, M. and Mulholland, P. (2005), 'Grassroots Initiated Networked Communities: A Study of Hybrid Physical/Virtual Communities', in Sprague, R. H. (ed.) *Proc. 38th Annual Hawaii International Conference on System Sciences*. (Big Island, Hawaii: IEEE Computer Society).

Hammersley, M. and Atkinson, P. (1983), *Ethnography: Principles in Practice* (London: Tavistock Publications).

Hampton, K. (2004), 'Neighbourhoods and New Technologies: Connecting in the Network Society', Keynote Speech at the workshop 'Neighbourhoods and New

Technologies: Connecting in the Network Society ', 26 April 2004 (London: The Work Foundation).

Hjorth, L. (2006), 'Being Mobile: In between the Reel and the Real', *Asia Culture Forum: Mobile and Popular Culture, 29* October (Gwangju, Korea).

Hughes, T. P. (1983), *Networks of Power: Electrification in Western Society, 1880–1930.* (Baltimore: Johns Hopkins University Press).

Ishida, T. (2002), 'Digital City Kyoto' *Communications of the ACM* 45:7, 76-81.

Leadbetter, C. and Miller, P. (2004), *The Pro-Am Revolution: How Enthusiasts Are Changing Our Economy and Society* (London: Demos).

Lievrouw, L. A. (2000), 'The Information Environment and Universal Service', *The Information Society* 16:2, 155 -159.

Mitchell, W. (1996), *City of Bits: Space, Place and the Infobahn.* (Cambridge: MIT Press).

National Information Infrastructure (1992), *Technology Position Paper for 1992 Clinton Presidential Campaign (Washington D.C.: National Information Infrastructure).*

Oldenburg, R. (1989), *The Great Good Place: Cafés, Coffee Shops, Community Centers, Beauty Parlors, General Stores, Bars, Hangouts and How They Get You through the Day* (New York: Paragon Press).

Putnam, R. (2000), *Bowling Alone: The Collapse and Revival of American Community* (New York: Simon and Schuster).

Ritas, C. (2003), 'Speaking Truth, Creating Power: A Guide to Policy Work for Community-Based Participatory Research Practitioners' (published online July 2003), http://depts.washington.edu/ccph/pdf_files/ritas.pdf, accessed 07 March 2006.

Sandvig, C. (2003), 'Assessing Cooperative Action in 802.11 Networks', *31st Conference on Communication, Information, and Internet Policy* (Washington, DC, USA).

Sandvig, C. (2004), 'An Initial Assessment of Cooperative Action in Wi-Fi Networking', *Telecommunications Policy* 28:7 - 8, 579-602.

Steel, L. (2004), 'The Autotelic Principle', in Iida, F., Pfeifer, R., Steels L. and Kuniyoshi Y., (eds.), *Embodied Artificial Intelligence: International Seminar,* Dagstuhl Castle, Germany, 7-11 July (Berlin/ Heidelberg: Springer) 231-242.

Tanabe, M., van den Besselaar, P., and Ishida, T. (eds.) (2002), *Digital Cities II, Computational and Sociological Approaches*, Lecture Notes in Computer Science 2362 (Berlin: Springer Verlag).

Verma, S. and Beckman, P. (2002), 'A Framework for Comparing Wireless Internet Service Providers with Neighborhood Area Networks', in *Proc. 8th Americas Conference of Information Systems (Amcis)* (Dallas, Texas, USA).

Ward, K. (2003), 'The Bald Guy Just Ate an Orange: Working at Home and the Organisation of the Domestic Internet', in *Proc. of New Media, Technology and Everyday Life in Europe Conference, 23-26 April,* London School of Economics (London: European Media Technology and Everyday Life Network).

Warschauer, M. (2002), 'Reconceptualizing the Digital Divide', *First Monday* 7:7.

Wellman, B. (2002), 'Little Boxes, Glocalization, and Networked Individualism', in Tanabe, M., van den Besselaar, P., and Ishida, T. (eds.), *Digital Cities II,*

Computational and Sociological Approaches, Lecture Notes in Computer Science 2362 (Berlin: Springer Verlag). 10 -25.

Wright, P. (2005), *A Community Intranet: Factors Affecting the Establishment of Information Communications Technologies at the Neighbourhood Level*, Ph.D. University of Melbourne.

Chapter 12

Augmenting Communities with Knowledge Resources: The Case of the Knowledge Commons in Public Libraries

Natalie Pang, Tom Denison, Kirsty Williamson, Graeme Johanson and Don Schauder

Introduction

> I have learned that excellence in cultural institutions means constant engagement in the problems of public knowledge. As proliferating themes of education and the culture of information remind us, what a culture knows, how it finds knowledge, and how it conducts its conversations about ethics, imagination, and policy are vital to its character and actions. These themes are also germane to understanding the promise of cultural institutions, where we can live out the unspoken parts of our experiences. Consequently, over the past several years when I have met audiences…I have done so as an advocate for the idea that we create and sustain these institutions because we want to become different people. We go to these places out of hope and will, and out of the desire for self-rescue. (Carr 2003, xix)

In August 2005 Hurricane Katrina, one of the costliest and strongest hurricanes ever recorded, struck the northern-central Gulf coast in the United States, with New Orleans and Mississippi being the worst-hit cities. It had many detrimental effects: economic, environmental, political and, often-forgotten, social and psychological damage (Karkosi 2006). StoryCorps, a national oral history project designed to help 'instruct and inspire Americans to record one another's stories in sound' facilitated the creation of stories shared by survivors, rescuers, and other Americans and contributed to the level of support and cohesion in the aftermath of the disaster (Flattery and Baronofsky 2003). These stories are now preserved as archives in the American Folklife Centre at the Library of Congress.

In another example of building of community cohesion, the State Library of Queensland in Australia has successfully brought together a series of resources to create a special collection: The Garage (Fielding 2006), featuring technical information pertaining to older cars, Australian motoring history, and a rich archive of digital photographs presenting vehicles dating from 1900. The creation of the special collection and the process of taking it online have had a fundamental impact on the outreach of the library to the user community: engagement with resources in the collection has grown from the conventional library user to the inclusion of "hard-to-reach" users in a dynamic way.

These are only some of the stories from the everyday life of cultural institutions, such as museums, libraries, national archives, public broadcasters, and public news agencies. As Carr (2003) noted in the opening quote, the role of cultural institutions in community building and the larger purpose of building and defending knowledge resources to be held in common by local communities is essential and implicit. Whether in time of disaster or during everyday life, communities are energised and sustained by cultural institutions – in formal and informal, direct and remote ways.

This chapter is focused on exploring the work of public libraries in communities. The term "library" has generally come to refer to places where collections of 'information resources in print or in other forms that are organised and made accessible for reading or study' (Britannica 2007). In today's media environment, the production and storage of collections by libraries for access by the communities they serve has become fragmented; which arguably shapes the library as a place and augments local communities. Regardless of their forms, however, libraries provide information to their users and use knowledge resources to connect to the cognitive, imaginative, personal and cultural energies and processes of local communities.

In particular, this chapter explores the concept of the knowledge commons, which is derived from the historical commons, and which refers to knowledge resources that are made freely available for all in society to build relationships, culture, and democracy. Such knowledge resources in public libraries could include information resources, community information, databases, even stories unique to the local communities. In practice, the impact of the knowledge commons can be conceptualised as the creation of an institutional space where communities can interact and exchange information, free from the constraints normally accepted as preconditions to markets.

Using two case studies, the chapter will explore how public library networks are being created and transformed into a knowledge commons that both serves and is defined by the communities it seeks to serve. It will also explore the role of physical spaces and information and communication technologies (ICT). Public libraries are seen as particularly appropriate in this context, being generally conceived as institutions that nurture and protect such spaces. Traditionally they have been used as an instrument that contributes to the creation of social capital and to the preservation of democracy by providing safe and open access to knowledge resources. This is changing with the contemporary idea of competitive advantage to be gained only through efficient information transfer (Lesk 2005).

The resources used are important because, both in form and content, they reflect the interaction of people within these communities and are integral to how people define their collective memories and their communal identities. It will be argued that, by contesting and reclaiming resources for use within the knowledge commons, the production of local knowledge is enabled, thereby building and sustaining the local communities.

This chapter will also bring into focus the partnerships public libraries build with other institutions, such as museums, educational institutions, community organisations, and local government, exploring how they work together to foster communities in local and public spaces, and some of the boundaries that act to constrain members while potentially excluding others. It will be argued, however, that not only does the interplay of the physical and virtual dimensions of the commons sometimes create these constraints, but it also can provide mechanisms to overcome them.

The Knowledge Commons

For some years now, writers and researchers have been passionately arguing for the importance of producing and protecting what is referred to as the commons, a space of "common" property produced by communities of people (Lessig 2004; Bollier 2003; Benkler 2006; Rheingold 2002; Liang 2004; Pang and Schauder 2007; Ostrom 1990). The idea is hardly new: since the first case of collective action for the common good of humankind, the commons has prevailed, rooted in communities of trust and cooperation (Bollier 2004). The knowledge commons stems from this historical concept, referring to resources made freely available for all in a community for knowledge production, building relationships and cultural democracy in the process. Such understandings contribute to the role of cultural institutions like the library, both in the conception of its mission and the perception of its role by the community. As Benkler (2003, 6) defines it, the knowledge commons comprises:

> Institutional spaces, in which we can practice a particular type of freedom – freedom from the constraints we normally accept as necessary preconditions to functional markets.

Thus resources in the knowledge commons are thought to be freely available or at least fairly accessible; the knowledge commons is distinct from the market.

Recent popularity of the concept has led to the development of several definitions, the more popular of which refer to the commons as 'a generic term which embraces all creations of nature and society that we inherit jointly and freely, and hold in trust for future generations' (Friends of the Commons 2004, 3). Levine (2002) describes the commons by the property residing in it – that is not possessed or controlled by any one individual, company, or government.

Such definitions imply that resources in the commons are not owned by any single entity, be it a person, organisation, or government. The Romans in the ancient world identified three types of property (Digeser 2003): "res privatae", "res publicæ" and "res communes". "Res privatae" identified property that is possessed by an individual, family or a company. "Res publicæ" is associated with things that are governed by the state, such as public parks, roads, and buildings. "Res communes" recognized resources that are common to all, such as resources in the natural world (such as water and air). In today's society there is frequently a lack of distinction between "res publicæ" and "res communes"; but this chapter asserts a difference between these two types of property. What is state-owned may not necessarily be common property. For brevity this chapter uses the term knowledge commons to refer to resources in the domain of "res communes", but although it does not include state-owned property or knowledge, it should be recognised that these domains do influence each other.

One implication is that the resources of "res communes" are also free and available to be used, borrowed, imitated, or altered by all, an aspect currently reflected in open content licenses such as those of the Creative Commons, which are often used as a means of placing resources in the public domain (Pang, forthcoming). As the focus of the study reported in this chapter rests on public knowledge as created, encouraged, cultivated, and defended by cultural institutions, this chapter refers to the commons as the knowledge commons.

The development of such commons holds many implications for communities and their management of resources, some of which include: design principles of systems, communication structures, ownership, institutional partnerships, and shared technologies. For example, participatory design principles may be relevant to guide community participation in the building of knowledge resources and technological applications, such as software created within the open source developers' community.

Private appropriations of the commons, referred to as enclosures (Bollier, 2003) are also often found within communities. With the knowledge commons, there can be boundaries around intellectual property contained in shared resources. For example, although there are many works, such as folktales, that are spread by storytellers across generations, there are also many literary works are re-interpretations of an original work, such as that of *Faustus* by Christopher Marlowe, amongst others, that could be considered private appropriations (Liang 2004).

There are varying degrees to these "enclosures", and Pang (forthcoming) quotes many contemporary examples. It is sufficient to note , however, that while boundaries exist between resources in the private sphere and the non-state public sphere, such boundaries are being renegotiated in the contemporary media environment. This will be discussed in the later part of the chapter.

The Creation of Knowledge Resources

The community is the centrepiece around which all resources, spaces, instruments, and the knowledge contained in them evolve. That being said, it must also be emphasised that the community is continually being shaped by the very things it is shaping. This perspective is provided by Giddens, who recognised through structuration theory that 'man actively shapes the world he lives in at the same time as it shapes him' (Giddens 1986, 21).

The concept of the knowledge commons purposes communities for collective action and enables them to become co-producers and participants in the commons, instead of mere end consumers. In fact, it is this construct of the knowledge commons that leads people in communities away from being passive consumers: in order for public knowledge to be used meaningfully within the commons they have to co-create and co-produce. The consumer of knowledge resources thus becomes a producer of knowledge as well.

Another unique characteristic of the community in the commons lies in the continuum of knowledge processes, in which boundaries between individual and collective knowledge are blurred. Self-knowledge is shared with the peer and the community and at the same time is constructed by the knowledge received from the peer and the community (Castells 2003).

Resources in the knowledge commons, representing such knowledge, are brought in by members of the community and shared collectively and equally with others. This is characteristic of research groups, where members come together with their disciplinary knowledge for the purpose of collaborating towards research goals. Other resources are produced as a result of the collective action of the community. They are owned collectively, although some community-oriented licenses may apply to external use. This is evident in open source projects, where resources are created and owned collectively by groups of developers, and open content licenses govern their use.

With the emphasis on the use and production of knowledge as a public resource, public libraries can fulfil their purpose as centres of knowledge that inspire, support and sustain communities. Hellstrom (2003) highlighted the benefits to be gained: knowledge as a resource increases in value the more it is used. Unlike physical resources which may deplete with use (such as water and grazing land), the more knowledge that is created and used in a community, the more useful it becomes. Collective wisdom is acquired in this fashion, providing the basis for emergent and shared understandings in the community. Wikipedia is an excellent example in that it provides open access to resources and allows anyone to edit pages. Recent evidence has also shown how the quality of resources in Wikipedia increases the more they are edited (Wilkinson and Huberman 2007).

The Contemporary Media Environment

The Internet has significantly changed the use and impact of technology on everyday lives. Shifts in mediating technologies have profoundly changed the way communities interact. Collective wisdom and the way it is created, used and disseminated in communities has been transformed as a result of, and in response to the contemporary media environment.

Private and public boundaries

Boundaries differentiating private and public domains existed to differentiate property that was private, state-owned, and common public property. Property is distributed within respective domains, and collective action comes about as people in communities move their action from one domain to another (Olson 1965). Such boundaries also determine whether an action lies within the commons or the public domain. As conceptualised in traditional collective action theory, in order to participate or contribute, there must be conscious, coordinated, and known intentions amongst people in a community and no one member or individual may be able to contribute to the public or communal good alone for the benefit of others (Olson 1965; Olson and Zeckhauser 1966; Oliver and Marwell 1988).

With the contemporary media environment, however, the world is now witnessing radical shifts in the way people communicate with one another. With digital media presenting old and new information in various ways, communication technologies overcoming the distance between people, non-commercial publishing penetrating everyday media, and easily available information any time of the day, boundaries between what used to be private knowledge and public knowledge are now breaking down on many levels. Participation in communities can now happen without necessarily requiring knowledge of others in the same community, or needing a conscious intention of contributing to the communal or public good. Bimber, Flanagin, and Stohl (2005) refer to this as second-order communality. There are many examples, from unconscious actions (such as the sharing of a bookmark with others) to conscious activities (such as adding value through the posting of a review on a web page or community blog). Value can be derived from the communal or public good through such sharing and retrieval of useful information that has also been value-added by the contributions of others. Contemporary technologies and the media environment, along with the communities of people resulting from the latter, provide such functionalities and make it possible for knowledge to increase in value the more it is used and shared.

Another implication of the breaking down of boundaries is that knowledge now continually and easily moves back and forth between the private and the public realms. The nature of production within communities has been amplified with the new media environment.

The nature of production in communities

With the advent of the Internet and the increasing availability of communication technologies, there has been an enormous growth in non-commercial publishing, reflecting a paradigm shift in the ways that people are creating knowledge, communicating, collaborating, interacting, and distributing knowledge (Lessig 2004). One story inspires another, and one's knowledge is the amassed wisdom of a few others (Boynton 2004). Knowledge is produced collectively and interactively within communities. Even blogging, for example, is mediated and networked by track-backs and thoughts from interested readers. According to Castells (2004), self-knowledge has never been a separable construct, and as communities come together they inevitably construct self-knowledge, which may also contribute to the collective wisdom of the community. Such collective wisdom has the potential to be included as common property in the knowledge commons. The ability of communities to generate such collective wisdom is helped by the Internet and communication technologies, which amplifies social networks and enables broader, faster, more transparent co-ordination of activities, at lower cost (Schauder and Pang 2006).

This model is efficient for a number of reasons. Knowledge appreciates in its use and context because of the dialogue contributed by participants in a community. But the production of knowledge in a community is never an isolated process. The use and growth of the number of linked pages on the Internet has certainly illustrated this interdependence. In this context, public libraries have always been one of the custodians of community memory, which is crucial to that the sustainability of communities. They frequently select information resources, including artefacts, from amongst their vast collections to represent identities and unique information needs of communities, and which then form the basis of an ongoing dialogue between the public library and the community.

Augmenting Communities: Public Libraries in Victoria (Australia)

Public libraries are a part of the larger social structure. They create, promote, and sustain knowledge resources for the local communities which they serve, regardless of economic class, literacy levels, education or ethnicity. They provide spaces – traditionally physical ones – in which communities can come safely together in the pursuit of acquiring and producing knowledge. They act as an integrated facility providing access to both physical and digital resources, most of the latter being acquired through the negotiation of licenses permitting use by local communities. Public libraries perform a number of profound roles in the larger context of the knowledge commons: allowing communities to come together for the purpose of knowledge production within safe spaces, enabling collective action in the process, engaging people in a continuum of knowledge processes from the construction of self-knowledge to the sharing of it.

Using the knowledge commons as the construct, a study of public libraries in Victoria (Australia) was undertaken to examine the ways in which they perform these roles.

Multiple case studies were undertaken within an interpretivist framework, but due to space limitations, only two can be discussed here. One is that of a regional public library serving connected shires and towns on both sides of the border between Victoria and New South Wales, the other a public library service that forms part of a learning centre in a relatively newly established community.

Interviews of one to two hours were carried out with the library managers from each library service. These interviews took place at venues within each library service, giving project team members opportunities to observe the location and many of the physical spaces. Written documents for each of the services provided additional information. To gain further insights on the meanings allocated to the public library services, interviews were also undertaken with individuals likely to have deep understandings of the roles of public libraries. These included two university librarians, one local government employee, an employee of the Wodonga Learning Centre, and three representatives of community organisations who are active in their communities. The interviews focused on a wide range of questions related to the role of public libraries in community development and in nurturing shared knowledge resources.

Public libraries in Victoria have faced a number of challenges in recent years. For example, the restructuring of the state's local governmental sector in 1993-1995 introduced compulsory competitive tendering to libraries, leading to new service agreements with their relevant governing councils, and forcing them to review their operating environment (Dudley 1998).

There has also been an increase in co-operation amongst public libraries (Dudley 1998) in Victoria and they have also significantly increased their engagement with the broader community in order to overcome problems in the last ten years. The issues vary but include information access and equity (Williamson et al 2000), lifelong learning (Bundy 2001), increasing free public access to the Internet (Hardy and Johanson 2003) and improving children and adult literacy (Gibbs 1990). This has been part of a more general increase in direct engagement with their communities, which has brought public libraries to a higher level of involvement in constructing memories and making meaning: an inevitable process of forming communities. Sometimes this has been the result of direct action, such as collecting and publishing local history about key figures in the community.

At the same time, global trends have been at work. Numerous examples of changes in libraries reflect a larger movement of cultural institutions away from traditional deliveries of services. Spaces are being transformed and new outreach services to communities designed (Axtman 2005; Elsevier 2005). In recent years, public libraries have found themselves confronting the idea of the "learning commons", where support services, resources, spaces, and technologies are brought together in comfortable workspaces to facilitate learning. As a result, hybrid spaces involving both physical and digital dimensions have been created. The communities functioning within such hybrid spaces in public libraries are finding themselves having to adapt quickly to new ways of communicating and constructing knowledge while remaining cognisant of previous practice.

A significant finding from the case studies is the emergence of partnerships pursued by public libraries. While public libraries have long recognised that they need to be more tightly integrated within their communities, no two institutions have adopted the same strategies. This reflects the fact that needs differ between communities and public libraries are driven to engage with the needs of their particular communities.

The two libraries discussed here have attempted to engage with educational institutions, community organisations and local businesses, one seeking partnerships with cultural institutions such as museums and art galleries, and another co-locating public libraries with social services such as child care and health centres. Regardless of the manner in which public libraries engage with the communities they serve, they seek creative partnerships with other institutions to allow them to improve services and enhance their engagements. As one participant noted: 'Isolation is deadly'.

In the face of such efforts it is necessary to address the core role of public libraries in the cultivation and construction of resources in the knowledge commons. But while the knowledge contained in resources is critical for the sustainability and growth of communities, the move to digital, as opposed to physical, resources is forcing change. Digital resources have mediated levels of access, and the ways knowledge is acquired and shared commonly amongst communities often now requires the use of media technologies and Internet applications.

Yet access and use are not the only issues with regard to resources in public libraries. With adequate funding, a public library may be able to provide abundant access to the finest resources and ensure that the appropriate infrastructure is in place to support access and use – but there are other issues involving communities and the way resources are nurtured in the knowledge commons.

Resources in the knowledge commons, now evolving in the new media environment, are emergent and cannot be established overnight by any community. They involve the construction and re-construction of meanings by individuals, and sharing such meanings with others in the community. Such shared understandings lead to the accumulation of knowledge resources, shared openly by the community that defines them. Many processes support this formation – significations of resources, structures and agencies to facilitate operations, and the founding of communicative facilities (Giddens 1986).

More so than ever, resources in the knowledge commons are constantly challenged with market enclosures such as global privatisation seeking to govern resources that public libraries have, and seek to build. One example of this is the use of licensing frameworks for electronic journals which potentially result in fragmented access and inequality in literacy, democracy, and power (Hardy & Johanson 2003). Such challenges, however, are balanced by the rapid generation of public knowledge through the Internet, communicative technologies, and new community relationships nurtured by public libraries. The public libraries in the study, while different from each other in many ways, are united in the role of creating and defending knowledge resources for communities in a range of ways.

The Hume Global Learning Village

The Hume Global Learning Village (HGLV), which incorporates six public libraries in the city of Hume, is located about 20 kilometres from the centre of Melbourne. Although not all parts of the city are economically or socially disadvantaged, there are high levels of unemployment and one-third of the city's population has a low level of family income, and there are low levels of tertiary education. Because the library service is highly integrated in the community, the library manager (LM) was interviewed together with the General Manager of the Learning Community Department, Hume City Council (GM). Not surprisingly, the HGLV has a strong focus on social justice.

> But we're (I think) we're the only council to deal with a social justice charter and a citizens' bill of rights, and that sort of is the overarching philosophical kind of document for us, and then sitting underneath that is this desire to create this learning community as the vehicle through which social justice will be achieved (GM).

Both managers recognised ongoing partnerships as their major achievement. This was further affirmed through a local government National Award for Excellence (Australian Government 2006). While the growth in partnerships has been central to the growth of the public library, GM also noted that the library has become more aggressive in going out to the community, and not all activities needed to be confined within the walls of the library:

> If it's not held in the library, it doesn't matter because in fact we've integrated it (GM).

With such partnerships, the boundaries of the public library as a local and public space are broken down and diffused. With contemporary media also influencing the way knowledge is created in communities, the spaces in which public knowledge is created, value-added, and defended are now very much decentralised and integrated in the communities themselves. There are various examples from HGLV, such as:

the use of a digital database as one medium to provide access to local community information (such as local organisations and their activities), providing Internet access and technical support so that resources on the World Wide Web are available to the community, providing suitable study spaces in the library for people in the community, storytelling program taking the library out to play groups, kindergartens and neighbourhood houses, and the use of a volunteers program

> As one of 56 strategies for creating a learning community (GM).

Using information and communication technologies, the knowledge commons found in the library exists not merely in the digital databases set up to provide access to local community information and resources, but also in the soliciting and distribution of inspiring stories from people in the community via the Imagine, Explore, and Discover (IED) project. As LM recounted:

> Just being able to bring everybody together and find that common thread, and the nice thing [is that] people contribute because they have a passion for making a difference in the community…so what we've done is simply connect people, and we often talk about how we connect them across silos (LM).

Integrated spaces providing access to information and communication technologies, as well as digital resources were also perceived as important to the community. This point was reiterated by a community representative (CR1):

> The library's got very strong relevance, especially in making resources that are otherwise unaffordable available to the community with a lot of disadvantaged people and families on really low incomes. It's not just books – but also services like assisting with technology and reading disabilities, spaces, and so on. Everything in the library is a community resource (CR1).

In the case of HGLV, the participation of people in the local community through the IED project and volunteer programs was also instrumental in strengthening social ties in the community in ways similar to those mentioned early in the chapter. Such programs utilises information and communication technologies to coordinate collective efforts as well as solicit, build, and distribute relevant knowledge resources to the community. In HGLV's view, the library service is integrated with the community in the current media environment. As such, the media environment, together with its information and communication technologies, supports the knowledge commons of the library. In this sense, the library service is aligned with the social charter discussed earlier. The changing roles and job scopes of people working in the library service are also important.

> It's the broadening out of the role of the librarian out into the community and we're calling it learning advocate or learning facilitator (LM).

This action has helped in understanding the emphasis on institutional partnerships within HGLV. To sustain the knowledge commons, it is crucial that public services in the community work together, frequently exchanging skills and communicating shared visions and goals. For HGLV this vision has been manifested by having internal staff members work with institutional partners, such as community representatives, learning organisations, and other public services in the community. The online environment and communication technologies have provided the opportunity for such effective partnerships.

> It's about using the connections with the community (…) and this has a lot to do with the shift from being a stand-alone service to being a member of the local community network and technologies have definitely helped us in such work (GM, LM).

The Upper Murray Regional Library Service – Albury Wodonga

The second public library, the Upper Murray Regional Library Service – Albury Wodonga (UMRLS), uniquely serves two states (Victoria and New South Wales) in Australia. It caters to a population of 130,000 over an area of 28,211 sq km, and has a mixture of regional centres, smaller towns and rural areas. Other than twelve branch libraries, the UMRL also has mobile libraries providing a range of materials and Internet access at approximately 72 locations throughout the service area (Local Government Victoria 2006).

As with the HGLV, the UMRLS initiates and maintains close involvement with communities through working relationships with other community organisations and activities reaching out to the local communities with the likes of the mobile library services. The Chief Librarian (CL) explains the popularity of the public library as a community learning space:

> They're going to want to know how to update their hardware, how to update their software, but they're not going to go and ask somebody who wants to sell them that. They just want to know how to do it. And some of the feedback, the initial feedback, was that they thought libraries would be a good place to come and sit and have that demonstrated, so that they're in a non-threatening place, which is probably a learning space but it hasn't got any expectations (CL).

The public library as a non-threatening space has been integral to the redesign of services and spaces at UMRLS, and extended to other cultural institutions in the local community. The public library fulfils the primary purpose of the knowledge commons in this way, providing an institutional space where communities are free to create, use, and exchange knowledge with freedom from market constraints.

The redesigned spaces also provide opportunities for the collection and presentation of public knowledge unique to various communities. Interestingly, there are mixed definitions as to what constitutes the community:

> There are some small communities that have things that they feel are very much for their community, that they do have significance and they want to hang onto them (…) [which will] restrict access. [But] if they were in, say, a museum or in a bigger collection somewhere, more people would have access to them – particularly an issue in regional and small communities, because they want to hang onto things that make their community, or their town, unique (CL).

Such reflections are congruent with earlier discussions on the renegotiations of private and public boundaries. As a public space, the public library is even more challenged today in negotiating these boundaries with communities: especially with the imposition of both contraints and empowerments. These challenges aside, the library is perceived as active in strengthening ties in the community. A community representative (CR5) praised the mobile libraries in the fulfilment of such a role.

> Travelling to central towns and areas can be quite difficult for those of us who live in remote areas – this is where the mobile libraries come in. They allow people to pick up books and drop them off at these libraries, access IT services which gives them a wealth of knowledge, and the librarians that come with them are absolutely wonderful (…) I know it is expensive but it's such a good service for the community, for reasons of community and capacity building (CR5).

As articulated by CR5, people in the community of this library service are generally separated by vast distances; with a few key towns between them. Mobile libraries were then found to be very helpful in connecting people in the community – and the integration of information and communication technologies and networked databases providing access to local information and resources is therefore crucial. How this helps to enhance ties in the community and foster social cohesion was evident in the Chief Librarian's explanation of the growth of technological services in the mobile library.

> We have two mobile libraries, one with network access and computers which stops in key towns for at least four hours, because that's where people congregate and want access to the IT services and databases. On the mobile libraries, people come from everywhere and for the few hours, they become social hubs – people exchange recipes, use emails, access information on the Internet and databases, everyone chats and you hear it in the small space within the mobile – and with the IT services people are also using digital resources to interact with one another (CL).

For the community, UMRLS functions almost as a network, connecting people in the community and providing them with resources that they would otherwise have no access to. Such sentiments were expressed by the community representative:

> I liken the library to the roads connecting our towns, and people in those towns. The library provides bridges for people to cross and reach one another…and the provision of computers and access to information on the Internet is important because otherwise, they would not have access at all (CR5).

The availability of the Internet was found to be advantageous in providing access to public resources that are enhanced by the expertise of librarians. Such access is crucial in ensuring that resources are maintained in the knowledge commons, and to inspire further knowledge to be acquired in local communities. As CR5 put it, disconnected communities that are often found in rural areas will stay disconnected if there is no public library. In such cases, mobile libraries that come with computers furnished with networked databases and Internet access to help overcome the limitations of physical resources are especially welcoming.

Conclusion

This chapter has explored the concept of resources in the knowledge commons in the contemporary media environment, particularly as it applies to public libraries as local and public spaces.

Public libraries offer an effective space to protect the creation and exchange of knowledge resources in the community. Such spaces form the very basis and purpose of public libraries, as they interact and engage communities towards making resources significant, establishing communicative facilities, using structures and agencies to help communities create and shape shared knowledge resources. The contemporary media environment poses both opportunities and challenges in the development of resources in the knowledge commons. As central and integrated facilities in the community, public libraries provide local public spaces and knowledge, possessing the potential to engage communities proactively. Institutional partnerships form an essential part of this picture, being equally matched by parallel trends in interdependencies between institutions in the context of knowledge production in the contemporary media environment. These partnerships can be harnessed for the betterment and empowerment of the broader community.

The chapter has discussed two case studies of public library services from Victoria, Australia, using pragmatic examples to demonstrate how information and communication technologies support library services and help enhance social ties and foster connectedness in local communities. As demonstrated by the variety of methods and programs used, public libraries are highly localised spaces, which can harness opportunities presented by the media environment to engage and enhance local communities in new and exciting ways.

References

Axtman, K. *Bye, Bye, Library*, <http://www.cbsnews.com/stories/2005/08/23/tech/main791462.shtml>, accessed 30 September 2005.

Australian Government. *National Winners of the 2005 Awards,* <http://www.dotars.gov.au/local/awards/winners.aspx>, accessed 10 June 2006.

Benkler, Y. (2003), 'The political economy of Commons', *Upgrade* 4:3, 6-9.

Benkler, Y. (2006), *The wealth of networks: how social production transforms markets and freedom* (London: Yale University Press).

Bimber, B., Flanagin, A. J. and Stohl, C. (2005), 'Reconceptualizing Collective Action in the Contemporary Media Environment', *Communication Theory* 15:4, 365-388.

Bollier, D. (2003), 'The rediscovery of the Commons', *Upgrade* 4:3, 10-12.

Bollier, D. Is the Commons a movement? (published online 12 June 2004) <http://www.bollier.org/pdf/BerlinWizardsofOS3speechJune2004.pdf>, accessed 22 June 2005.

Boynton, R.S. 'The Tyranny of Copyright?', *The New York Times,* Magazine section (published online 25 January 2004) <http://www.nytimes.com/2004/01/25/magazine/25COPYRIGHT.html?ei=5007&en=9eb265b1f26e8b14&ex=13903

66800&partner=USERLAND&pagewanted=all&position>, accessed 22 June 2006.

Britannica. *Library,* < http://www.britannica.com/eb/article-9106477/library>, accessed 22 January 2007.

Bundy, A. (2001), 'Essential connections: school and public libraries for lifelong learning'. in *Proc. of the Forging future directions the Seventeenth conference of the Australian School Library Association* (Queensland, Sunshine Coast).

Carr, D. (2003), *The promise of cultural institutions* (Walnut Creek: Alta Mira Press).

Castells, M. (2003), *The power of identity* (Blackwell: Blackwell Publishing).

Castells, M. (2004), *The network society : a cross-cultural perspective* (Cheltenham: Edward Elgar).

Digeser, E. D. (2003), 'Citizenship and the roman res publica: Cicero and a Christian corollary', *Critical Review of International Social and Political Philosophy* 6:1, 5-21.

Dudley, G. (1998), 'Victorian public libraries: facing the future', *Australasian Public Libraries and Information Services* 11:4, 161.

Elsevier. (2005), 'Embracing Change at Singapore's Newest Polytechnic Library', *Library Connect,* 3:2, http://www.elsevier.com/wps/find/librariansinfo.print/lc030203?avoidEmail=true

Fielding, E. (2006), 'Unlocking the garage: A web portal for car enthusiasts', *Aplis* 19:4, 145-154.

Flattery, J and Baronofsky, G. 'American Folklife Center at the Library of Congress to house the StoryCorps archive', *News from the Library of Congress,* News Releases section (published online 30 September 2003) <http://www.loc.gov/today/pr/2003/03-168.html>, accessed 20 February 2007.

Friends of the Commons (2004), *State of the Commons* (New York: Friends of the Commons).

Gibbs, R. (1990), 'Libraries and Literacy: The Role of Australian Libraries', *Australasian Public Libraries and Information Services* 3:3, 123-128.

Giddens, A. (1986), *The constitution of society : outline of the theory of structuration* (Berkeley: University of California Press).

Hardy, G. and Johanson, G. (2003), 'Characteristics and choices of Internet access users in Victorian public libraries', *Online Information Review* 27:5, 344-358.

Hellstrom, T. (2003), 'Governing the virtual academic commons', *Research Policy* 32, 391-401.

Karkosi, K. (2006), 'Hurricane Katrina changed people in uncommon and unknown ways', *Associated Content*, Health and Wellness section, (published online 25 September 2006) <http://www.associatedcontent.com/article/59042/hurricane_katrina_changed_people_in.html?page=2>, accessed 22 April 2007.

Lesk, M. (2005), *Understanding digital libraries* (San Francisco: Elsevier Inc).

Lessig, L. (2004), *Free Culture: the nature and future of creativity* (New York: Penguin Books).

Levine, P. (2002), 'Symposium: Democracy in the electronic era', *The Good Society* 11:3, 3-9.

Liang, L. (2004), *A guide to open content licenses* (Netherlands: Piet Zwart Institute).

Oliver, P. and Marwell, G. (1988), 'The paradox of group size in collective action: A theory of the critical mass', *American Sociological Review* 53, 1-8

Olson, M. (1965), *The logic of collective action* (Cambridge: Harvard University Press).

Olson, M. and Zeckhauser, R. (1966), 'An economic theory of alliances', *Review of Economics and Statistics* 48, 266-279.

Ostrom, E. (1990), *Governing the Commons: the evolution of institutions for collective action* (New York: Cambridge University Press).

Pang, N. (forthcoming), 'Cultivating communities through the knowledge commons: The case of open content licenses', in Sasaki, H. (ed.), *Intellectual Property Protection for Multimedia Information Technology* (Hershey,PA: Idea Group Publishing).

Pang, N. and Schauder, D. (2007), 'The Culture of Information Systems in Knowledge-Creating Contexts: The Role of User-Centred Design', *Informing Science: The International Journal of an Emerging Discipline,* 10, 203-235.

Pang, N., Linger, H. and Schauder, D. (forthcoming), 'Community-based partnerships in the design of information systems: the case of the knowledge commons', in *Proc. of the 15th International Conference on Information Systems Development: Methods and Tools, Theory and Practice* (Budapest, Hungary: Budapest University of Technology and Economics).

Rheingold, H. (2002), *Smart mobs: the next social revolution* (Cambridge: Perseus Books Group).

Schauder, D. and Pang, N. (2006), 'Keynote presentation: Digital storytelling', in *Proc. of the Storytelling through emerging technologies* (Victoria, Australia: Museums Australia).

Wilkinson, D.M., and Huberman, B.A. (2007), 'Assessing the value of cooperation in Wikipedia', *HP Labs* (published online 20 February 2007) <http://www.hpl.hp.com/research/idl/papers/wikipedia/wikipedia.pdf>, accessed 12 February 2007.

Williamson, K., Schauder, D., Stockfield, L., Wright, S. and Bow, A. (2000), 'The role of the internet for people with disabilities: issues of access and equity for public libraries', *The Australian Library Journal* 50:2, http://alianet.alia.org.au/publishing/alj/50.2/full.text/access.equity.html

Chapter 13

City Information Architecture: A Case Study of OTIS (Opening the Information Society Project) in Sheffield, UK

Mike Powell and Adrian Millward

Introduction

This case study tells the story and reflects on the implications of the OTIS (Opening The Information Society) Project undertaken by the Sheffield First partnership between 1999 and 2001.

The project provided a focus for action for those who wished to create a city-wide information architecture that was not simply the outcome of unstructured competition between private providers or of public sector dictat, but one that actively sought participation in the process and design from the citizens and communities served.

The OTIS project team had been directly involved in and inspired by community level informatics activities in the years preceding the project. However they were suspicious of treating "community" as a separate silo of information exchange, excluded from the more powerful systems of politics and commerce. Community informatics should be strengthened by access to information from these other domains and by the influence it is able to exert on them through the content it in turn produces and makes available. The project team believed that structures of local governance have a duty of "well being" that extends beyond the creation of physical urban landscapes and the provision of high quality public service. This duty extends to the world of the net and also includes ensuring virtual urban space for local social and economic interaction, which includes appropriate e-Governance and space for a wider more participative e-Society.

We believe this has many parallels with physical place making in urban settings, and can draw on good practice in urban design and town planning, which allows a city to build and thrive in a managed fashion.

A city's virtual urban space is important to citizens, particularly in those cities seeking to transform themselves from producer to knowledge based economies. It provides an opportunity to build a creative space that aids a city's competitiveness by generating a unique high quality "city asset". This is a hard asset in terms of a civic amenity and enhanced connectivity between business, public providers and the wider community. It is also a soft asset by encouraging and retaining talented and skilled people in its creation, by creating an enhanced city image and by utilising the self organising capacity of its people and its organisations. It is a key enabler of innovation, economic diversity, higher education and skills. Linking these places and people are the "new public spaces" that offer intangible improvements to citizens' quality of life through connectivity, access to information and new places to rest and relax.

The OTIS project sought to help Sheffield gain an understanding of its information architecture and how it could evolve; to promote information exchange as a critical success factor for partnership working; and to demonstrate how information production, exchange and use could be used to unleash the power of knowledge already present at community level.

OTIS was established under the umbrella of the Sheffield First Partnership, the formally structured multi-sectoral partnership charged with driving the city's regeneration. OTIS itself was also a partnership of those in civil society and in the public sector who saw the potential of collaboratively shaped information space for everyone – a new meeting place to share information and ideas – to refashion the relationships between local governance, public services and the communities they were supposed to serve.

Context

Sheffield is a city of some half million people located in the North of England. Famous as a producer of steel and surrounded by the Yorkshire and Derbyshire coalfields, the city suffered significant job losses during the 1980s and 1990s, with attendant economic, political and cultural consequences.

Due to the need to restructure the local economy, the city was eligible for a wide range of EU and central government regeneration funds. Given that the purposeful direction of external resources to reverse deep rooted social and economic trends was a relatively new activity, there was much argument, even within those charged with managing the process, as to how these resources should be applied and managed. This led, for a time, to the acceptance of a diversity of approaches and created some room for innovation.

Nowhere was this more the case than in the area of the emerging "information society", about which most senior figures had a very limited understanding. Consequently, information initiatives, in all sectors, tended to be promoted by "information activists", who, in keeping with the culture they sought to promote, were considerably better networked than the institutions they sought to change. There were a variety of initiatives in all sectors and at all levels but, to the extent they can be retrospectively seen as a whole, three features are remarkable. One was the pioneering role of the community and voluntary sector in what was happening – a phenomenon which ranged from groundbreaking work by local activists on rundown housing estates like the Manor (Manor Training and Enterprise Centre) or the nearby mining village of Grimesthorpe (Grimesthorpe Electronic Village Hall) to the hosting of the first national UK Communities On-line conference by the Department of Information Studies at Sheffield University. The second was an acted-on awareness, amongst all the activists but far less amongst the institutions, that the change processes underway involved social and political relationships and were not simply technological in nature. The third was the considerable overlap and collaboration between sectors. OTIS itself emerged from the "social strand" of a three pronged Sheffield ICT Strategy (the other two were enterprise and learning) led by the Hallam Group, a high level local private sector network, in 1998.

The OTIS Project

OTIS was developed primarily by the authors of this article who had both been active in several of these earlier initiatives - one as an officer of the city council with responsibilities for regeneration work in a number of geographic communities and one who had been working in voluntary and community sector networks. Using the networks which had been developed, we created, consulted and reported to an informal body of information aware professionals and activists from the community and voluntary sector, Sheffield First Partnership, the city council, health authority, police and universities, which held well attended meetings two to three times a year before and during the three years of the project's life. This body supported the development of an application for European Regional Development Funds which formally located the project within Sheffield First. The small co-ordination team reported formally to a small multi-sectoral steering group whilst continuing to report to the wider body.

The project team was acutely aware of the general failure of top down approaches to the development of virtual information spaces and of the old-fashioned political suspicions which existed of any unilateral attempt to define and enforce what an information society might consist of. It therefore adopted a highly distributed approach to its work, seeking wherever possible to support and involve project stakeholders in informational activities of direct relevance to their current concerns, confident that the experience of such work, and the opportunity created to reflect on it, would lead to greater interest in wider information exchange and greater collaboration. It thus supported groups and organisations in a number of small pilot projects, local level research activities and external learning which ranged from exploring how information about mother and child services was communicated within the Pakistani community, support for a Women's Forum ICT conference, the evaluation of both council and community information projects, visits to both local and international programmes and events to study what was being done elsewhere.

Key themes and responses

The strategy of supporting the ideas and explorations of stakeholders meant that the project became involved in a wide, some argued too wide, range of activities. We would emphasise here those aspects which relate to the notion of information space in the context of the city and the challenges it faces.

Sharing information space One fundamental question is the extent to which information which has been created by the public sector using public money should, subject to protocols relating to privacy and confidentiality, be public information. Although usually discussed in terms of rights, this question also encompasses ideological issues in the approach to information, whether it is something to be used for a limited purpose or shared as a public good. One extreme example was provided by an adult education programme in a neighbouring city which trawled all records available to public sector agencies in order to provide an incredibly, one might say frighteningly, complete record of the social problems of each household on a particular estate – single parent families, recipients of unemployment or disability benefit, interactions with the criminal justice system, data on truancy etc. This information was then used to develop a plan to increase the participation of community members in adult education initiatives. Although in this case driven by benign intent, such an approach should ring serious political alarm bells. By using information which was far too confidential to be made available for public discussion and entirely asymmetric, the officers concerned were imposing a subject – object relationship on the process in which all space for participation and partnership was denied. Public information space is not therefore simply a human right, but also a necessary precondition for citizen engagement in the functioning of a democracy.

Nor is this simply a theoretical point. If such engagement is to take place and be of value, the public information space must enable citizens to find, rely on and use information relating to their particular concerns. In Sheffield, probably not untypical of similar UK cities, the management of information was fragmented and confused. Similar data was held by an array of different agencies using a host of different procedures and standards, even if funded through similar Government agencies. This made it hard for anyone, even the public authorities, to have reliable or adequate data readily to hand and promoted a silo mentality. It was also inefficient both in terms of the cost of maintaining data in such a way and the time it took users to find, access and verify it.

For example, the project coincided with a major re-organisation and expansion of services for young children in the city whose managers were keen to work with OTIS and others to identify the information needed for their new tasks. Their mapping exercise identified 33 separately administered, often overlapping data sets to which, despite being the public agency directly responsible, the apparent approval of relevant managers elsewhere and a clear authority from the city's chief executive, it only proved possible to access 18. The difficulties for a community based organisation, seeking data on which to develop their own ideas or services can only be imagined. The situation uncovered was one where

- potential information users had to compete for the attention of a bewildering number and not always visible information holders
- information holders were overwhelmed by requests for information
- multiple versions of the same data existed in different places
- standards of quality, format and update frequency varied
- information access procedures were inconsistent
- receiving updates of any data set usually required repeating the whole application process

Although such a number of data sets was excessive, centralising data collection and holding did not appear to be offer a feasible solution. Apart from anything else, a number of agencies had a legal obligation to maintain core data about the people with whom they work. Instead the young children's service, with help from OTIS, sought to develop a structured framework in which

- there were common standards agreed for maintenance, format and quality of data
- each information holder was responsible for establishing and making known rules for access to their data at operational, management and public access levels.

It was felt that if the many obstacles to creating such a framework could be overcome, accessing information in the city would be faster, easier, cheaper and, above all inclusive of all who might need it.

Tracking qualitative information Many significant decisions are taken on the basis of qualitative material which emerge from public consultations, consultancy assignments or reports produced by public service managers or by voluntary or community organisations. Often this material has limited circulation outside those immediately concerned with a particular decision and even then is often quickly forgotten. As a result

- people (communities and organisations) are often unaware of work done on issues of concern to them in other (similar) geographic or thematic areas
- relevant historical records and perspectives are under-used in current analysis and planning
- public sector organisations are often unaware of research and analysis carried out by voluntary and community sector organisations
- public bodies working on overlapping multi-sectoral issues can find themselves launching similar consultation exercises with the same populations, leading to cynicism and frustration with the whole process

OTIS constructed and piloted a simple database for such material which sought to list, provide short abstracts, and offer hyperlinks to or detail of the physical location of each item. The pilot contained only limited detail, but the intention was to create a city-wide mechanism whereby the producers of such material could easily add it to the database and thus render it permanently visible and accessible to those in the city who might find it relevant. Items were both listed for browsing or could be searched for under both thematic and geographic descriptors.

Navigation/visualisation of information space Each service presents information according to the logical structure of the service, which may or may not coincide with how different groups of user interpret the value or linkages of the information. OTIS sought to explore the potential of Java-based, user-controlled dynamic navigation tools but, outside the internal processes of the project, was unable to test such methods with any large data sets. It also, building on work done by the University of Sheffield with the Young Children's Service, explored the potential of GIS/spatial models to make information available in a geographic context, related to neighbourhood concerns. Learning from pilot projects elsewhere, OTIS demonstrated that it was technologically possible to create such an asset, and by using existing information sources could be done at modest cost. The barriers found were not technological but again the processes through which public information sources were managed.

Governance of information space OTIS also tried to address issues of governance – both for itself as a specialist city-wide partnership and for the work of its diverse stakeholders in shaping the information environment of the city. From the beginning it was clear that the issues which OTIS sought to address were of fundamental importance to future social and political relationships in the city. This relates far less to party politics than to the structures through which citizens can participate in producing, examining and debating evidence on which decisions which affect their lives are taken. In the case of virtual structures, it was clear that there was a need for more detailed and more informed analysis, reflection and debate than could be provided either by the traditional representative structures of local government or by any combination of ICT managers.

The solution piloted by OTIS was to set up a series of user-groups, each reflecting a particular perspective. One consisted of middle level public service managers (in this case an IT Champions group within the City Council), one of community and voluntary sector activists and one of professional information service workers. Membership consisted of individuals, well known within their respective communities and both aware of and interested in what Hamelink has described as "informational developments" (Hamelink 2003, 123). These individuals were selected but, at least in the information professional and community groups, in an open way. Names were published on the web pages of each group along with open invitations for others to express their interest. The groups could not be regarded as representative, but their output was in the public domain and open to challenge by peers.

The purpose of the groups was to advise the project on how the issues it was addressing were perceived from their particular perspective. They had limited budgets (under their control) which allowed their members to report on what they saw as interesting and relevant work being done elsewhere. They were also a resource for others to use as peer review groups: to pass draft reports or plans to them for their comments. This function, however, was little used. There is no culture of seeking and using constructive criticism in local public discourse. Only a few more progressive individuals were prepared to submit their work to such scrutiny and those then felt hurt and upset when the response, provided enthusiastically by groups for whom this was an unprecedented opportunity to have their say, was critical.

Working in partnership OTIS sought to identify and support the informational components of the new models of working on community regeneration issues. Over the past fifteen years "partnerships" have been strongly promoted by local and central government in the UK as a means of encouraging diverse stakeholders in a set of issues to work together to develop and deliver common solutions. For this dynamic to function partners must be able to share data and ideas in a collaborative and iterative process, rather than seek the compliance of others for fixed positions. For such work to enhance citizen participation and democracy, it must also be as transparent as possible. OTIS worked with two local area partnerships providing ICT tools and training – internet connected computers, discussion lists, extranets, web pages – intended to facilitate the working of the partnerships, especially for their community members, and the communication of their activities to the communities their work affects. It soon became clear, again, that providing and using the technology was the easy part. A more fundamental issue was the lack of any culture of openness with information, originating in the often confrontational history of relationships between community activists and local public service managers. OTIS sought to negotiate new protocols for the sharing of draft and confidential documents to facilitate genuine partnership work but this met with some resistance. Public sector bodies in particular felt uneasy sharing their internal debates even with their colleagues on the partnership boards.

Project end

By the time OTIS neared the end of its formal funding, its approach, based on developing the existing interests and concerns of stakeholders, had led to much greater interest and awareness of the issues it sought to address. Seventy people, mainly from the community and voluntary sectors, attended a half day conference, which reported on the project and engaged the audience in planning what they wanted to see happen next. The audience was divided into six groups and asked to imagine an ideal information environment ten years hence and the various issues and events which had taken place leading up to that point. All the groups saw the huge potential of addressing these issues for developing an effective and vibrant community life in the City. Each group, through their own processes, produced a chart shaped like an old fashioned pharmacists bottle, representing a mass of frustrations and problems which needed to be addressed straight away, after which it was imagined that progress would be smooth and easy. OTIS and Sheffield First, who hosted the conference, were given a clear mandate to develop new plans to tackle these bottlenecks.

However from the perspective of the project team the continuation of the project posed a number of problems, in particular the attitude of the City Council. This offered lip service in praise of the project, but showed little real understanding of it, far less any commitment to implementing the changes in its own informational behaviour that the project implied. Thus although plans were developed for a second stage, the team, in the absence of that commitment, decided not to pursue them.

Main lessons

Project timeliness In a cross-sectoral initiative, what is timely for one set of stakeholders might not be so for another. In Sheffield, the community and voluntary sector had pioneered aspects of an "information society" and there was demand, motivation and capacity to build on what had been started. By contrast, the City Council in 1998 was, for all its expressed interest in innovation and change, facing some very real crises to do with budgets and basic services, which demanded its attention. It is interesting to note however that when, in 2006, the council did get round to submitting an (unsuccessful) bid to the central governments "Digital Challenge", it could only do so in partnership with some larger voluntary sector organisations with little history of interest or innovation in informational initiatives. The initial flowering of experimentation had withered in the face of neglect and the expropriation of its ideas by better funded and more established institutions.

A second issue of timing related to the needs of a "developmental approach" in a culture of targets, measurements and a demand for quick results. If the original message was unclear, it was because the project team, as much as anyone else, needed to learn from and reflect on how the various strands the project was seeking to bring together could best work in practice. In retrospect it was also unrealistic to expect very entrenched and often antagonistic working relationships to change quickly. It was not enough to develop new tools and procedures and expect their value to be obvious and their adoption inevitable. Like the development of a successful city, the construction of successful virtual space in a locality, is a task for the long haul.

The concept of city information architecture The biggest lessons, however, related to the importance of well designed information architecture to the efficiency and effectiveness of social and economic interaction by people in a given locality.

The reality in Sheffield was, and to a considerable extent still is, one of information confusion. Most public agencies make information available to the public in their own way and according to their own priorities and standards. As we have seen, they do not hold their data in a way which is efficiently available to other public services, let alone to other sectors. Much qualitative information gets lost. Spaces for exchange are sites of frustration and dispute rather than collaboration and creativity.

As in its physical reality, the virtual space of a city needs to reflect the interests and needs of its citizens and businesses. Virtual urban space is not something that can be effectively delivered through a national "e-government" template, as has been attempted by the central government in the UK with such limited success. A successful City has to have the ambition, confidence and capacity to be entrepreneurial, an approach is unlikely to be achievable from the Public sector acting alone. Social inclusion and local design creativity are important mechanisms to avoid the "clone city" that would be provided by reliance simply on national or international private providers. Inclusion is a mechanism to add vitality, creativity and distinctiveness to a City and to its external image, at a cost considerable lower than creating physical distinctiveness through iconic buildings.

Building Blocks of Information Architecture in Practice

We think the metaphor of "architecture" is appropriate for two reasons. First, as we will discuss below, it is an apt metaphor for describing the processes, including but not restricted to the technological, by which information can and should be structured across an organisation or, in this case, city (Davenport 1997; Powell 1999). Second, it provides a narrative with which to explain what needs to be done to an audience which perhaps lacks confidence in managing information flows – and want excuses to leave it to their ICT teams – but which is at home with the concepts of physical urban planning. It is recognized in the physical environment that good design is good business. However there are no "best practice design guides" for city level information architecture, and no equivalent to the Commission for Architecture and the Built Environment (CABE) to offer advice and to collate and disseminate good practice. We believe that OTIS offers some indicators on the principles upon which such guidelines should be based, consistent with the principles of information architecture in development organisations described by Powell (1999). We do not suggest that this is the only approach. Christopher Alexander developed a more reflective guide to thinking about physical space in 'A Pattern Language' (Alexander 1977) and this method has been explored in relation to organisation and communication for social progress by a virtual community led by Doug Schuler.[1] It suggests a method which could be developed further as a guide to constructing virtual spaces.

In our view, a city information architecture provides a framework for managing a city's information assets and exchanges. It should seek to

- Provide a framework for systems, information and procedures at the city level needed to support growth in the new information economy.
- Set shared governance and decision making structures and standards for the shared use of information as a city resource.
- Co-ordinate, integrate and harmonises information practice by using an exploratory, inclusive but planned approach across a city.
- Assist individuals, companies and public bodies to make better use of information already in the city to make better decisions.
- Identify additional unmet information needs.

The principles upon which such an architecture needs to be based include:

User-led design Good town planning design processes include participation by the community to provide places and buildings which are better tailored to need, engender a sense of ownership and reduce crime. Design processes must cater for all including people of different age, gender, ethnic background and disabled people. This is to create and retain people-friendly places that are well-used and well-loved.

OTIS sought to involve local people in shaping their local information architecture. This is not just because this is a "good thing", or because it leads to better, more accessible design, although these are valid reasons. More fundamentally this is because at a strategic level, if city information architecture is to be source of lasting competitive advantage, it requires the dynamic involvement of local people to create its distinctive identity.

1 See http://www.publicsphereproject.org/patterns.

Capacity for alteration As a city continues to evolve, it has to be capable of adaptation and change, to reflect the changing external environment. Information architecture should be a critical tool to assist in this ability to make the city more "future proof" by making it more adaptive and not create another barrier to change. This mirrors recent changes to the Town Planning process in the UK which aim to increase flexibility and continuous revision within the strategic context of a Local Development Framework and area plans.

OTIS provided mechanisms whereby ideas for change could be circulated and commented on by user-groups to ensure that changes identified and met the changing needs and views of diverse user communities.

A thought-out pattern for the whole It is recognised that effective town planning requires a long term structure plan and more localised "Neighbourhood master-plans" to facilitate and co-ordinate development. Consider what our cities would look like in a world with no planning structure plan to shape development, left entirely to the private sector.

In the same way the information needs and resources for the city must be understood, but this does not mean that every last detail must be rigidly or centrally defined. Good planning should create a space in which people can act with considerable freedom, but without detriment to their neighbours. OTIS demonstrated that there was little recognition that city information architecture requires active strategic planning or design or that it was up to civic leaders to take the responsibility to facilitate this.

Multiple perspectives Defining and mapping the information needs of a city is problematic, given their size and complexity. It needs to be understood that similar information is used by various user-groups in different ways for different purposes. For example, the links between road pollution and childhood asthma might be researched by traffic planners, environmental groups, health policy agents, doctors, schools and parents each with a distinct set of information needs and a particular – although given that it is possible to be a transport planner with an asthmatic child – potentially overlapping perspective.

An inclusive urban information architecture has to encompass and be shaped by the knowledge and perspectives of all sectors to be effective – business, public and community. In OTIS, the city's universities were core participants and local colleges and schools were involved through the community networks. However, the sources, types and application of knowledge in a large urban space go far beyond the traditional purveyors of formal knowledge and the architecture needs to allow accommodation and exchange between differing perspectives.[2]

The idea of a series of user groups giving their information-literate perspective on how the information environment meets and reflects their perspectives was one way OTIS sought to build a multiple perspective approach into the development of

2 This is carefully illustrated and discussed with reference to competing "knowledges" in an Australian mining town by Valerie Browne (2006).

the city's information architecture. Dynamic, user-controlled navigation might, if developed, have offered another.

Consciousness of scale The project recognised the need for information to be collected and made accessible at in an appropriate manner for each level at which it needs to be used. To return to our asthma example, the level of detail required at a case level is far more detailed (and more confidential) than that required by a healthcare planner, which in turn is different from the sort of generalised health statistics which might be used by environmental campaigners in the city. It should be possible to collect the information once and then organise it for its various uses, rather than build the data from scratch for each potential use.

Making links with information The way in which information flows, or does not flow, around a city has parallels in city traffic management. There are barriers to prevent us reaching our destination and insufficient space for parking (storage) when we arrive. How much does the cost of ICT, inaccessibility of data, navigation of poorly linked sites act as a deterrent for participation or a barrier to entry for small business and low income individuals or communities?

It is recognised in Town Planning that a city poorly connected internally and externally, is a poor place to live and a poor location to do business and will not flourish. In information terms what is required is an understanding of

- what information can be made use of, where, and by whom?
- where is the information created and best stored?
- what links are required between supply and demand?
- is there equality of opportunity and access?

Governance, security and freedom of information Political discourse, local democracy and civic rights and duties are core elements of vibrant and successful cities. They need to be as visible and as participative in the virtual plane as in the physical one. Virtual space designed by bureaucrats or ICT professionals is likely to be dead space.

Implications for Policy

UK Governments have launched a series of policy initiatives over the last 25 years to combat poverty and exclusion and create conditions for social and economic development. They have also invested heavily in e-government initiatives as well as spending £billions in service level ICT projects. All such initiatives talk of improving economic competitiveness, of strengthening partnerships, of enhancing the participation of citizens and communities.

At national level and locally in Sheffield, which in terms of economic and educational achievement continues to lag behind most of the country, it is hard to see signs of the promised transformational change. As everywhere else in the world, many more people are connected to the Internet but its use for collective economic, social or political engagement is limited. There are many fewer grass roots experiments now, compared to ten years ago. The pioneering community projects, much visited by ministers and academics, have become increasingly lost in mega-strategies and a broader policy drift towards expropriating community experience and "doing it to" citizens rather than learning from and building on community foundations. This mirrors the similar processes of centralisation in central government, restricting the freedom of public agencies to engage in informational developments with their direct stakeholders at local level.

However the issues which prompted these initiatives – social dysfunction, areas of poor economic performance, alienation from politics, increasing competition for resources in delivery of public services – still remain and so solutions are still needed. Our experience is that without real transformational change both in thinking and approach, "solutions" will continue to lead to recurring economic underperformance and disappointment in terms of public-community relationships.

Action research is by its very nature exploratory. OTIS, at the beginning, was not able to fully articulate what it wanted to do. However, in the final project document which considered future needs, the entire team, with its community and public sector representatives, was able to articulate a vision for a local information society which we offer as a starting point and perhaps a clearer route map for future initiatives;

> We believe that the information society will offer many opportunities for people to participate in society in new ways. This is especially the case in the areas of regeneration, social development, public service and governance with which OTIS is concerned.
>
> Such opportunities will not be created just by using the technology and financial resources, which are likely to be available. They depend on the active engagement of people, building on the multiple experiments already forged by people and organisations in all walks of life, owning, shaping and confidently developing a new set of practices in which

- it is easy for people and groups to find out and participate in what is going on (transparency, efficiency)
- the development of information strategies, local information architectures and new services is the responsibility of open and transparent partnerships committed to quality, collaborative working and user-led solutions (accountability, joined-up government, partnership working)
- all information systems are interactive allowing citizens and civil society organisations to, both publicly and privately, create new content themselves as well as respond to the information/services they receive
- people as citizens have access to information and services, capable of responding to the need of the user – content, form, time – rather than that of the supplier (citizen focused services)
- citizens, civil society organisations (such as community groups) and public service managers have, subject only to justifiable policies of confidentiality, access to all relevant information on their areas/subjects of interest from all relevant sources (freedom of information, evidenced based working)
- people and organisations can find, in the freer exchange of information and in the acquisition of new skills in the handling and use of information, new

opportunities for personal, social and economic advancement

We believe that such a society will be

- more democratic in that all actors in political and social debate will have access to all relevant information
- more effective because decisions will be based on better evidence and are likely to take into account more perspectives
- more efficient because the implementation of such a vision will reduce duplication of effort, mistakes based on wrong information and wild goose chases
- more inclusive both because the information on which policies are based is open and channels exist to challenge either the information or the policies
- more entrepreneurial as people gain the skills and confidence to develop their on-line creativity into commercial ideas (OTIS 2000).

Disclaimer

The views in this article are those of both authors only and do not represent the views of Sheffield City Council, the Sheffield First Partnership or any other organisation.

References

Alexander, C. (1977), *A Pattern Language* (New York: Oxford University Press).

Browne, V. (2006), 'Towards the next renaissance? Making collective decisions combining community, expert and organisational knowledge', *International Journal of Knowledge, Culture and Change Management* 6:3, 43-48.

Davenport, T.H. (1997), *Information Ecology* (New York: Oxford University Press).

Hamelink, C.J. (2003), 'Human Rights for the information society', in Girard, B. and Siochrú, Ó. (eds), *Communicating in the Information Society* (Geneva: UNRISD).

OTIS (2000), 'Work in Progress', Documentation for Opening the Information Society conference 23 November 2000 (Sheffield, Sheffield First Partnership).

Powell, M. (1999), *Information Management for Development Organisations* (Oxford: Oxfam).

PART 3
Planning Challenges in the Augmented City

Alessandro Aurigi

The previous two sections of the book have considered different aspects – and indeed viewpoints – that characterise how the city gets "augmented" by ICT; how both its physical spaces and communities gain new dimensions – or maybe old dimensions get extended. Comparing those sections and their papers will certainly highlight how interdependent all of these aspects – artificially pigeonholed into categories for the book's economy – are. It will also highlight how crucial it can be to look at urban space, community and ICT from two perspectives. On the one hand, the more "formal" one of what goes on in terms of purpose-made initiatives aimed at enhancing space and community through high technologies. On the other, the "informal" – though certainly design-rich in its background – dimension of the "spontaneous", unplanned use of ICT devices and networks by individuals and small groups, and how this changes our perception of city space, its usage and our "place" within it. And, even more crucially, it becomes easy – and somehow worrying and exciting at the same time for researchers in the field – to conclude that all of this "augmentation", rather than being approachable and readable in very rational and rigorous ways, thanks to the blurred boundary between formal and informal layering of space and ICT, adds further complexity to urban problems.

It can seem a paradox – at least conceptually – that the impact of computing-related technologies might end up complicating – rather than simplifying – the life of those who are in place-making jobs and positions. But once the debate gets out of the "comfort zone" of ICT used as a set of tools for urban analysis, and starts considering high technologies as part of everyday city life, the added complexity of a relatively uncharted and unfamiliar inter-disciplinary "territory" to look at seems to be the price to pay, and a major challenge to face.

This discomfort and need for a stronger awareness is something that is highlighted by all authors in this section, who have been looking from different viewpoints at what city-makers and planners have to face – or indeed may fail to face – when it comes to dealing with places, people, jobs and lives augmented by ICT.

But should planners bother with augmented space at all, the first question could be? Most papers, in different ways, seem to embed the tension between the traditional "formality" and physicality of planning practice, aimed at controlling land use and development, and the informality and fluidity of augmented spaces. But the latter does not just constitute an anomaly. It also carries with it huge opportunities as well as potential inequality threats which should not be underestimated or left on their own. As Odendaal puts it when referring to the informal street-based, ICT-boosted economy in Durban's public and semi-public spaces, and the spatial "networking" resulting from it 'This is difficult to manage from a land use perspective, but requires a flexibility in approach without compromising the public good', though the prize for this will be fostering those 'less obvious "spaces" for empowerment and mobility'.

This points straight at the fact that planners do have a role here, and although this involves a flexible, innovative and possibly more participative approach, it is still important that the "public good" is kept well at the centre of things.

Townsend critically describes some aspects of what a proactive series of interventions which aim to design and build augmented neighbourhoods can lead to, by looking at the ground-breaking developments in Seoul.

Other authors also highlight opportunities which arise from an ICT-enhanced use of space, and these tend to be the result of a combination of social/community and micro-economic improvements, and the interplay between community, economy and successful planning.

Fistola also tells us about the role of an augmented place – the "telematics square" concept and one of its applications – in helping regenerating one of the most deprived and crime-ridden neighbourhoods in Naples. Di Maria and Micelli focus on another Italian, yet diametrically opposite example describing how professional communities can be boosted by ICT and ensure that the rich and well-organised Veneto region can keep its strength through networking its otherwise fragmented SMEs fabric.

But positive results in planning augmented urbanity do not come automatically by simply implementing new technologies in cities and regions. Beyond the experiences' results, the writings in this section highlight a series of tensions which should be in the planner's (and urban designer's) mind when considering an integrated approach towards proactively shaping the augmented city, and these are obviously the most valuable and transferable pieces of knowledge available here.

A potential problem for planners is taking a shortcut, by trying to oversimplify augmentation. Both Firmino and Odendaal note how planners and local authorities – when it comes to ICT – tend to focus on what is easy to understand and deal with by their traditional practices. So, a lot of attention might be dedicated to the placement and dimensioning of physical telecommunication infrastructure[1] – as they know how to deal with controlling the location of tangible elements of space. Similarly, efforts have been made in the past fifteen years to define the characteristics of civic websites and discrete initiatives that can usually be categorised as part of the e-government wave of innovation. However, the far more blurred area of how these former – and more formal – elements of augmented space, combine with changing lifestyles, relationships, spontaneous networking and what Odendaal calls "implicit manifestations", is what can really be hiding the most interesting and beneficial potential – both economically and socially – for cities. Townsend shows among other things how – especially in the case of Korean youth culture – ICT can "remediate", somehow regenerate but also problematise traditional instances of "third place" in

1 See for instance Aurigi A. (2006) "New Technologies, yet same dilemmas? Policy and Design Issues for the Augmented City", *Journal of Urban Technology*, 13(3), pp.5-28.

the city, affecting heavily – and in an un-planned way – urban life and the public sphere.

Being able to 'plan' and design within this reality requires mainly two resources which are too often taken for granted: knowledge and strategy. Knowledge needs to be questioned and – as a consequence – expanded, as you can control and plan only what you know. Strategy needs to rely on this expanded knowledge, and be drawn so that regeneration initiatives, space and place, ICT developments can be thought and planned as one, as the combination of all elements of the space-society continuum is what really counts in the end.

Examples of possible challenges to knowledge and consequent strategies come again from the section's papers. Pelizza for instance questions the notion of community and highlights the tension – and very different outcomes – between conceiving community as a fixed entity to protect through the augmentation of space used as a tool for control and surveillance, or looking at it as a fluid, organic, process to foster through initiatives aimed at 'multiplying potential narratives' and therefore linking – rather than dividing – people. Odendaal herself implicitly suggests that renouncing – or relying less on – a logic of control is essential to becoming able to engage in the shaping of multi-dimensional and complex augmented spaces. Townsend also highlights how planners and urban designers should be wary of falling into the trap of utopian "new state efforts", of thinking that they can – and should – re-shape urban reality. Although 'compelling new urban environments' that combine physical and digital can be thought of, an attention to the values proper to the place, local communities their culture and traditions is an essential ingredient to increase chances of relevance and sustainability of innovative urbanity.

Fistola on the other hand takes a more 'modernist' viewpoint, and envisages a strategy for defining a sort of sophisticated "augmented development control" process which he calls a "digital plan", applicable only in regions and countries where local authorities have a considerable amount of power and independence. But despite his proposal is about control, he acknowleges how crucial.

Partnership is certainly another keyword, as it is seen as the only way to acquire more and better knowledge and draw effective strategies. Building institutional capacity and working within wider arenas is not a particularly new concept, but certainly looks like needing to be extended into the augmented city debate.

Di Maria and Micelli also remark how good results are "the joint outcome of deliberate policy and of bottom-up grassroots effort" and see the role of planning digitally-enhanced space as a strategic process of brokering and facilitation. Knowledge has to come from the communities involved, and a one-size-fits-all approach in shaping augmented spaces is rejected by these authors, who see as important the ability to join people, but also to maintain fragments – discrete augmented spaces for specific communities – in order for the potential of diversity to emerge.

What basically the contributors' analyses raise are issues which are already proper to contemporary planning and urban design debates. As Bolter and Gruisin (1999) would argue, the emergence of digital lifestyles and ubiquitous computing ends up being a "re-mediation" of older practices, facilities and problems.[2] As the introductory chapter of this book tries to suggest, what is needed is "updating" current theories and practices to reflect the extended possibilities and relationships made available within the physical-digital spatial nexus. In other words, we might not need new theories for planning the city at all, but we should "augment" the ones we already know.

"Spatial" planning, for instance, should acknowledge space as augmented, and promote research, new knowledge and ideas about it as its own. Working on anything involving the notion of "community" without considering its ICT extensions, for instance, is bound to make increasingly less sense. Similarly, the courage to look at "community" as something fluid and dynamic rather than as a fairly static group of people or manifestation of a culture, is needed regardless of any considerations of ICT and space, but it is certainly "augmented" by them.

"Augmented" planning will have to operate within a yet more strongly interdisciplinary and multi-actor arena, something which is more and more requested to planners anyway.

What can it provide? A much-needed platform for strategising, but also gradually understand the digital/physical nexus. It could indeed be argued that planners should not just get involved, but that they should represent the central characters in this effort, the pivot point.

Planning should then still be itself, configuring and organising space to ensure public good, but with new knowledge and the will to explore new conceptions of space.

2 Bolter J.D. and Grusin R. (1999) *Remediation: Understanding New Media*, MIT Press: Cambridge, MA.

Chapter 14

Public Space in the Broadband Metropolis: Lessons from Seoul

Anthony Townsend[1]

Introduction

The pervasive deployment of telecommunications networks is a defining characteristic of contemporary cities. Wristwatches and televisions organized the industrial and consumer transformation of cities in earlier eras – mobile phones are now more numerous than both, are subsuming many of the same functions, and providing new forms of personal augmentation through synchronization and communication. However, over the last decade as the Internet and mobile communications have transformed economic, political and social communications, it has been difficult to identify clear impacts on public space that stem from this technology. This stands in stark contrast to the visible disassembly of American cities following the rise of the private automobile, and the utter transformation of public space that followed. Unlike that experience, in this socio-technical encounter, public space is proving far less flexible than consumer technologies. In many cases, therefore it makes more sense to flip the question around – what about public space has changed that led to the rise of these technologies?

Several other challenges complicate any attempt in urban studies to understand the implications of advanced telecommunications on public spaces. First, urban scholars rarely possess a nuanced understanding of information and communications technologies – study of these industries, their markets and regulation, and the infrastructures they create are curiously neglected in our curricula and texts. In fact, some of the most creative and active research is coming from researchers in the field of human computer interaction (HCI), trying to understand how their field will change as computing is liberated from the desktop.[2]

Yet we know something powerful is going on, as two important trends unfold.[3] The first is the rise of ubiquitous mobile communications, which among highly-

---

1 This research was made possible by a research fellowship through the Fulbright Scholar Program and the Korean-American Educational Commission, with assistance from the Seoul Development Institute.

2 See for instance, the work of Eric Paulos' Urban Atmospheres project at the Intel Research co-loaboratory at the University of California, Berkeley.

3 I assume that most readers will agree, even though prominent scholars and critics in architecture and urbanism continue to decry the banality of digital telecommunications.

augmented young people is starting to take on some of the characteristics of telepathy. These functional telepathic capabilities – the ability to rapidly and seamlessly communicate both functional and emotional information – provide great flexibility in how groups choreograph their activities in urban public space. The second key trend is the rapid deployment of material sensing in urban space, such as video surveillance to GPS, which is driving a whole new set of feedback loops that govern the management and operation of public space. Together, these social and material sensing networks are driving a thorough integration of virtual and physical spaces.[4]

One way to study the future is by living in it, and in terms of advanced broadband communications, the Seoul metropolitan region of South Korea is the world's most networked city. In the decade since the devastating financial crisis of 1997, the Seoul metropolitan area was transformed into the world's most wired metropolis. By 2007, in Seoul the broadband penetration rate surpassed 100 percent, suggesting that some homes are using multiple broadband lines, a practice not seen elsewhere.[5] Thus, Seoul offers an ideal case for understanding how other cities may experience this urban communications revolution, as its meteoric growth during the late 20th Century so resembles today's urban expansion in China, India, Latin America and Africa, driven by rapid industrialization and urban-rural migration. With a long and rich tradition of dynamic and fine-grained urban public space, Korea also provides a window onto the future street-level implications of these new networks and the interactions they enable.

The Forgotten Megacity

Few cities in history have grown as large and as rapidly as Seoul in the years following the Korean conflict of 1950-1953. Between 1950 and 1975 alone, the city's population doubled approximately every 9 years, growing from 1 million in 1950 to 7 million in 1975. By the 1990s, 'Seoul was no longer an independent city but was rather the central city of a rapidly expanding metropolitan region of 20 million' (Seoul Development Institute, 2003).

Seoul's postwar expansion compressed into a space of 50 years growth that had taken centuries in cities such as London and Paris. A chronic housing shortage produced a harsh, semi-permanent urban landscape, most notably characterized by clusters of identical high-rise apartment blocks, quickly constructed beginning in

---

For instance, as recently as 2003 *New York Times* architecture critic Paul Goldberger argued that "[w]hen you walk along the street and talk on a cell phone, you are not on the street sharing the communal experience of urban life. You are in some other place—someplace at the other end of your phone conversation", an observation utterly ignorant of what people are usually talking about on mobiles in the city – meetings, encounters, rendezvous, and current events – a networked urbanism operating in an augmented public space. See Goldberg P. 2003. "Disconnected urbanism", *Metropolis*, < http://www.metropolismag.com/cda/story.php?artid=254>

4 See Alex Pang's End of Cyberspace blog for an extended discussion of this trend. <http://www.endofcyberspace.com>

5 Ministry of Information and Communication, Republic of Korea. 2007.

the late 1960s, and rarely built to last more than 30 years. Large tracts of these buildings are now being demolished and the sites redeveloped featuring much larger apartments.

The frenetic pace of Seoul's growth over the 1950-2000 period left as its legacy a chaotic, multi-nodal metropolitan landscape that is one of the most disorienting cities in Asia. Nudged into decentralization by government policy, and amplified by rapid economic growth of the 1980s and 1990s, Seoul seems at times to be on the verge of disintegration as it struggles to employ, house, feed and move its increasingly affluent millions on a daily basis (Seoul Development Institute 2003, 6). Highways, office towers, and apartment blocks intersect at unpredictable intervals, and the newness of the physical landscape lends the whole a uniform appearance that is at once shocking and confusing.

The coda to Seoul's 20th century development was the fallout from the 1997 economic crisis, which resulted in an extended period of economic uncertainty and stagnation. During the decades leading up to the crisis, Korea's massive industrial conglomerates (*chaebol*) had become highly leveraged due to a belief that the government would never allow them to fail. While the *chaebol* had been instrumental in implementing the national governments urban development policies through their enormous housing and construction arms, by 1997 world financial markets demanded a reconciliation of the financial imbalance caused by the heavily indebted corporations. While the IMF stepped in with an emergency loan to stabilize the economy, the nation and city were humbled.

The Korean Broadband Miracle

Despite the setback of the 1997 financial crisis, Korean society is remarkably resilient – as a history of surviving frequent foreign occupations has demonstrated. In the wake of the 1997 financial crisis, the physical limits of land and road capacity were largely exhausted, and the modernist industrial development model of the *chaebol* was widely discredited. In a bold but visionary move, the political leadership, entrepreneurs and even housewives staked their future on the Internet as a new platform for development.

Pursuing the Internet was a risky national development strategy. Under the military dictatorships of the 1970s and 1980s, the telecommunications industry was a tightly regulated monopoly. By 1997, telecommunications reforms had led to the emergence of several new competitive companies in the residential telecommunications market. A variety of government programs spurred investment in a national fiber optic backbone linking 144 cities and towns. Korea's population density and relatively new network infrastructure[6] was a boon to rapid buildout, as shorter distances and the economies in wiring large apartment complexes greatly reduced the cost and speed up deployment.

6 Essentially, every piece of urban infrastructure in South Korea was less than 50 years old, having been built after the near-total devastation of the Korean War.

Broadband truly caught on with consumers in 2000. Competition, government investment, the structure of the housing market, and Koreans' growing exposure to broadband through cybercafes created a surge in demand for residential broadband services. Between 1995 and 1999, the Korean online population doubled every year, and nearly tripled between 1998 and 1999. From 1999 to 2004, the online population nearly tripled to over 29,220,000 users. By then, dialup connections had become an anachronism, with 95 percent of Internet users using broadband (Ministry of Information and Communications 2004).

The Case of Seoul

Studies of Korea's unique broadband marketplace have identified many factors that contributed to the country's rapid embrace of broadband: favorable regulation, strong consumer demand, and urban density (DTI/Brunel University 2002; Yun *et al.* 2002). However, this paper argues that the Seoul metropolitan region itself played a crucial role in providing an environment for success. As technopoles and "network cities" provided a seedbed for the early development of the Internet as an academic research network (Townsend 2001), Seoul was a crucial incubator for the Korean

Table 14.1. Korea's Share of OECD Broadband Population, December 2002

	Broadband subscribers (millions)	% of OECD total
United States	19.8	35.5 %
South Korea	10.1	18.1
Seoul	(4.5)	(8.1)
Rest of Korea	(5.6)	(10)
Japan	7.8	14.0
Canada	3.7	6.6
Germany	3.3	5.9
U.K	1.4	2.5
Others	9.7	17.4
OECD	55.8	

Source: OECD Communications Outlook, 2004 CIA World Fact Book.

Internet miracle, not just as the seat of government policymaking and industrial power, but also as the center of Korean fashion, culture and consumerism.

The swiftness with which Seoul became a dominant global trend-setter in urban broadband is difficult to overstate. In 1997, the greater Seoul region, which accounts for approximately 45 percent of the national population, was home to an estimated 700,000 of the nation's Internet users. Since broadband services were not yet deployed, these were almost exclusively dialup users. By 2002, just five years later, Seoul's Internet population had grown to 4.5 million households, almost universally using broadband connections. Due to multiple users per household, the actually online population was much higher.

As a result of this rapid growth, by 2002 one in every twelve broadband Internet users (8.1 percent) in OECD countries was in Seoul, and one in six was Korean[7] (Table 14.1). There were more broadband homes in Seoul than in the entire nation of Canada, or Germany, or the United Kingdom. Compared to other international cities for which statistics are available, Seoul's leadership in residential broadband is even more remarkable (Table 14.2).

Table 14.2. Household Broadband Penetration, Selected World Cities, 2004

Seoul (2002)	75 %
Hong Kong	54
Singapore (2002)	40
New York	38
Los Angeles	35
San Francisco	35
Barcelona	26
Washington, DC (central city only)	22
Rotterdam	20
Sydney (2002)	7

Sources: KRNIC, Singapore Infocomm Development Authority, comScore Network, European Union, Western Australia Technology & Industry Advisory Council, Hong Kong Office of the Telecommunications Authority.

At the time of writing, Seoul remains the world's leading broadband metropolis, and is likely to maintain a significant lead. Household broadband penetration in Korea is nearly complete, with most 90-95 percent of households subscribing to broadband service in major cities. In 2004, the date of the last major metropolitan-level survey, the leading American cities only had a 52 percent penetration rate. (A reasonable projection for 2008, given strong recent market expansion in broadband, would place the figure closer to 65 percent.) As a result, most other developed countries are somewhere between 5-10 years behind Seoul in residential uptake of broadband.[8] This fact is evident on the usage side as well – Internet video, social

7 There are a negligible number of broadband households outside the OECD region

8 Author's calculation based on data from Leichtman Research and the Fourth Section 706 report on broadband deployment of the United States Federal Communications Commission <http://www.fcc.gov/broadband/706.html>

networking, and multiplayer online gaming, seen as cutting edge in the US today, have been established markets in Korea since 2002.

With a per capita GDP of $10,000 (close to that of less developed EU nations like Greece and Portugal), Korea remains the outlier in the global broadband race. The question thus remains – what was it about Seoul that made it such a powerful driver of broadband adoption? The remainder of this chapter analyzes three dimensions of Seoul's urban character that were powerful drivers of the expansion of broadband infrastructure: neighborhood cybercafes, wireless networks, and urban cyberculture.

Retail Bandwidth: Seeding Demand for Broadband at the Neighbourhood Level

A seminal event in Korea's rapid adoption of the Internet, and a highly visible manifestation of Korean cyberculture, was the opening of over 20,000 PC "bangs" throughout Korea in the late 1990s. *Bang* – literally "room" in Korean – refers to a category of retail establishment found throughout Korean cities where small spaces are rented on an hourly basis for some kind of recreation. Typically small and family-run, PC bangs tended to locate near schools and subways stations, nestling themselves into sub-prime commercial spaces on the upper floors or basements of older commercial buildings. The typical PC bang offered Internet and LAN gaming, basic office software, and snacks for about $0.50 per hour (1000 won). Quickly, these establishments became so widespread that inexpensive access to broadband Internet was a reality for most Korean households. As a result, millions of Koreans were introduced to high-speed Internet through PC bangs, making them a key factor in seeding demand for residential broadband in Korea.

While cybercafés have flourished in many cities around the world, the Korean version differs significantly in both its cultural origins as well as its role in urban social life. In most parts of the world, public Internet access has emerged as a recombination of Internet terminals with existing "third places" such as cafés, libraries or transit terminals. PC bangs in Korea are no different, yet the venue is uniquely Korean. For over 1000 years, a broad variety of bangs have provided privacy and diversion to Seoul's densely packed residents. Before the 20th century, tea bangs and marriage bangs provided social and courtship spaces. In the 1970s and 1980s, it was karaoke bangs, billiard bangs, and board game bangs that dominated the urban public life of Seoul. In the 1990s as digital media technology and global teen culture are increasingly popular, these are being supplanted by PC bangs, comic book bangs, DVD bangs, and video game bangs.[9]

Beyond the provision of affordable, prevalent access to high-speed Internet services, PC bangs also act as a unique kind of third place, bridging the divide between physical space and virtual space. One consequence of the exploding popularity of the Internet in Korea has been the development of massively multiplayer online games

9 Aldridge G."A Bang of One's Own". Seoul Metropolitan Government. <http://www.visitseoul.net/english_new/discoveringseoul/theme04/menu_01.htm>

(MMOGs), in which hundreds of thousands of players compete in simulated fantasy worlds of knights, wizards, monsters, and treasure. Mimicking Korea's Confucian social order, these games often encourage players to organize into clans of several dozen players, who coordinate their online activities with military precision. Yet, despite the prevalence of broadband access and PCs in Korean homes, many of these clans will meet in PC bangs and venture online together, and mix face-to-face interaction with their clan, with online interaction with their clanmates and their enemies (Herz 2002). At these times PC bangs represent one of the most seamless, rich interstitial spaces between online and offline worlds, while also serving the needs of Korean youths (mostly male) for a transitional step between work/school and home. For Korean males age 15 to 35, the extremely popular network game *Starcraft* has replaced the board games played by their parents' generation. As one young Korean described it, "*Starcraft* is chess for my generation – it's what you do to relax with your friends after work."

As residential broadband services reached into over 70 percent of Korean households by 2001-2002, the PC bang phenomenon peaked and began to decline, having introduced the nation to the broadband Internet that was increasingly being brought into homes (Table 14.3). 'PC bangs provided a solution in Korea to the 'Chicken and Egg' problem of how to stimulate user growth and the development of compelling content simultaneously' (DTI/Brunel University 2002).

Table 14.3. Growth of PC Bangs in Korea

Year	Number of PC Bangs
1998	3,000
1999	15,150
2000	21,460
2001	22,548
2002	21,123

Source: Korea Game Development and Promotion Institute

Today, PC bangs are undergoing a transformation. They continue to serve a large customer base of teens that seek someplace outside the home to congregate for this unique form of online/offline collaborative gaming. They are also an important part of introducing new games and gaming technologies to the youth market, such as high end game consoles. Through both formal and informal arrangements with video game developers, PC bangs have become institutionalized distributors of new games, and play a crucial role in trend- and hit-making in that industry. Finally – these *de facto* public spaces are being transformed through the addition of amenities such as non-smoking sections to maintain their attractiveness.

Untethered Access and the Last Mile

The most recent phase of digital network building is the deployment of wireless infrastructure for mobile data communications. Again, Seoul has led Korea in deploying and utilizing these new technologies for broadband communications. It is essential for urbanists to understand the evolution of wireless technologies because they are a classic case of technology being adapted to the realities of urban public space – wireless devices put a flexible face on the end of heavy, fixed telecommunications infrastructure that acknowledges our human need for mobility and intimacy (Townsend 2001b).

As wired broadband infrastructure has spread, wireless networks have been deployed as edge extensions to further leverage the value of information and data services outside traditional fixed access locations such as the home and office. Public spaces, once places for disconnection, now are venues of intense telecommunications activity. The sidewalk café, the park and the public plaza have been given new vitality by being reconnected to the conduits of modern urban social, political and economic exchange. Additionally, wireless technologies are far more adaptable to the architectural legacy of existing cities – connectivity can be provisioned incrementally within existing envelopes, rather than requiring renovation and reconfiguration of utilities.

As with wired infrastructure, the high population density of Korean cities has made wireless deployment less expensive and more rapid. Furthermore, a widely shared recognition of the value of pervasive access to cellular service has hastened the deployment of network infrastructure into subway tunnels and stations, and the interiors of large buildings.

Korea is a leading experimental site for future wireless technologies that may encourage new kinds of behavior in public space.

Telecommunications carriers worldwide have deployed. Korean firms have invested heavily in Wi-Fi "hotspots" at popular public venues, which allow mobile subscribers to access the Internet at high speed. Unlike in most countries, where Wi-Fi instrastructure is scattershot and fragmented, in Korea the national telecom firms provide integrated, comprehensive, dependable and affordable hotspot access. SK Telecom's NESpot alone offering service at over 23,000 locations nationwide by the end of 2004. By comparison, Tmobile, North America's largest commercial WLAN operator only offered 5,600 hotspots in the US and Canada. In the latest round of investment, Seoul is again leaping ahead in the deployment of Wi-Max technologies, which provide broadband access to mobile users outside of hotspot areas. Today, Wi-Max (Korea's local flavor is marketed as WiBro) can be used to watch streaming real-time video from the Internet in public spaces across Seoul – even while traveling underground between stations on the Seoul subway.

Mobile TV was launched in Korea in 2004, and is already transforming the way people use public space, especially interstitial, compulsory public spaces like subways and buses. Digital Mobile Broadcasting is a service that provides satellite television and radio programming to mobile devices such as telephones and PDAs, moving at up to 130 km/hr. The service is provided from a dedicated satellite in geosynchronous orbit over Northeast Asia, launched as part of a joint venture between SK Telecom and NTT DoCoMo in Japan. The government mandates that a set of free channels be provided to all subscribers, in addition to the premium content paid for by subscribers. Potentially, this technology could be more disruptive to public spaces, as it displaces the two-way communication of mobile voice or Internet usage.

Urban Cyberculture and Daily Life

In a major report on the future of broadband, the United States' National Research Council (2002) argued that 'broadband commands attention because it enables dramatically different patterns of use that offer the potential for significant changes in lifestyle and business'. Journalists have documented some early signs of this transformation in homes in the broadband suburbs of American cities such as San Diego.

Meanwhile, Seoul's substantial early lead in deploying broadband infrastructure has led to the emergence of a full-blown urban cyberculture that has been largely overlooked by the foreign media, industry, and academy. Through an increasingly seamless integration of digital networks in the daily lives of its inhabitants, Seoul has become a leading laboratory for the development of new social, economic, political and cultural uses of network technologies.

Pervasive access to digital networks means that Koreans spend more time online than citizens of any other country. The Internet is widely used for a variety of urban functions – including banking, gaming, media, and socializing. As a result, data traffic on Korean networks is nearly five times the average of other countries (Korea Network Information Center 2004). The universal presence of broadband Internet has come to define Korean homes, and is a major component of the national government and industry's view of the future of domestic life (Chosuniblo 2004).

Perhaps the most visible manifestation of Seoul's urban cyberculture is the considerable amount of time, space, and energy devoted to PC gaming. In addition to the 20,000 neighborhood PC bangs described previously, several television channels are devoted exclusively to videocasting of important competitions. Game designers also seek to break down the barriers between virtual and physical worlds by adding interfaces between games and mobile communications devices. Many online role-playing games, for instance, allow a character that is threatened to summon help from teammates by short text message (SMS). The teammates receive these distress messages on their mobile phones and can rush to a terminal to log in and join the battle.[10]

Seoul's massive regional subway system offers another glimpse into this emerging cyberculture. Like many other Asian and European transit systems, mobile voice and data services are available throughout Seoul's system. The synergy between transportation infrastructure and mobile communications is a key defining characteristic of urban cyberculture. In combination, these two networks allows

10 Oh S J. 2004. Personal interview, Game Portal Team Manager, Hanbitsoft. Seoul,. South Korea. August 16.

for unprecedented mobility and freedom even for youngsters, because parents can monitor their activity. Rheingold's "smart mobs" swarm the back alleys of Seoul's trendy Shinchon, Hongdae and Apgujeong neighborhoods coordinated on the fly by mobile communications.

The rise of urban cyberculture has already reaped significant economic benefits for Seoul. Local firms dominate the global market for large multi-player online games, and have developed a lucrative export industry. Electronics conglomerates Samsung and LG have used the rapid domestic product cycle as a laboratory for rapid innovation. (The average shelf-life of mobile phone models in Korea is approximately 18 months) Enormous amounts of retailing and business-to-business commerce have moved to the Internet, introducing new efficiencies into the urban economy. Seoul boasts one of the nation's highest online shopping rates, with 1 in 5 Internet households regularly shopping online (Ministry of Information and Communication 2002).

However, there is growing concern about the negative consequences of the digitally networked society. Korean newspapers abound with stories of teens who cocoon themselves into wired bedrooms for hours, days, and weeks on end playing games and socializing on the Internet, no longer drawn out by the need to visit the PC bang's wellspring of bandwidth. Such Internet addiction, fueled by unsupervised access to residential broadband, is also becoming a major source of inter-generational conflict.

Seoul's Lessons for Augmented Urban Public Space

Seoul is perhaps the best laboratory in the world to see what happens when digital technologies, especially broadband telecommunications, are introduced into a thriving, complex urban society.

Perhaps the most important lesson of all is about how designers should think about the integration of technology into urban public space. All too often in history, urban planners, urban designers and architects have sought to use technology to transform urban society into some ideal form or state. Such technologically deterministic experiments usually end disastrously, and do no good to public confidence in any of these professions. While the trend is now clearly towards community-driven planning, with the possibilities raised by each new wave of technology we again must confront this techno-utopian temptation.

What's so remarkable about the experience of Seoul with broadband, however, is that by fully embracing its transformative potential, Korean urban society thoroughly domesticated this technology. While a public debate on the impacts of broadband in Korean society rages to this day, to the itinerant observer the Korean Internet appears to have augmented the ability of Korean society to be Korean. Urban cyberculture in Seoul is so distinctly Korean as to be unrecognizable to outsiders, from the succession of PC bangs in a long lineage of hybrid public/private spaces to the ad hoc clan society and real-virtual synthesis of massively multi-player games like *Lineage*. The lesson is simple – the integration of broadband technology into public space can, and

Figure 14.1 Typical Seoul Streetscape, Credit: Peter Hessedahl

Figure 14.2 Korea's Web Aesthetic Reflects the Streets

should, reflect deeply held values and social norms about public behavior, the built environment, social relationships, and privacy.

The roots of Korea's cyberculture become more evident the deeper we look. Like the more familiar (to Western cultures) streets of Tokyo, Seoul streetscapes are notable for the density of visual information they present at every turn. (Figure 14.1) Bewildering arrays of signs sprout from every surface, employing every possible technique to stand out – neon, Day-Glo, and flashing lights. Should we be surprised that Korean web portals employ a similar design aesthetic, packing large amounts of information onto each screen? (Figure 14.2) Content modules provide news about key groups that the visitor is a member of - group affiliation being the sole important determinant of social status in Korean society. While broadband has certainly enabled Korean portals to incorporate rich multimedia, these portals reflect deeper needs of Seoul's urban society – titillating entertainment and group identification. In contrast, the generally dry and sparse appearance of California-based Internet giants like Yahoo! and Google seems to reflect the American landscape of wide-open spaces and suburban sprawl. Even the Korean version of Yahoo! has adapted itself to this urban visual literacy.

Finally, the enthusiastic embrace of mobile communications reflects a number of cultural and spatial practices that combine in a unique way in Seoul. The challenge of living in a large Asian metropolis is eased through the convenience and flexibility provided by mobile phones. Particularly for residents of Seoul, flexible communications provides a way of organizing a modern life across the many public and private rooms – for moving, working, eating, playing, and resting – that define life in the Korean metropolis. The all-important social ties to groups can be frequently reinforced through conversation and messages, while ornamentation of devices and the freedom from parental supervision provides some outlet for individuality to manifest itself. Whereas American teens put aside savings for private automobiles, high school and college students in Korea frequently replace their mobiles at enormous cost to maintain their social status.

Seoul's final lesson is that pervasive deployment of broadband technologies inevitably brings with it a spatial reconfiguration of activities at various spatial scales. These changes reflect the exploitation of opportunities broadband presents to firms, individuals, and the government to reduce cost, increase convenience, or meet other needs in novel ways.

The Seoul Digital Media City (DMC) project, initiated in the late 1990s under former Mayor Goh Kun, exemplifies the opportunities presented by broadband at the neighborhood scale for rethinking one aspect of the urban realm, the street. Envisioned as a hub for Seoul's booming digital media industry, DMC proposed a number of innovative broadband fixtures that sought to carefully weave broadband technologies into new solutions to timeless urban design challenges.

Media Board used urban design regulations to require the use of integrated digital signage across building facades, presenting the opportunity to create a unified citywide digital canvas for use by artists during special events. The Digital Odometer, to be located at the entrance to DMC, is a large obelisk that functions as a micro-economic stock ticker by displaying real-time data about DMC and its tenant companies and residents. Finally, intelligent street lighting remotely controlled by broadband illustrates the way in which even mundane urban management tasks can be improved through networking.

Projects like Digital Media City suggest a future of experimentation and possible breakthroughs in creating compelling new urban environments through the use of broadband technology as a tool. However, Seoul's broadband experience also suggests a number of challenges to existing urban environments. Foremost among these challenges is what might be called "virtual opportunity cost". "Virtual opportunity cost" is a concept to describe the process by which widespread and frequent use of broadband devices by city dwellers may interfere with their ability to interact with each other and the built environment. In Seoul, as in other cities, this usually takes two forms.

The first form involves escape from unpleasant or inflexible urban environments into virtual worlds that provide a more desirable or customized experience. The widespread appeal of network gaming and "virtual sitting rooms" on Korean community web portals suggests that this has become a popular means of escapism. The negative consequences for urban public space are reduced use and participation. To some extent, it appears that a generation of Koreans is in the process of cocooning itself into its bedrooms and virtual parlors.

The second form of "virtual opportunity cost" occurs through the use of mobile communications devices as a means of coping with social anxiety in public space. Mobile users caught in an unfamiliar or uncomfortable space are far more likely to reach out to their familiar social network than engage strangers in conversation – a necessary step to integrating themselves into society. The negative consequence may be an overall reduction in serendipitous encounters between individuals with complementary interests, the lifeblood of the city's magic. While it could also be argued that mobile communications allows individuals to manage greater use of public space because they are not confined to an office and often use third spaces as a substitute for one, these encounters still remain confined to existing professional and social circles and do not contribute to the "mixing" function of urban public space.

Both of these forms of virtual opportunity cost also suggest a third possible outcome, which is reduced attention to the negative aspects of poorly designed urban space and a corresponding loss of interest in improving it. This can occur through disengagement by one of the means described above, or through the improved ability to overcome poor design presented by way finding and navigation capabilities increasingly being integrated into mobile communications devices. For example, the widespread use of phone-based maps in Tokyo has effectively erased any need to rationalize that city's archaic and ineffective street address system.

What can urban designers do in the face of these challenges? Unfortunately, Seoul provides few suggestions, as Korean society is just beginning to struggle with these questions. However, given its significant lead in exposure to these challenges, it is the logical place to watch in the coming years for insight. As with Digital Media City, Seoul will continue to be a site of experiments that seek to accentuate the positive effects of broadband on urban life, while mitigating the negative ones.

Figue 14.3 Rendering of Digital Media City Design Proposals, Credit: Dennis Frenchman, MIT Program in City Design and Development; Donyun Kim, Archiplan

Acknowledgments

This chapter is an adapted version of a research article originally published as Townsend A M. (2007), "Seoul: Birth of a broadband metropolis". *Environment and Planning B*. 34(3) 396–413, Pion Limited, London.

References

Chosuniblo (2004), MOIC Unveils Ubiquitous Korea Plan <http://english.chosun.com/w21data/html/news/200406/200406100024.html>.

DTI/Brunel University (2002), Investigating Broadband Deployment in South Korea. October.

Hafner, K. (2004), 'Living the Broadband Life', *New York Times*, July 15

Korea Network Information Center (2004), Internet statistics <http://isis.nic.or.kr/english/index.html>.

Herz, J.C. (2002), 'The broadband capital of the world' *Wired* 10.08 <http://www.wired.com/wired/archive/10.08/korea.html>.

Kanellos, M. (2004), 'Korea's KT to have Earth's largest Wi-Fi network?' *CNET*, May 20.

Ministry of Information and Communication (2002), *Informatization Statistics 2002* (Republic of Korea).

Ministry of Information and Communication (2004), *Korea Internet White Paper 2004*. (Republic of Korea).

National Research Council (2002), *Broadband: Bringing Home the Bits. Computer Science and Telecommunications Board* (Washington, DC: National Academies Press).

Seoul Development Institute (2003), *Seoul, 20th Century: Growth & Change of the Last 100 Years*. (Seoul, South Korea).

Townsend, A.M. (2001), 'The Internet and the rise of the new network cities: 1969-1999', *Environment and Planning B*, special issue on "Cybergeography". 28(1):39-58.

Townsend, A. (2001b), 'Mobile communications in the 21st century city', in Brown B. (Ed.) *The Wireless World: Social and Interactional Aspects of the Mobile Age* (Berlin: Springer-Verlag).

Yun, Heejin Lee, and So-Hye Lim (2002), 'The Growth of Broadband Internet Connections in South Korea: Contributing Factors' Kyounglim (Stanford University, Asia/Pacific Research Center) <http://iis-db.stanford.edu/pubs/20032/Yun.pdf>.

Chapter 15

Stretching the Line into a Borderland of Potentiality: Communication Technologies Between Security Tactics and Cultural Practices

Annalisa Pelizza

If politics is defined as the progressive composition of collective life, some sociologists, tired of the revolutionary period, found a way to shortcut the slow and painful process of composition and decided to sort out by themselves what were the most relevant units of society. The simplest way was to get rid of the most extravagant and unpredictable ways in which actors themselves defined their own 'social context'.[1]

Introduction

In the Western "exploded" city, where residential, entertainment, commercial, technological and administrative poles were born in order to satisfy previously integrated functions, there seem to be two opposite attitudes towards physical public spaces. On one hand, the growing claim for security and standardization of social practices under the umbrella term of "civility" or "proper behaviour" have been addressed by scholars since the publication of the "Broken Window" theory (Wilson and Kelling 1982); on the other hand, artists' initiatives emphasizing the unpredictability of different socio-cultural realms in cities have been reclaiming public domain open to heterogeneous groups (Broeckmann 2004).

With the emergence of the terrorist threat, war, insecurity and cities have been redefining each other in new ways (Graham 2004). Domestic places of civil society have emerged as geopolitically charged spaces, 'battlegrounds on which global powers and stubbornly local meanings and identities meet, clash, struggle and seek a satisfactory, or just bearable, settlement – a mode of cohabitation that is hoped to be a lasting peace' (Bauman 2003, 21; quoted in Graham 2004, 8).

As a consequence, while historically cities were born in order to assure protection for the inner population from external threats, with the urbanization of insecurity and violence the binary construction underpinning the notion of protection – namely, the basic construction of an "us" and a "them" – gets charged with new political and social implications. Trying to separate the "inside" from the "outside" in order to identify the potential menace is becoming the dominant pursuit of the new urban warfare not only for public and private security operators, but also in the everyday experience of common people. However, such an activity requires skills and

1 Latour 2005: 40-1.

information that often are not at the citizens' direct disposal (Molotch and Mcclain 2003). Therefore, the long standing construction of "us" and "them" – once updated – becomes a relevant source of information, able to assure a bearable degree of risk. In other words, the state of perpetual emergency (Agamben 1998) assigns a new political relevance to mechanisms of construction of an "us vs. them" binary and to the information sources fuelling this construction. Furthermore, it should be noticed that local and national mainstream media play a crucial role in creating a "phenomenology of fear" by stretching particular groups and social behaviours into a dichotomized, potentially dangerous and unpredictable "otherness".

One of the dangers is that the perception of increasing insecurity is leading to the inclusion into the logic of warfare of social phenomena and behaviours that under other circumstances would not be considered criminal, and therefore to the rise of (geographical, social, cultural) borders *inside* cities (Graham and Marvin 2001) and to the general acceptance of the privatization of public spaces as a necessary evil. It is to show the limits of this trend and in order to enable an open "borderland" where potential identities can be constructed through communicational processes and "us vs. them" binaries can be enriched that transnational artist and media-activist networks are designing politics of cultural practices. The embedding of installations, video works and performances in contested public spaces can be conceived of as an attempt to blur and subvert the usual framing of the politics of security and fear.

ICT themselves are to a great extent involved in shaping the responses to both the requests for security and openness, as a sort of interface between the familiar and the *unheimlich* of urban public environments. On the one hand, computer code automatizes the sorting of individuals and groups into categories of risk and degrees of "otherness" (Lyon 2002). On the other hand, artist networks activate public domains and create conditions for agency by pushing invisible interactions out on to the visibility of city streets. Here, ICT are seen as an additional communicational layer providing opportunities in order to substitute binary narratives of conflict with grassroots or just multi-faceted ones.

This paper, on the one hand, investigates how security policies led by local authorities tend to resort to ICT according to a reactive control attitude, as tools to analyse traditional spatial and social problems. On the other hand, it explores how projects aiming at experimenting in the urban domain the potentialities provided by ICT have been developed by artists and media-activists worldwide. Notably, this paper questions how information technologies can be used towards the "inflation" of shrunk narratives and the hybridization of dominant images of what the city should be.

These reflections draw on an analysis of a series of case studies embracing interviews to policy makers and public officers operating at some local governments of the Emilia-Romagna Region[2] (Italy) as well as participant observation of artistic

2 In the last years the Emilia-Romagna regional government has been highly committed to the design of experimental projects aiming at reducing the perception of insecurity by citizens. See <http://www.regione.emilia-romagna.it/sicurezza/> (web site of the Department for the Advancement and the Development of Security and Local Police Policies of the Regional Government).

performances and installations held at various venues worldwide from 2004 to 2006. These case studies revealed two different attitudes toward the employment of ICT in urban spaces, sketching two different regeneration strategies for cities and communities. The first strategy implies a traditional idea of community as *gemeinschaft* to be preserved. Case studies exemplifying the second strategy, instead, consider also unstable communities where the use of ICT is not supposed to help identify pre-built subjects, but creates them through the same process of communication.

In the first paragraph urban disorder as one source of insecurity that – although not being concerned with exclusively criminal behaviours – acts as a powerful mechanism for the rise of borders inside cities is introduced through the presentation of two research projects. The first survey, governmentally led, aims at defining an interstate common taxonomy of phenomena producing an intense demand in security by citizens. The second research project, led by a group of artists, explores the nature of urban disorder's perceptions by means of a qualitative approach and realizes an intervention aiming at the public sharing of the research's results.

In the second, more theoretical, section 'urban disorder' is defined as a discourse aiming to constitute mono-dimensional identities and to shrink narratives. This theoretical perspective thus allows the comparison of government-driven ICT initiatives concerning urban disorder with artist/activist projects addressing issues like the blurring of the dominant images of a city and the hybridization of identity. Within this comparison, the paper will seek to figure out how different attitudes towards spatial and social problems imply different attitudes towards the situated application of communication technologies.

In the third section, wider political aspects of the discourse on security are taken into consideration. Here, three problematic aspects arising from an intense use of ICT as surveillance means to assure the smooth running of urban life are discussed. Notably, the paper focuses on the relationship between the communicational model implied by ICT applications in security policies, citizens' agency, the political pattern underpinning the discourse on urban disorder and the process of privatization of public spaces.

Finally, the fourth section compares some artist and activist projects stressing the significant potential provided by ICT to foster democratic life, on condition that their use is seen as an opportunity to introduce new elements in established images of what the city is and what it should be.

Introducing Urban Disorder

'The urban impulse is an impulse toward community – an impulse toward being together, and toward accepting the idea that however different we may be, something unites us.' (Goldberger 2001).

This statement, pronounced in 2001 by Paul Goldberger during the spring semester opening speech at the University of California in Berkeley, could sound either far-sighted or provocative. It seems, in fact, to posit some "sense of the Urban" that would characterize all those living in the city as a sort of tolerant community.

Far-sighted because, on the one hand, it is commonplace that one of the most noteworthy phenomena taking place in contemporary cities is the coexistence of different ways of life, behaviours and informal normative systems. On the other hand, with the thematisation of the urban experience and the imperative of competition among places in a global system increasingly oriented to resource delocalization, the capability of attracting city users becomes a paramount factor in assuring the wellbeing of a territory (Martinotti 1993). Therefore, many metropolitan areas as well as medium-size cities around the world evaluate the presence of diverse lifestyles as a tool to foster an image of vibrant, cosmopolitan city.

However, on the other side of the social imaginary spectrum, diverse behaviours have also shown their potential to give rise to conflict dynamics, especially when it comes to public opinion, as often reported by local mass media. In a growing number of cases in European cities, especially in town centres, coexistence turned out to be all but easy. Worried reactions against phenomena of incivility and urban disorder articulate the more or less explicit refuse by long time dwellers and city users to share public spaces with populations or individuals expressing different identities and lifestyles as far as social time, behaviour and use of public places are concerned (see, among others, Cooper 1998). Very often these reactions are accompanied by feelings of insecurity and perceptions of a decrease in urban safety, so that local authorities are requested to intervene and take into account these demands when designing political strategies or security tactics.

Even if feelings of insecurity towards different lifestyles and identities have always been part of complex forms of urban life, two aspects are totally novel when it comes to current phenomena of urban disorder and urban policies emphasizing "quality of life" and security of the urban experience.[3] These aspects – strictly related – are the 'virtualization of insecurity' and the new political relevance attributed to fears towards uncommon – yet often not criminal – behaviours by the same political sphere as well as the mediascape, also as a consequence of the current state of emergency threatening cities worldwide.

It is because of these two aspects that focalizing on the ways according to which ICT are used to address urban disorder (or counteract the attitude arising from this concern) reveals to be a strategic research choice that could lead to a wider understanding of the challenges to be faced by augmented urban planning.

3 In the UK urban renewal linked to neo-liberal urbanism and concerns about social sustainability of the urban domain was subsumed under the term 'urban renaissance' (see Rogers and Coaffee 2005). On a European level, the way social, economical and security aspects of urban management are interlaced sometimes still depends upon specific national contexts. The European Commission includes such aspects into the wider 'multilevel governance' framework. According to this, urban governance is characterised by three main strategies: 'decentralisation of the responsibilities and resources of local authorities, encouragement of the participation of civil society, and the creation of [public-private] partnerships with the aim of realising common objectives' (United Nations cited by European Forum on Urban Safety 2006a). For an analysis of neoliberalism as a new mode of political optimization and of agency and participation as technologies of subjectivity in Foucault's terms, see Ong 2006.

Virtualization of Insecurity

Very often, urban disorder deals with phenomena which are not predominantly criminal or violent. Mass media, local officers and citizens groups are increasingly using the terms "incivility" and "urban disorder" to indicate behaviours taking place in public places which – even if not subjected to penal law *strictu sensu* – are nonetheless perceived by the majority of citizens as explicit acts of rupture with traditional, informal codes of behaving, cohabiting and taking care of common urban spaces (Barbagli 2002).

In order to better understand the citizens' sense of insecurity and map a potentially very heterogeneous range of phenomena, some initiatives have been undertaken. One of these, the *Survey on Urban Disorder and Feelings of Insecurity* (*S.U.D.*),[4] led jointly by the city councils of Birmingham (UK), Malmö (S) and Bologna (IT) and developed with the support of the European Commission, Directorate-General for Justice, Freedom and Security, aimed at outlining an interstate common taxonomy and measurement method starting from the definition of urban disorder events as phenomena affecting urban public places and producing an intense demand in security by citizens (Nobili 2003a).

The survey was split into two phases. During the first part the CATI (Computer Assisted Telephone Interviewing) system was used to detect perceptions associated with their neighbourhoods by inhabitants of the three partner cities. According to the survey, instances of "urban disorder", as named by dwellers, span from the perceptible[5] presence of beggars, prostitutes and punks, to traces left in public spaces by young people hanging out during night hours, drug dealers, students' social rituals, dogs, street musicians: 'These persons are considered alien to the decorum of public spaces and even dangerous since they are unpredictable, even able to commit a crime' (Nobili 2003b, translation by the author). But even disrupted roads, traffic jams, damaged phone boxes, car pollution and decaying facades are seen as occurrences of incivility.

What emerges clearly from the survey is a growing tendency in the public opinion of the partner cities towards associating very heterogeneous social and environmental phenomena under the umbrella term "urban disorder". Even if often not harmful in themselves – *S.U.D.* argues – these phenomena nonetheless induce a cumulative effect: if a single sign of incivility is rarely considered serious, a high frequency in uncivil manifestations becomes critical (see Roché 2002). Being usually more visible than criminal events, urban incivilities are perceived as signals of abandonment of urban spaces, thus involving feelings of isolation and insecurity.

This latter consideration partly casts light on one of the most significant results of the survey. Direct observation by researchers compared to citizens' perceptions demonstrated that feelings of insecurity occur even when the probability of being victimized is very low (Nobili 2003b). This phenomenon could be named "virtualization of insecurity".

4 Hippocrates project 2001/HIP/043.

5 Either by sense of smell, hearing or sight.

Towards Politics of Cultural Practices. Sezione Zero's Visual Intervention in Contested Spaces

In order to investigate more deeply the nature of urban insecurity as emerged from the previous survey and its relation with the notion of visibility, an artistic neighbourhood-based group in Bologna realized an intervention in public perceptions using visual images. During the fist stage of the project, *Sezione Zero*'s video crew conducted a video-based ethnographic documentation involving the heterogeneous social groups implicated in the conflict for the definition of the identity of the city's allegedly most degraded spaces. The study revealed that traces and presences recognized as occurrences of urban disorder by citizens, although corresponding to very different social behaviours, had in common a semantic reference to dirtiness: being 'out of the right place' (disordered) was strictly associated to being 'dirty' (Sezione Zero 2006).[6]

This qualitative result shed also light on the "virtualization of insecurity" phenomenon. In the context elected as case study, references to urban disorder could be interpreted as a discourse aiming at defining "us vs. them" binaries and therefore identities for both human and physical subjects (Pelizza 2006).[7]

This discourse was then made visible and was embedded in the same contested space. During the second stage of the project, in fact, people's interviews and scenes of daily life in contested areas were digitally edited by *Sezione Zero* according to a dialogic pattern, so that self-representations by students, punks, street artists, migrants and other subjects involved in the rupture of informal normative systems alternated with accounts by shopkeepers and inhabitants about their discomfort and feelings of insecurity. The resulting video was then beamed over a video wall placed in one of the same contested squares, thus making the terms of the disputes among space users accessible by communities and individuals involved in the struggle as well as passers-by.

Setting Up the Borderland. A First Theoretical Perspective

From an analytical perspective, data from *S.U.D.* and *Sezione Zero*'s video ethnographic research show that the perception of urban disorder deals with social phenomena (feeling of insecurity, isolation, fear of the "other") deeply embedded in an image of the city made of multiple spatial-temporal layers. On the one hand, the process of accumulation of traces of "dirtiness" – through direct or mediated experience – could be seen as acting as a temporal link, similarly to a back-forward technique, and as a mechanism recalling other spatial orders. On the other hand, what is striking in citizens' accounts on urban disorder is also the repetitiveness and scarcity of interpretative patterns used to describe assorted behaviours and traces.

---

6 We cannot account here for the vast literature dealing with pollution and otherness. A classic reference study for this literature is Douglas 1970.

7 By 'subject' we do not imply only human actors. Rather, we refer to any agent – be it a human being, a natural element or an artifact – involved in a process of creation of meaning. See Latour 1987; Hayles 1999; Whatmore 2002.

Both cumulative effect and repetitiveness are related to the virtualization of insecurity phenomenon: insecurity is not necessarily associated to a high probability of being victimized, while the presence of particular groups or individuals is a sufficient condition for the perception of risk. It could be helpful, here, to recall what Amartya Sen points out: that violence and social insecurity are fuelled by the imposition of univocal identities on individuals through the extraction of a single affiliation becoming the only relevant feature in identity construction (Sen 2006). In our case study, the relevant feature would be, from time to time, race, nationality, socio-economic status, patterns of consumption, entertainment behaviours.[8]

The claim for civility could therefore be seen as a discourse where public spaces, communities, citizens and social practices, images of the 'real' and of the desired city, feeling of insecurity and sense of the uncanny constitute elements of *relatively closed hypertexts*.[9] All these elements, in fact, are sorted – among all the virtual narratives enabled by the rupture of informal normative systems – into only one potential discourse associated with a high level of risk and implying opposing, binary identities for individuals and spaces.

If the discourse on urban disorder aims at shrinking narratives and set up boundaries, it could therefore be of some interest comparing initiatives using ICT to address urban disorder with digital projects aiming at multiplying potential narratives by activating a public domain open to heterogeneous kinds of publics. In recent years, a wide range of artistic and media-activist projects have in fact been emerging as attempts to claim public domains as sites of development of democratic forms of agency coherent with the new nature of virtual and physical public environments. The work of these artists can be seen as "stretching the thin excluding boundary line into a thick borderland" where identities can be negotiated as part of discursive processes kept accessible. By deploying "tactical" technologies enabling the insertion of multimedia contents into contested public spaces, they attempt to blur and subvert the dominant image of the city and allow the emergence of multi-faceted narratives.

The *Memoria Histórica de la Alameda* project, for example, shows an effort towards using ICT to recall that urban space is not only the place of a vision, but also of relation and judgement.

Challenging the dominant image of the city: the Memoria Histórica de la Alameda Project

Memoria Histórica de la Alameda uses a locative media platform to create an augmented space where historical information provided by a geo-referenced system counteracts with the empirical vision of the *Avenida Alameda* in Santiago (Chile).

8 It should be stressed that direct observation (Sezione Zero 2006) suggests a correlation between a low quality level of consumption, the belonging to a low-income class and the use of public spaces for recreational purposes.

9 Here, 'hypertexts' are conceived of as discourses where elements placed on different layers are linked in a not linear way.

Developed by the workgroup at the *Estética y Tecnología suave* course of the Facultad de Artes, Universidad de Chile, in collaboration with the artistic network *Netzfunk.org*, the project aims at recovering the recent painful memory of Chile by means of multimedia contents embedded into Santiago's public spaces. *Alameda* boulevard is Santiago's main street. It houses many institutional buildings and, among these, *la Moneda*, the Presidential Palace bombed on 11 September 1973 during General Pinochet's coup d'état. With the rapid normalization of the country and the amnesty, traces of the recent past had been erased and the boulevard's name was changed into *Avenida Libertador Bernardo O'Higgins*, even if most people still keep using the old name.

The project team developed a system enabling the association between physical positioning, thanks to a GPS device, and multimedia geo-referenced contents recovered from public and private archives as well as personal memory. The *Cultural Luggage* platform, a cross media wearable system consisting of a GPS receiver, a laptop and a PDA, was used. The GPS receiver gets data describing the space crossed by the body wearing the system from GPS satellites, data are then transmitted to the laptop via a Bluetooth serial port. The laptop acts as a multimedia content server: data are decoded and stored into a real-time updated database. When the software detects the proximity of the mobile system to an area previously associated with geo-referenced multimedia contents, the corresponding multimedia file is automatically pushed to the PDA via the USB port.

The multimedia open repository contains video, photos, interviews and texts which interact with the user's direct experience within the urban area. Multimedia contents have been collected through research in official archives as well as personal micro-stories. Audio recording of Salvador Allende's last declaration, video footage showing the bombings of *La Moneda*, family photos and real time mobile-phone-generated personal memories of the same places under the military regime are all made visible and audible on headphones and on the PDA's screen along the boulevard's surroundings. Multimedia contents thus counteract the dominant image of the city centre conveyed by neo-liberal urban renewal.

In *Memoria Histórica de la Alameda* ubiquitous computing contributes to the setting up of an open system where heterogeneous and unpredictable memories take part in an ongoing discourse where identities of human beings and places get constantly renegotiated. 'Each word detaining for too long the same meaning loses its strength, its potential for action. We want to grant memory other multiple meanings in order to produce a constellation of partial and simultaneous perspectives able to create a domain for dialogue more than a definitive statement'. (Memoria Histórica de la Alameda workgroup 2005).

Patterns of Communication, Agency and the Privatization of Public Space

As previously recalled, the second reason why phenomena of urban disorder can be considered as novel conflicts is related with the new political relevance attributed to fears towards uncommon behaviours not only by local media, but the same political sphere.

Over the past few years, European cities have been involved in developing projects to face citizens' security expectations and to prevent phenomena of urban degradation on the basis of strategies linking urban regeneration with the use of public spaces as vibrant places for social contacts. These strategies undoubtedly bring with them a much more democratic flavour than securitization tactics. Yet, while it is widely known that vibrant communities are able to get rid of alien or unwanted behaviours, ruptures in informal normative systems are often seen by individuals as signals of the absence of institutions (Barbagli 1999). Distrusting, isolated, regarding themselves as potential crime targets, citizens ask local authorities to return safety to public urban spaces, which are perceived as hostile.

Very often, the demand for the re-establishment of a sense of familiarity within urban spaces is so pressing and charged with political implications that local authorities commit many political and economic resources in undertaking policies aiming at regulating a wide range of assorted – yet often not illegal – social practices taking place in public spaces. Instances, spanning from soft to hard measures, range from the Municipality of Barcelona asking people for *civismo* through site specific adverts, to the traditionally left-wing Municipality of Bologna deploying very restrictive strategies in order to restore a sense of legality and adopting preventive measures in order to limit disturbing behaviours (see Pavarini 2005).

Traditionally, local authorities' response to the growing feeling of insecurity has regarded information technologies as useful tools to get a real-time knowledge of what has been going on around the city. Especially video-surveillance systems have been viewed as technological substitutes of the "witness", in a certain way projecting its presence over urban spaces. A growing number of public and private CCTV systems has been recognized either as the ultimate deterrent in the toolbox of crime preventive strategies or as the first instrument of security tactics.

However, many doubts of different nature accompany the intense use of surveillance systems as means to assure the smooth running of everyday life. Some scholars pointed out that video-surveillance systems have usually been implemented according to a deterministic approach towards social problems, since the strengthening of surveillance is supposed to favour risks prevention (Pecaud 2002).

Others stress the incoherence between political declarations in favour of citizens' agency and the asymmetry implied by patterns of information transmission enabled by digital control systems. As an example, the final *Manifesto on Urban Safety and Democracy*, adopted at the *European Forum on Urban Safety* held in Saragossa (E) from 2 to 4 November 2006 by representatives of several hundred European cities, pays great attention not only to the prevention of the social causes leading to urban insecurity, but also to the role of citizens' agency in avoiding strategies of fear.

> 1. Safety is an essential public interest, closely linked to other public goods such as social inclusion and the right to work, to health care, education and culture. Every strategy using fear is to be rejected in favour of policies furthering active citizenship, an appropriation of the city's territory and the development of collective life. Access to other rights also favours the right to security. […]
> 9. By means of urban regeneration and reconstruction strategies, by providing basic services in the areas of education, social security and culture, cities

> have the ability to act on the causes and effects of insecurity. By developing integrated, multisectorial approaches, and with the support of regional, national and European authorities, urban policies are innovative if they do not put security solely in the hands of justice and the police. (European Forum on Urban Safety 2006b)

However, when it comes to the topic of ICT, the *Manifesto* does not make any reference to the role of digital technologies in 'creating links between the city's various spaces' (*Ibidem*), but focuses on exposing opportunities ('It is unavoidable for cities to resort to technological means for ensuring the smooth running of facilities accessible to the public and video surveillance systems in particular'), and constraints ('although the results remain mixed') of CCTV systems (*Ibid*). Nor, during the three-day-long meeting, panels approached alternative possible applications of ICT in relation to urban security.

As this latter case suggests, it could be said that, even if supported by cutting-edge sociological and criminological paradigms on a theory level, current urban security management policies – when it comes to implementing the use of ICT – are interested in communication technologies considered as tools to analyse traditional spatial and social problems, rather than as means enabling communicational processes whereby new subjects can get constituted.

Since video-surveillance systems are based on an asymmetric pattern of information transmission, our hypothesis is that a strong correlation between the CCTV communicational model – and more generally ICT applications in security management policies – citizens' (lack of) agency, the political pattern underpinning the discourse on urban disorder and the ongoing process of privatization of public spaces can be envisioned.

As the asymmetric communicational model distinguishes between passive subjects (the ones under surveillance as well as the protected) and proactive ones (police, private guards), in a similar vein the political logic underpinning the discourse on urban disorder implies a lack of agency on the side of citizens. The interviews carried out by the author show that local officers are usually aware of the political perils a deterministic application of ICT to urban disorder problems could lead to.

> The project we are developing asks citizens to signal episodes of urban disorder by means of cell phones and allows the setting up of a priority agenda. However, there is a great risk in this: the project could produce a passive attitude, if citizens are not directly involved in the solution of the problem. ICT cannot only stimulate authorities, but should also activate citizens themselves.

According to the discourse on urban disorder, local governments are asked to act as grantors of the access to public spaces perceived as emptied, insecure and abandoned. Moreover, especially for highly symbolic places like city centres, inhabitants often keep looking at themselves as the legitimate users, whose legitimacy comes from the repetition of practices of everyday life (*Sezione Zero* 2006).[10] Unable to directly control access to "their" public spaces, individuals thus ask local authorities to act as an access regulator agent.

However, spaces where access control is introduced can be seen as undergoing a process of privatization. As architect Manar Hammad points out, spaces 'are not public insofar as access to them is controlled in favour of their "legitimate owner"' (Hammad 1989, 24 – translation by the author). Besides generating boundaries, the discourse on urban disorder can therefore be conceived of as a claim for the privatization of public space enacted by local authorities in favour of long time inhabitants and retailers. In contrast with what many e-participation projects assume, privatization of public spaces is thus a process implying the partial renounce to agency by individuals. Together with this political model, the logic underpinning the application of communicational technologies to face the growing feeling of insecurity bears this asymmetry too.

Digital cartography as closed form of social knowledge

The reactive control approach towards ICT so far drawn also underestimates a third point highlighted by many scholars. By assigning individuals and social groups into categories of risk on the basis of one single feature, computer code perpetuates and reinforces many social exclusions of our time. In other words, according to these scholars, inequalities are already embedded in electronic systems of 'software sorting' (see Lyon 2002; Graham 2005).

An initiative developed during the second phase of *S.U.D.* and then continued as part of the *Ril.Fe.De.Ur* project (Survey on Urban Disorder Phenomena) constitutes a good example of this.[11] The project made use of PDAs and databases to geo-reference non-orthodox behaviours whose presence ends up increasing the feeling of insecurity in citizens. The experiment was tested in ten small and medium cities in the Emilia-Romagna area of Italy from 2003 to 2006. Resorting to PDAs, GIS and GPS systems as means to investigate the relationship between urban spaces and social behaviours, the project aimed at developing a digital evolution of crime-mapping systems.

The test was divided into three phases. During the first one, direct observation around neighbourhoods was carried on by foot by eight trained operators equipped with palm PCs. The operators conducted a census of traces of disorder observed in the assigned areas using their PDAs as a sort of notebook (De Marco 2003). A software application was developed *ad hoc*, thus enabling the geo-referencing of data related to occurrences of urban disorder. When detecting an event of incivility, operators selected the interested area on the digital vector cartography appearing on

---

10 *Sezione Zero*'s ethnographic research showed a profound attachment to an apparently univocal endangered identity of the city. Such attachment was expressed by statements like 'I can not recognized my beloved city anymore'. (Sezione Zero 2006)

11 The latter project was developed by the Emilia-Romagna Region in partnership with ten municipalities and co-funded by the Italian Ministry for Technological Innovation under the national e-government call for proposals.

the PDA's screen. The data entry process was piloted by a grid structured by type of phenomenon and metadata about the survey (date and time of detection, street, closest house number). The urban disorder event could be classified by three macro categories (physical, social, mobility and traffic-related phenomena) and by a sub-list of fifty-one more specific items related to behaviours, damaged objects, undesired presences, littering and dirt, abandoned or occupied areas, types of pollution, etc.

At the end of every detecting session, data stored in the PDA were synchronized with the police station server via the PDA's dock. Here, a database was constantly updated with geo-referenced data coming from the operators' observations as well as from citizens' signalling by phone and web forms.

Thirdly, the system allowed queries by type of phenomenon, period of time or area and the drawing of land register maps populated with occurrences of urban disorder. These maps were then used as analytical tools for security policies design of given urban zones.

In this experiment, PDAs were explicitly used as notebooks, thanks to the acceleration they can provide to the data entry process. However, it can be noticed that the use of PDAs to represents social life lacks the flexibility of the flâneur's paper notebook. In order to be entered into the system, events of incivility needed in fact to be encoded into pre-determined categories. It should also be stressed that this taxonomy was based on citizens' accounts collected during the first phase of *S.U.D.*

This fact could testify for the most advanced form of political participation and agency: the shaping of the same categorical organization underpinning powerful forms of social knowledge on the basis of descriptions provided by citizens.[12] However, this argument overlooks the fact that the experiment was led on the basis of "relatively closed hypertexts": mobile technologies have been used to reinforce a discourse whose aim is the setting up of a distinction between "us" and "them", civilization and savagery, normality and anomaly. In other words, ICT have been used as tools to crystallize mono-dimensional identities rather than to enrich them.

Of course, the process of identity definition is a fundamental need of societies. At the heart of the issue is not so much contesting the search for identities but ensuring the open access to a public borderland where the freedom to choose what priority should be given to the different affiliations every human being detains at the same time is assured. Especially when it comes to phenomena like urban disorder and the virtualization of insecurity, the ability – also for "uncivil" minorities – to question what features are more critical for the constitution of complex identities could turn out to be more effective than the attempt to reinforce strong dichotomized identities.

On the contrary, the digital cartography from the latter example is a form of social knowledge whose production process is not accessible. Maps give a representation of public spaces only as containers of social practices. The "disordered other" is seen as a passive element of a discourse developed elsewhere: on mass media, on the server at the police station, on the software running on the PDAs.

---

12 Maps are powerful forms of social knowledge because the encoding of aspects of social life into maps, statistics, charts has historically been linked to the same process of state formation (see Foucault 1977).

More generally, the logic often underpinning applications of ICT to manage feelings of insecurity implies a traditional vision of community as stable *gemeischaft* to be preserved and does not recognize that, as for the *Memoria Histórica de la Alameda* project, augmented spaces do not only enable a point-to-point encoding of social practices into a digital substance. They also possess the potential to shift the line of what is possible or "normal" a little forward. In the case of urban disorder this could possibly lead to a decrease in feelings of insecurity and in expectations towards local authorities.

Experiments Towards Re-Writing the Hypertext of the City

An approach to ICT and urban space focused on the *making* is shared by many artists' and activists' initiatives which stress the significant potential provided by ICT to foster democratic life, on condition that their use is seen as an opportunity to introduce new elements in established images of what the city is and what it should be.

Many of these projects focus on the elaboration of critical forms of representation of space like cartographies of contested areas.[13] Others integrate actions aimed at activating public spaces and re-writing in a participative way some of the elements that constitute the hypertext of the city. *Geograffiti,*[14] for instance, enables the spontaneous recording of the points of an itinerary ("waypoints") by means of a GPS device. Waypoints are then associated with accessory information like text, images, audio and video. Every online user can upload contents related to a particular place, tag accessory information or retrieve contents left by other users.

Other projects go beyond individual forms of agency to set up an investigation on connective, transient relations emerging during participation in a communicational process rather than from well-established identities. Two examples of this are the *Meta^7-Medium* project realized by *Orfeo TV-Telestreet*[15] in Rotterdam (NL) during January and February 2006, and the *Khirkeeyaan* project developed by artist Shaina Anand in different neighbourhoods in and around Khirkee Extension, New Delhi over three weeks in April 2006.

---

13 Some examples are *Cartografia resistente* (<http://www.cartografiaresistente.org>) and *Transacciones/Fadaiat* (<http://mcs.hackitectura.net/tiki-browse_image.php?imageId=593> and <http://mcs.hackitectura.net/tiki-browse_image.php?imageId=592>)

14 <http://www.gpster.net/geograffiti.html>

15 Telestreet is a self-organized network of TV stations producing and broadcasting on a neighbourhood area. Single street televisions are widespread all over Italy and, since the stations are open to everybody, allow the short-circuit between the creation and the consumption of media. They broadcast on narrow areas using "shadow cones", frequencies granted to commercial networks but unusable because of territorial obstacles <http://www.radioalice.org/nuovatelestreet/index.php>. Since the beginning, while single TV stations used to air-broadcast on a specific urban area, the networked got constituted on line, for example using peer-to-peer systems and bit torrent files to mutually exchange videos. By so doing, the networked combined low and high-technologies for video distribution and social networking. Orfeo TV is the first born of the Telestreets.

The *Meta^7-Medium* project set as a starting point the overcoming of an urban sociality seen as a localized *gemeinschaft* to be preserved and recognized the city as the domain of intermittent communities. *Meta^7-Medium* was basically an integrated media infrastructure making use of Wi-Fi, internet cable, radio transmission and air-broadcast to link different public spaces in the city of Rotterdam: *Tent – Witte de With Centre for Contemporary Arts*,[16] Rotterdam's *Blaak* library, the main market

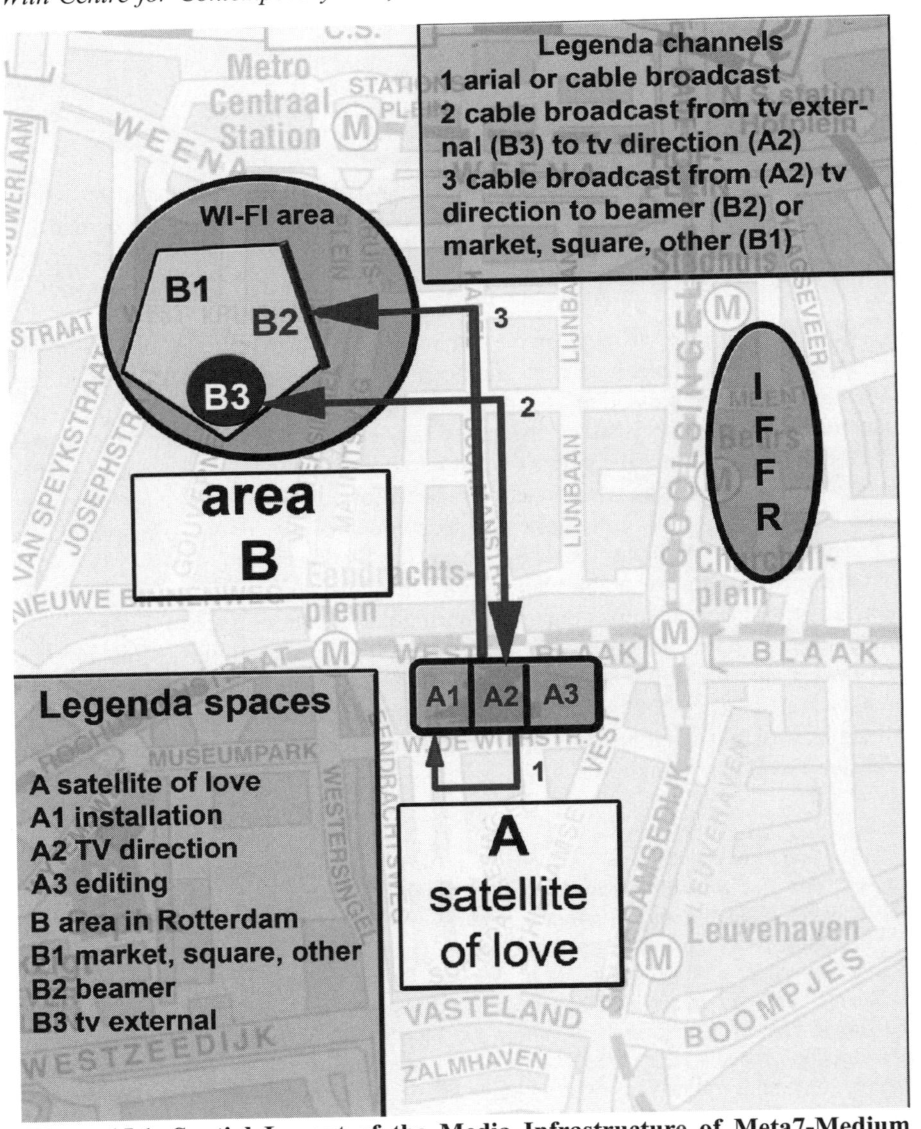

Figure 15.1 Spatial Layout of the Media Infrastructure of Meta7-Medium Project. Credit: Orfeo TV-Telestreet

16 The Centre hosted the installation and the media centre as part of the wider *Exploding Television* section of the Rotterdam International Film Festival.

square, a dedicated website, two nomadic video-equipped crews acting on streets and an indefinite number of locations crossed by people spontaneously involved in sharing their images of the city through contents produced using their cell phones and uploaded on the website.

In the project every virtual or physical location was designed to act both as a receiving and as a transmitting hub. The project tried to let the process of documentation of city life open by allowing a short-circuit between the creation and consumption of media. For some days, the sites involved (and their notion of community) were "staked out" as openly accessible. As a result, passers-by, sellers and buyers at the market, library users, visitors at the *Centre for Contemporary Art* as well as internet users and nomadic mobile users enacted symbolic forms of appropriation of urban spaces by publicly sharing their images of the city. At the same time, these images set reciprocal connections and contributed to the re-writing of the meanings associated with specific places.

KhirkeeYaan[17] is another experiment where the TV image was transformed from the box of passive viewing into an immediate self-reflecting device which could be looked into and looked out of. The aim of the project was the exploration and employment of an "open circuit TV system" as a local area network, feedback and micro-media generation device within Khirkee Extension in New Delhi.

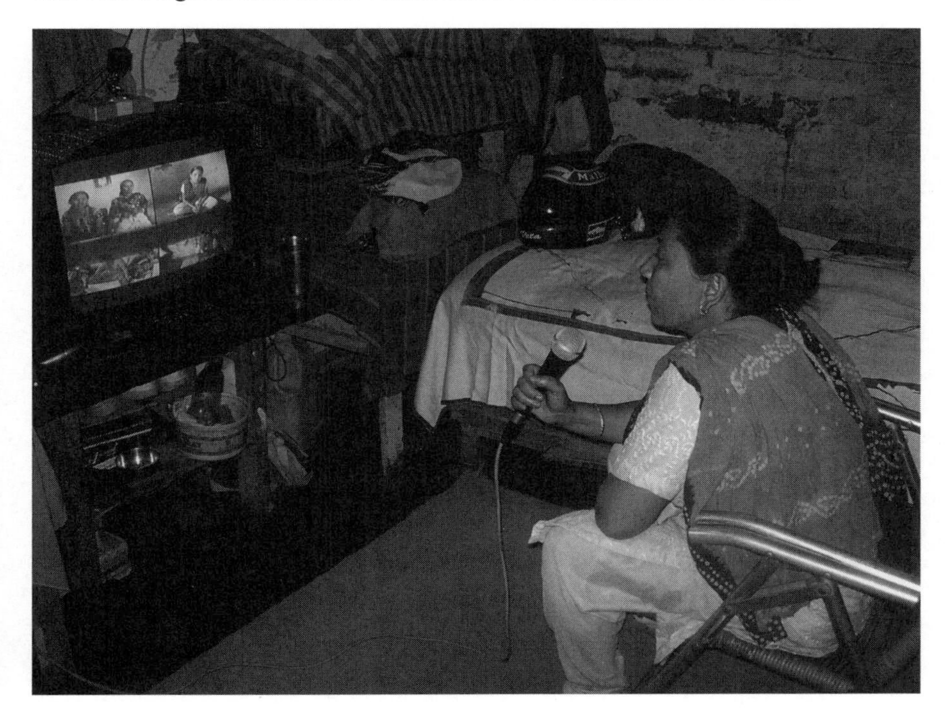

Figure 15.2 Woman Interacting in Front of the TV Quadrant. KhirkeeYaan Project. Credit: Shaina Anand

17 <http://www.chitrakarkhana.net/khirkeeyaan.htm>

Cheap CCTV and cable TV equipment was laid out to set up temporary communication systems for the use of the communities of the neighbourhood. Four sets of cameras, TVs and microphones were positioned within a 200-metre range of each other. The audio-video from the four venues were connected to a quad processor and audio mixer. This quadrant comprising of sound and image from all four locations was then fed back to the TV sets, allowing viewer/performer to interact with others in the frame.

As in surveillance systems operated for security purposes, here the process of filmmaking was automatic, made possible through a local network of mirror views, eye-level communication and the absence of cameraman and editor. However, unlike traditional use of CCTV, here the process of information transmission was visible and kept accessible: negotiations were unpredictable, made possible by immediate micro-contacts, while the intervention was unplanned and collaboratively organized.

Conclusions

Although ICT driven security policies can result in an efficient answer to citizens' need for institutions' visibility, they seem to fail in addressing more general issues regarding the complex nature of hybrid public urban spaces.

First, the implementation of ICT in urban disorder management policies endorses a political model according to which citizens are seen as passive subjects delegating local authorities the role of granting access to the public domain. If we assume that physical and virtual spaces are private in so far as access to them is controlled in favour of the legitimate user, initiatives using mobile devices to control access policies reveal a role in the process of privatization of public realms, not very differently from the one played by video surveillance systems.

Secondly, information technologies are often seen as mere tools to analyze traditional spatial problems, thus reinforcing many social exclusions of our time. By assigning individuals and social groups into categories of risk on the basis of a point-to-point digital encoding of social practices, digital crime mapping and CCTV systems perpetuate an idea of urban space as mere container for social behaviours and prevent minorities from affecting the overall image of the city. This approach shows a reactive control attitude, rather than the proactive commitment to create ICT-mediated opportunities for the negotiation of adversarial interpretations of urban environments, especially when it comes to phenomena of insecurity not strictly associated with crime events.

In this regard, it is worthwhile to recall what remembered by Bowker and Star:

> Classifications are powerful technologies. Embedded in working infrastructures they become relatively invisible without losing any of that power. [...] *classifications should be recognized as the significant site of political and ethical work*. [...] In the past 100 years, people in all lines of work have *jointly* constructed an incredible, interlocking set of categories, standards, and means for interoperating infrastructural technologies. We hardly know what we have built. No one is in control of infrastructure; no one has the power *centrally* to change it. To the extent that we live in, on, and around this new infrastructure, it helps form the shape of our moral, scientific, and esthetic [*sic*] choices. Infrastructure

> is now the great inner space. [...] The moral questions arise when the categories of the powerful become the taken for granted; when policy decisions are layered into inaccessible technological structures; when one group's visibility comes at the expense of another's suffering. (Bowker and Star 1999. *Author's emphasis*)

However, the evidence that finding mediation between different lifestyles is becoming an increasingly critical issue for urban planners as well as for local officers and social operators cannot be underestimated. This paper argued that planning with "lines" shapes subjects placed on one of two opponent sides of the line: that is to say, antithetical subjects involved in a relation of mutual exclusion. Planning with "borderlands" means on the contrary thinking of thick, open zones of potentiality where urban subjects can get constituted through a communicational process.

By examining suggestions on how artists are experimenting with ICT as means to try and activate the public domain, it can be argued that such attempts to create new conditions for agency in common spaces should not be relegated to intermittent demonstrations. The challenge for urban designers and planners, as well as for local authorities, deals with the developing of imaginative urban policies which will ensure that ICT will play an important role not only in assuring safety for "capsularized" zones, but also in inclusively design the image of everyday cityscapes.

References

Agamben, G. (1998), *Homo Sacer: Sovereign Power and Bare Life*. (Stanford: Stanford University Press).

Amin, A. and Thrift, N. (2001), *Cities. Reimagining the Urban*. (Cambridge: Polity Press).

Barbagli, M. (ed.) (1999), *Egregio signor sindaco. Lettere dei cittadini e risposta dell'istituzione sui problemi della sicurezza*. (Bologna: Il Mulino).

— (2002), 'La paura della criminalità', in Barbagli and Gatti (ed.).

Barbagli, M. and Gatti, U. (2002), *La criminalità in Italia*. (Bologna: Il Mulino).

Bauman, Z. (2003), 'City of fears, city of hopes'. (London: Goldsmith's College).

Bowker, G.C. and Star, S.L. (1999), *Sorting Things Out. Classification and Its Consequences*. (Cambridge: MIT Press).

Broeckmann, A. (2004), 'Public Spheres and Network Interfaces', in Graham, S. (ed.), *The Cybercities Reader*. (London: Routledge).

Cooper, D. (1998), 'Regard between strangers: diversity, equality and the reconstruction of public space', *Critical Social Policy* 18(4), 465-492.

De Marco, A. (2003), 'Metodologie di rilevamento e tecniche di visualizzazione dei dati sul degrado urbano', *Proceedings of the workshop on the European project S.U.D. 'Survey on urban Disorder and Feelings of Insecurity'*. Bologna, 26 March 2003.

Douglas, M. (1970), *Purity and Danger: An Analysis of Concepts of Pollution and Taboo*. (Harmondworth: Penguin).

European Forum on Urban Safety (2006a), 'The governance of security. Objectives'. Available at <http://zaragoza2006.fesu.org/article.php?id_article=216&lang=en>, accessed 15 February 2007.

European Forum on Urban Safety (2006b), *Saragossa Manifesto on Urban Safety and Democracy*. Available at <http://zaragoza2006.fesu.org/article_pied_page.php?id_article=267&lang=en>, accessed 15 February 2007.

Foucault, M. (1977), *Discipline and Punish: The Birth of the Prison*. (New York: Vintage).

Goldberger, P. (2001), 'Cities, Place and Cyberspace'. Available at <http://www.paulgoldberger.com/speeches.php?speech=berkeley#articlestart>, accessed 15 February 2007.

Graham, S. (2004), 'Introduction: Cities, Warfare, and States of Emergency', in Graham, S. (ed.), *Cities, War and Terrorism. Towards an Urban Geopolitics*. (Malden, Mass.: Blackwell Publishing).

— (2005), 'Software-sorted geographies', *Progress in Human Geography*, 29: 5, 1-19.

Graham, S. and Marvin, S. (2001) *Splintering Urbanism: Networked Infrastructures, Technological Mobilities and the Urban Condition*. (London: Routledge).

Hammad, M. (1989), 'La Privatisation de l'Espace', *Nouveaux Actes Sémiotiques* 4(5), 1-81.

Hayles, K. (1999), *How We Become Posthuman*. (Chicago: University of Chicago Press).

Latour, B. (1987), *Science in Action*. (Cambridge: Harvard University Press).

— (2005), *Reassembling the Social. An Introduction to Actor-Network-Theory*. (Oxford: Oxford University Press).

Lyon, D. (2002), *Surveillance as Social Sorting: Privacy, Risk and Automated Discrimination*. (London: Routledge).

Martinotti, G. (1993), *Metropoli. La nuova morfologia sociale della città*. (Bologna: Il Mulino).

Memoria Histórica de la Alameda workgroup (2005), *Memoria Histórica de la Alameda project*. Available at <http://www.memorialameda.cl/>, accessed 15 January 2007.

Molotch, H. and Mcclain, N. (2003) 'Dealing with Urban Terror: Heritages of Control, Varieties of Intervention, Strategies of Research', *International Journal of Urban and Regional Research*, Volume 27.3, 679-98.

Nobili, G.G. (2003a), *Detailed report of S.U.D. 'Survey of Urban Disorder and Feelings of Insecurity'* (Hippokrates project 2001/HIP/043). Report submitted to the European Commission, unpublished.

— (2003b), 'Disordine urbano e insicurezza: una prima indagine su Bologna', *Quaderni Città Sicure* 28, 91-122.

Ong, A. (2006), *Neoliberalism as Exception. Mutations in Citizenship and Sovereignity*. (Durham and London: Duke University Press).

Orfeo TV (2006), *Meta^7-Medium project*. Unpublished.

Pavarini, M. (2005), *Il governo del bene pubblico della sicurezza a Bologna: analisi di fattibilità*. Report submitted to the Municipality of Bologna, unpublished.

Pecaud, D. (2002), *L'impact de la vidéosurveillance sur la sécurité*. (Paris: éditions IHESI).

Pelizza, A. (2006), 'Spazi pubblici a un bivio', *E/C. Proceedings of the XXXIII AISS Congress 'Per una semiotica della città. Spazi sociali e culture metropolitane'*. San

Marino, 28-30 October 2005. Available at <http://www.associazionesemiotica.it/ec/pdf/pelizza_15_03_06.pdf>, accessed 9 February 2007.

Roché, S. (2002), *Tolérance Zéro? Incivilités et Insécurité*. (Paris: Odile Jacob).

Rogers, P. and Coaffee, J. (2005), 'Moral panics and urban renaissance. Policy, tactics and youth in public space', *City* 9(3), 321-340.

Sassen, S. (2006), *Territory, Authority, Rights: from medieval to global assemblages*. (Princeton: Princeton University Press).

Sen, A. (2006), *Identity and Violence. The Illusion of Destiny*. (New York: Norton & Co).

Sezione Zero (2006), *Le Voci di Piazza Verdi* (video survey and installation recorded and beamed in Bologna) .

Whatmore, S.J. (2002), *Hybrid Geographies*. (London: Sage).

Wilson, J.Q. and Kelling, G.L. (1982) 'Broken Windows', *The Atlantic Monthly*, 279, 3, 29-38.

Chapter 16

(D)urban Space as the Site of Collective Actions: Towards a Conceptual Framework for Understanding the Digital City in Africa

Nancy Odendaal

Introduction

Durban, like many other cities in South Africa, and the rest of Africa, displays an interesting array of experiences to the onlooker: colonial architecture towers over informal markets where traders sell cheap imported goods to local passers-by. Street barbers offer squared haircuts in brightly colored tents at busy intersections while pedestrians make rushed phone calls to friends and family at makeshift phone "shops" on crowded sidewalks. The dominance of the informal economy is reflected in the high numbers of street vendors on the city's streets, at its transport nodes and amongst the shack settlements that sit on its urban fringes. The sale of Information and Communication Technology (ICT) "services" is available through the use of sidewalk phones, mobile technology in container shops and small internet cafes. Mobile phone use is ubiquitous from township to inner city as users log on using "pay-as-you-go" payment options – the only service available to those without bank accounts and credit.

These transactions, exchange of information and communication facilitated through technology use, occur in an urban environment typified by change. The city of Durban, or eThekwini as its metropolitan area is called, has undergone tremendous restructuring and change in the last ten to fifteen years. Not least has been the transition of the Central Business District (CBD). Once home to multi-national companies (most of these have moved to an affluent and private-sector development to the north of the city), it now consists of an agglomeration of subdivided offices, private colleges and public spaces dominated by informal trade. Central (D)urban spaces are not defined by the usual standard road reserves, building lines and street furniture but by energies and activities that defy formal parameters. The public realm may be regulated on plan but terrestrial activities often defy the uses and boundaries imposed by cadastre-based zoning plans.

Two dimensions of this phenomenon and its relation to ICT are explored in this chapter. The first refers to the *contrasts* that typify the city: imported motor vehicles are parked in parking lots that double as makeshift Sunday morning market sites whilst wide well-maintained road intersections double as the point of sale for cheap imported plastic coat hangers. This contrast implies inequalities in terms of income and access to resources, marginality and exclusion; what this paper seeks to understand is how ICT may enable a softening of this divide. Does ICT enable us to address the inequalities embedded in these contrasts? Does it potentially reduce marginality and exclusion, whilst providing the means for growth in income?

The second dimension refers to the *extent and diversity of informal trade* in South African urban spaces. These informal activities are often the domain of those at the margins of urban policy making and resource distribution. Such groups include the many immigrants – some official, others not – that rely on activities such as car guarding and hair cutting to earn a meager living. Related to both dimensions is the importance of networks to enable communication and information sharing. The use of ICT in maintaining and enhancing these networks provides a focus for understanding its role in maintaining relationships in Durban's public spaces. There are many functions that can be associated with ICT in relation to these networks; advocacy, information sharing and general negotiations. These may not be obvious, but provide an augmentation of the relations that determine how people function in urban space.

This chapter originates from the theoretical premise that information and communication technologies, their distribution, their use and their adoption for social, economic and cultural production processes are integral to the evolution of cities, urban life and urban spaces. The potential of ICT for development has been embraced by development agencies such as the World Bank and the United Nations Development Programme (UNDP); some note the potential divisive impacts of digital technologies (Graham and Marvin, 2001; Castells, 1996) whilst others have tried to understand their impact on urban space. Scant attention, however, has been paid to the notion of the "African Digital City", and in particular, the impact of ICT on urban space and how it may contribute to the reproduction of that space. Furthermore, Brown (2006) argues that the urban design literature is very limited in its understanding of public space in developing countries; what its form could be, the nature of its urban spaces and the dynamics underpinning them. By extension, little attention has been paid to the impact of ICT on urban space in developing countries. This chapter seeks to explore some of the issues related to the relationship between urban space and technology in cities such as Durban.

Are urban public spaces in African cities such as Durban different? One distinction that Brown (2006) makes is the importance of public space for economic activity, as displayed in the street trading activities that dominate many such areas. An understanding of augmented urban spaces (through ICT potentially) needs to take cognizance of that. Not only are these economic activities the lifeblood of many urban Africans but the public realm is an essential economic input into street trade since pedestrian movement, traffic and visibility are key factors in drawing customers. Thus, public space is a resource, as are the associational networks that street traders, for example, rely on to acquire information and gain access to resources. Social capital, Brown (2006) argues, is an important input in this regard. It is worth exploring then, how the added "layer" of digital technology could potentially impact upon the relationship between urban space and informality.

Conceptually, the city is considered a site of collective actions for which urban space provides the stage. This chapter seeks to build a conceptual frame for understanding the future digital African city through the following:

- An overview of Durban, challenges facing South African cities in general and the debates surrounding informality, and its impact on public space.
- An overview of the extent of ICT use and availability in the city.
- Discussion of the notion of associational networks with and the degree to which ICT use may provide the basis for virtual capital.
- A conceptual frame is posited for understanding digital layering in urban space in Durban and other (South) African cities with suggestions on possible interventions.

In addition to literature on South African cities, ICT and cities and community informatics, the paper draws on preliminary findings of empirical work that has been done in parts of Durban. It should be noted that this is work in progress that includes:

- The development of web sites for community networks in Inanda, Ntuzuma and KwaMashu, a government-funded urban renewal node in the city's northern metropolitan area that is home to half a million people, many living in informal settlements and marginal living conditions. The findings used here are based on participant observation during meetings, demonstrations and interviews as well as input into the design of these web sites.
- Preliminary field work, involving mainly mapping, has been done with Siyagunda, an association of street barbers (most of them from the Democratic Republic of Congo). An analysis was done of their operations in the Durban CBD, and two major metropolitan transport interchanges.

A literature review informs the conceptualization of digital spaces whilst the preliminary field work is used to inform examples of what probably is, and what could be. The context of Durban is considered useful for this type of enquiry. It is a multicultural urban centre with a diverse array of nodes. The city also displays the usual disparities between rich and poor that inform the use of space.

Durban: Spaces of Collective Actions and Experiences

Over half of South Africans live in urban areas, ranging from larger cities such as Johannesburg and Durban (with over three million inhabitants in each) to smaller urban areas such as Pietermaritzburg and East London (Borraine et al, 2004). Cities contain 26% of the country's total population on less than 2 percent of its land area (South African Cities Network, 2006). Urban development is considered a priority at national level, given that in 2001 the nine cities included in the country's 'State of the Cities' report gave employment to 51 percent of the country's total working population. Yet 1.2 million households continued to live in informal dwellings in

2001, accounting for 34 percent of all informal dwellings in the country (Ibid.). Durban ranks second after Johannesburg in terms of population size with just over three million inhabitants. It is a coastal city that relies heavily on its port for its economic activities, while the beaches and sub-tropical climate account for a large part of its tertiary economy in the form of tourism revenue. Durban is classified as a metropolitan area in terms of the country's local government legislation (RSA, 2000), run by the eThekwini Metropolitan Authority (EMA), the municipality in charge.

Informal trade in city spaces

The South African Cities Network (SACN) (2006: 2-25) notes the increase in informality in the country's cities as taking three forms:

- Use of land outside the parameters set by the regulatory environment such as informal housing;
- Unregulated small and micro enterprises
- Unregistered and informal labour – casual, unregistered, moonlighting and multiple jobs.

Small enterprises such as fruit selling, shoe repair, hair cutting and phone shops are abundant in the city, particularly in the CBD, and at major transport interchanges, such as Warwick Junction, a train, bus and minibus taxi hub where 1 billion South African Rand (about US $134 million) is turned over every year in the informal economy (Nomico and Sanders, in Brown, 2006). No formal statistics exist with regards to the number of informal traders in the city due mainly to the fact that traders are highly mobile given the narrow margins that determine their livelihoods.

The city administration's response to informal trade is an accommodation of trade within boundaries of acceptable public health standards. The informal economy is recognized as playing a significant role in the city's economy, given the income generated at household level and the countless jobs created (www.durban.gov.za). Formal and informal enterprises are considered to be co-inter-dependent, with the performance of the one seen as having an impact on the other. The intention to improve the circumstances within which informal traders work is contained in a number of strategies that seek to improve the functioning of urban space in the city. These include an inner city revitalization plan, the CBD Revitalisation Strategy, research into international success stories through the Best Practice City Commission, and initiatives seeking to create more defensible spaces through the Safer Cities Project. Practical interventions contained in the draft informal trade policy include the establishment of a unit for Informal Trade and Small Business Opportunities (ITSBO) attempted to support small businesses such as the establishment of numerous satellite market places. Furthermore, eThekwini's Health Department has in the last five years attempted to upgrade the skills, and improve the working environment of informal traders. Various projects such as the Warwick Junction Urban Renewal Project have been implemented to improve the inner city so that informal traders can benefit. Significant funds have been allocated to the upliftment of informal market infrastructure.

Despite these initiatives, there exists very little guidance on the actual location of trade in relation to other economic activities and in terms of traders' needs. Work with Siyagunda indicates that designated spaces allocated to street barbers by the municipality have been in inappropriate locations. Furthermore, they are deficient in number with designated spatial configurations not allowing for clustering with similar enterprises or allowing for maximum exposure to pedestrian traffic (Ralfe, Briginshaw and Odendaal, 2007). Barbers are continuously under threat of being moved whilst negotiations with local street committees (set up by the municipality to assist in local level governance) are tempered by xenophobia and suspicion (Bikombo, personal communication).

Figure 16.1 Street Barber in Durban

The management of street trade activity is a complex endeavor that layers physical considerations of public health, pedestrian traffic management, maintenance of public transport infrastructure, public open space upkeep with difficult negotiations between formal and informal interests, local and foreign claims as well as between formal institutions and organs of civil society. The interplay between formal and informal, between official and unofficial results in fluid relationships between all actors involved. A complex web of contested claims to space emerges internationally. Using case study research from 4 developing cities (Dar es Salaam, Kumasi, Maseru and Kathmandu), Brown (2006) notes some of the constraints to informal trade in public spaces include continuous risk of harassment, being labeled as unofficial and illegal, being seen as temporary and transient, with inappropriate or limited policy responses. The use of public space is therefore contested for a number of reasons. Urban public space is essential physical capital where location is of utmost importance. Hence access and location are subject to bargaining and deal making between traders, between traders and gatekeepers as well as traders and officials. Tenure arrangements are complex where spaces are sublet and sometimes time-shared. Official licenses are not necessarily consistently rewarded, permits vary and initial access to space is often negotiated through kinship networks and deliberations with officials (that may involve bribery) (Ibid.). All of this is often done within a physical context of inadequate infrastructure where the maintenance of public surfaces does not necessarily keep up with demand for trading surfaces.

Informal trade in public space: typologies In Durban, the interface between street trading and public space manifests in a number of typologies that translate into variations of the distinction drawn by Brown (2006) between spaces used in the civic domain (such as streets, plazas and piazzas) and "edge" spaces – street corners, bus stations, vacant land etc. Perhaps a more useful classification is to conceptualize the relationship between formal and informal space as a continuum mediated by the economic, social and cultural relationships underpinning urban public space. On the one end of the spectrum is formal business; shops, commercial enterprises located in designated, physically enclosed spaces; secure, taxed and official. On the other end of the continuum is highly informal trade occurring in edge spaces such as street corners and on traffic islands; highly insecure, transient, temporary and essentially unsafe. Further complexity exists in the many CDB office buildings used for economic production (many of them illegal "sweatshops"). Between these two poles are the variations determined by context and informed by policy. In the Durban context these include the trader markets developed and maintained by the city administration where traders are engaged in management and governance, permits are paid and relationships maintained with formal traders (such as Warwick Junction in central Durban where a dedicated inner city development agency facilitates this). This type would sit closer to the "formal" end of a formal-informal space continuum. Closer to the "informal" end would be the negotiated spaces (such as that referred to in the discussion on street barbers above) where spaces are designated but are often inadequate and largely unused, permits are paid but markets dictate the use of public transport nodes, pedestrian sidewalks and public open spaces. Notwithstanding negotiation of permits and maintenance of relations with municipal officials, traders are nevertheless required to be mobile in the threat of removals. What determines a trader's position on this continuum? The literature and empirical investigation suggests a complex web of relationships, kinship networks and associational configurations of negotiations and bargaining.

The importance of associational networks Social capital, Brown (2006) argues is an essential input into the economic survival of street traders and others engaged in the informal sector. Kinship networks and familial ties may determine access to a particular space, which may be negotiated with a particular association of functionally grouped traders (such as the street barbers' association Siyagunda), looser networks may be used to alert traders in a particular space to danger of harassment or threats of removal while broader geographic networks may be used to access goods and credit. The networks are as complex as the spaces they negotiate, but they stretch beyond the geographic boundaries that define these areas. Urban public space is not frozen in time. Its physical definitions are merely that: physical boundaries that represent a dimension of the intentions, communications and movements exchanged between a multiplicity of actors making sense of their life worlds through negotiating, scheming and bargaining. Manifested in space, in particular with regards to informal trade, are the relations, deliberations, histories and meanings ascribed to urban life.

This understanding of public space sits comfortably with Amin and Thrift's (2002: 9) re-imagination of the city that builds on three metaphors: *transitivity* that reveals the permeability and porous nature of city processes, relations and interactions; *rhythms*, created through multiple movements, experiences and interactions; and finally, *footprints*, the evidence left by history, daily movement and outside networks. The use of metaphor is useful in creating the conceptual spaces for deeper understandings of the relations between city spaces, people and social, cultural and economic interactions. Building on this relational perspective, and their experiences in African cities, Swilling, Simone and Khan (2002) suggest that within the context of globalization and market liberalization, cities have become nodal points in international market networks and trade intentions that may be more adept at excluding, rather than including. Relationships between poor urban citizens in African spaces need to be constantly reconfigured and renegotiated within a context of ongoing precariousness and general "living on the edge". African urbanity is not the outcome of modernization processes that underpinned the North, but an organic renewal of 'behaviors, dynamics, activities and processes whose own logics are explicable in terms of the specificities of African cities.' (p. 313). Activities in cities may relate to survival in specific spaces, but enabling the engagement with the specifics of the local often entails the negotiation of social spaces across boundaries, markets and immediate spaces. African urban areas can therefore be linked in simultaneously different ways to 'national, regional, and global markets as well as different modes of production and spatial organization.' (Simone, 2004: 239). The 'highly mobile social formations' (p. 2) are fluid manifestations of a rich, yet often insecure, associational life that underpin survival networks. In the absence of the usual money, skills and educational resources available, much is made of the nature of associational life in African cities as a means towards resource exchange. These exchanges are elusive, dynamic, and unpredictable to the policy maker and are often manifested in informal modes of economic activity that is typical of African centers, reflected in its urban spaces.

Understanding informal networks is difficult for the policy maker argue Simone and Gotz (2003). There is little coherence in these processes, little that can be pinned down for long enough in order for meaningful city policy making to occur. While, 'the traditional tools have been directed at tying identified actors to preferable behaviors in approved *territories* (author's emphasis)', 'displacement is accelerating and progressively eroding the conditions for clarity and certainty' (p. 123). Not only are these transactional spaces unlikely to correspond with physical spaces, but Simone and Gotz are of the opinion that African identities also display a remarkable capacity *not to need fixed places* (emphasis in the original) (2003: 125). Local space is the locus from where transnational and global frameworks are tapped into for enhancing opportunity in the local. The reference point may be local space but the associational processes that enable sense-making of the local requires broader mobilization. These 'localized constellations of interests and urban practices' (Simone, 2001: 46) produce a 'complex topography of intersecting social networks, which simultaneously "dissolve" into each other but also often maintain rigid operational hierarchies, norms and criteria for participation', supplemented by 'strong personal, street, and face-to-face networks, which are important in residential areas of great density.' (pp. 52-53).

Approaching any form of intervention in African spaces, should therefore take cognizance of the workings of the social capital and associational networks. There are no absolutes, no models, and no fixes. Considering how digital technology may augment, enhance, change or deepen public urban spaces in an African city such as Durban requires consideration of the social processes that underpin urban life. It requires an engagement with the "other spaces" that transcend the physical in negotiating a sense of belonging. To facilitate a conceptual entry into this, the following section explores debates surrounding community appropriation of technology and the potential for "virtual capital".

Digital Urban Space: Manifestations of Virtual Capital?

The entry point for understanding the notion of augmented public space in the South African context (as perhaps in other African countries) is to understand the availability and use of ICT. This provides a base for understanding how technology is currently used to enhance and deepen the relations that underpin urban space. Furthermore, it provides an understanding of the potential through which digital technologies can contribute to improving the quality of urban places by reinforcing their functions as productive, cultural and social spaces.

If one is to consider ICT as consisting of cell phone technology as well as computer access to email and the Internet, and by extension telecommunications infrastructure, then the South African situation emerges as a patchy landscape of intentions and barriers. Perhaps one of the biggest barriers to internet access is a legislative environment that gives Telkom, the country's official telecommunications provider, a monopoly over telecommunications networks (although this is about to change through the long-awaited introduction of a second service provider). Broadband access is available through private service providers but under very tight legal and physical (the availability of cabling) constraints, leaving most households with the option of expensive and underperforming dial-up services. Many municipalities are now challenging this through provision of broadband internet services utilizing their own backbone networks. The eThekwini Council uses their fiber-optic network that connects municipal offices throughout the Durban metropolitan area as a backbone for extension of these services to all public libraries (Subban, personal communication). Free wireless access is also available now in libraries. Other cities are leasing spare infrastructural capacity to registered telecommunications operators; existing electricity infrastructure is being used to provide wireless access to poor communities. In the Western Cape Province, the small coastal town of Knysna aims to provide a blanket wireless network over the whole of its municipal area, providing broadband internet access and voice-over-Internet protocol (VOIP), thereby enabling free local calls from public phones. In Tshwane (Pretoria), power line communications are tested to provide residents with broadband internet access and VOIP (Gedye, 2005).

In a situation where there are 6.85 personal computers per 100 inhabitants in the country, and only 3.1 million South Africans had internet access in 2001 (ITU, cited in bridges.org, 2002) it is clear that the digital divide is about more than just infrastructure distribution. Where individual access is limited, some effort has been made to provide public service centers with computer access. The most notable of these are tele-centres or multi-purpose community centers established by the Universal Service Agency as well as private companies such as Microsoft and Hewlett Packard. Experiences of these have been patchy, with many suffering from lack of capacity to ensure sustainability and lack of initial community engagement (Benjamin, 2003). High internet connection bills, lack of staff for ongoing training and maintenance as well as lack of community buy-in have undermined a lot of these initiatives (bridges.org, 2002).

Internet access in township areas such as in Durban's Inanda-Ntuzuma-KwaMashu (INK) cluster, for example, is frustrated by lack of telecommunications infrastructure making dial-up services slow and expensive. Two digital hubs (tele-centres) have been established (through the Universal Access programme) in Ohlanga and Amaoti, two largely informal, very impoverished areas within INK. Both centers provide basic computer training, internet access and conference facilities with audio-visual equipment. Discussion with the center's manager (Zuma, personal communication) reveals frustration with the cost and limited availability of using dial-up services; connectivity is assured through satellite service using a (costly) private service provider.

In INK and other townships within Durban mobile phones have broadened access to telecommunications in remote and under-serviced areas. In general, South Africa has an extensive cellular network intended mainly for voice communications but nevertheless growing as a data transfer medium through Short Message Services (SMS) and picture messaging. There are 13 million cell phone subscribers in the country, according to a 2003 study (Mohlageng et al); covering most urban areas and national roads with more than 70% of the population served. The prepaid market is particularly important: accounting for more than 90% of new connections apparently (Ibid.). In addition to individual cell phone ownership, cell phone service providers have set up phone shops in disadvantaged communities whereby fixed-line services are provided in partnership with local entrepreneurs. Nearly 98% of South Africans live within two kilometers of a telephone (Benjamin, cited in Mohlageng et al, 2003), an improvement from the 87% that were within a 60 minute walk of a phone in 1998 (DFID, cited in Mohlageng et al, 2003).

Figure 16.2 Cellular Phone 'Shop' in Durban Central Business District

ICT-enabled networking

Statistics exist on infrastructure access and market penetration by cellular phone companies (4 exist in South Africa and services are being expanded to internet access through 3G and GPRS technology) but not much research has been done on the extent to which ICT is used for community networking in the developing

Figure 16.3 Trading Fruit and Phone Calls

world context. It is useful to distinguish between internet access and the use of mobile technology with regards to understanding digitally augmented public spaces in cities. Whilst technology convergence is causing the two technology streams to blend, the two are still used in different ways (and will be until palmtops become universally available).

Figure 16.4 Phone Services Advertised

Web-based networks are considered in the Community Informatics (CI) literature which explores the social appropriation of information technology. Day (2005) contends that CI emerged as a response to network society phenomena where some work has been done in understanding how technology may augment social capital. In a study of three types of UK community web sites (driven by local government, the private sector and communities respectively) Liff (2005) concludes that social networks, practices and organizational forms on-line are still very limited. Her study shows that community-based sites that were well maintained tended to express communities of interest such as for example environmental groups. Location-based sites that were developed by communities tended to have participation and ownership as operating principles whilst standardized sites (such as that maintained by a regional authority for example) often gave insufficient attention to communities to express themselves as part of a cohesive, distinctive community. It would appear that the pro-forma community web site developed by a private enterprise has limited ability to engage with the distinctiveness of communities. The focus on communities of interest that relate to particular places as shown by environmental sites reminds us that 'any particular virtual representation of place is inevitably partial – and in a healthy real and virtual community contested!' (p. 51). Liff concludes that high levels of social capital development are evident in these findings, but tend to be of the "bridging" type, characterized by 'looser relation-ships, which might for, example, provide a link between distinctive social groupings.' (2005: 42).

There is a subtle point to be clarified however, and it relates to virtual capital in African contexts. Liff (2005) quotes Putman in raising concerns that differential access to internet resources may, indeed, be destructive to social capital. The online tendency of people to interact with communities of interest may be destructive to community cohesion, although research on Canada's Netville experiment shows that it may indeed enhance further involvement (Hampton, 2002). Certainly, Liff quotes a number of other studies that suggest positive correlations between internet use and various forms of social capital (Katz et al; Wellman et al; in Liff, 2005). Kavanough and Patterson (in Liff, 2005) indicate that there is added potential for a community to build social capital locally by drawing in new participants through ICT. Liff (Ibid.) nevertheless concludes that the provision of a web site and easy means to upload and maintain information will not necessarily result in vibrant community web sites supportive of social capital. Clearly the relationship between virtual and social capital is a dynamic one; if not a reciprocal one as illustrated.

How then, do networks facilitated through mobile phones differ from that enhanced through the internet and email? In a study on mobile use amongst small entrepreneurs in Rwanda, Donner (2005) finds that the convenience of maintaining business and personal networks in real time is valued as important. Of interest would be to understand whether new networks and new ways of building associational relations are facilitated through mobile phone use. He notes the blurring of boundaries between personal and business use of mobile phones; 'users may be using the mobile technology to bring home to work.' (p. 16), which seems especially true for entrepreneurs at the lower end of the income scale. A further blurring occurs between mobility and connectivity effects of the mobile phone, especially in developing countries where landline use may be low (Ibid.). In the context of street trading, connectivity is perhaps more important in terms of social networks whilst mobility is essential given the transience of many street trader conditions. The relationship between public space and digital lifestyles that are facilitated through connectivity and mobility is reflected, Townsend (2002) notes, in the accelerated metabolism of cities as information transmission and responses are recorded in real time, 'on the move'.

Mobile phone communications (Kopomaa, 2002) increase efficiency given that the coordinates for everyday living can be determining at any time of the day – in any part of a social network. The mobile phone has become an icon for 'mobility, freedom and possibility' as also supported by Townsend (2002) in examining the marketing ploys of cell phone marketers. In the Rwandan case, cellular phones are also seen as important status symbols (Donner, 2005). Put differently – mobile phones may very well reflect aspirations to become part of a particular network or group.

Another aspect to be considered is that mobile phones offer instant connections; 'other people working and moving in synchrony with oneself are the new co-ordinates to which life is anchored.' (Townsend, 2002: 243). Mobile phones connect people to a wider social network (creating instant communities in some cases), a decentralized meeting place leading to a mobile neo-community of shared interests and constant dynamic maintenance. This communality is not manifested spatially, in the conventional sense, but phone anyone and gauge how often the question 'where are you?' emerges, and one is reminded that place is dynamic but nevertheless considered a reference point. The high degree of mobile phone use in public spaces as noted by Kopomaa (2002) and Carey (in Graham, 2004: 134) shows that 'urban spaces become no longer defined by landmarks, but by coverage' leading to a new sensation of space and the subtler separation between private and public.

The many technologies available to the urban dweller, therefore, avails a range of opportunities available for extension of current networks amongst urban citizens. A glance at the relationship between urban spaces and ICTs reveals a picture of inequalities but also possibilities whilst the way that people make sense of these spaces may very well be enhanced by electronic networking. The latter is not clear however, we know that ICT-enabled communication does not replace face-to-face contact, yet, we are not sure how it may contribute to associational life and the ongoing re-invention of social capital necessary to survive in cities. Evidence of virtual capital exists and there is some uncertainty as to whether social capital is actually enhanced through virtual capital. The final section of this chapter offers a number of dimensions through which the digital layering of urban space in the (South) African context can be understood.

Towards an Understanding of Digitally Augmented Urban Space in African Cities: Conclusions and Suggestions

Informality has been explored as a key informant into the experience and functional use of public spaces in cities such as Durban. Whether marginal or defined, public spaces give those at the economic margins access to markets represented by pedestrians and commuters. This chapter has posited a spectrum in understanding typologies of informality in city spaces, a continuum where the formal, secure and taxed shop owner sits at opposite ends to the highly mobile and insecure coat hanger salesman standing at street intersections. In between are the traditional healers selling medicines in designated spaces, the street barbers cutting hair outside the inappropriately areas allocated to them and the vegetable and fruit sellers at bus stops. Complex social and cultural relations determine where they are located in space, but also where they are located on this spectrum of formal to informal. Negotiations determine access to public places, networks inform the information required to do so whilst deliberations with officials and gatekeepers often determine the degree of permanence and continuity that can be expected. How then does a layer of digital technology affect these arrangements and their spatial manifestations?

There are two dimensions worth considering with regards to the digital layering of urban space in Durban. The first refers to explicit manifestations of this relationship; the land uses and functions that are specifically digital. The second refers to implicit inferences that are less obvious and not necessarily physical. Augmenting urban space given these two categories, it is argued here, requires a variation in approach, informed by where activities sit on an formal-informal continuum.

Explicit manifestations of the digital

The *explicit* translation of digital technology into space is represented by internet kiosks and cafes, phone shops and physical evidence of ICT such as mobile phone use in public places. This translates into nodal points on the continuum presented above and deserves its own empirical enquiry. Of interest would be their spatial configurations and interface with pedestrians and other uses, how ubiquitous they are, how different financial and institutional arrangements manifest spatially and what urban design measures may be required to accommodate them. Clearly the addition of digital technology constitutes new land uses and economic opportunities. Some of this economic activity is highly informal and entrepreneurial such as the makeshift phone tables on street corners whilst others are multi-functional spaces such as libraries where internet access is available.

Interventions should accommodate and encourage, not restrict. Much like other commercial land uses, location is particularly important, particularly in terms of capitalizing on passer-by traffic. The less formal phone shops are particularly vulnerable to changes in pedestrian circulation; the use of a telephone needs to be as convenient as a mobile phone – or purchase of airtime needs to be immediate. Location of these activities close to public transport nodes, on sidewalks and in areas where other traders are active is important. From a planning perspective this required urban design measures that allow for shade for traders, location on pedestrian routes without obstructing circulation and basic infrastructure such as tables, lock-up facilities and seating.

On the other hand, land use planning needs to allow for agglomeration and intensity. Economies of scale require that phone shops be located near stalls that sell airtime in a public environment with good cellular phone coverage. The sale of illegal "second-hand" (stolen) phones can be avoided through community policing measures and accommodating "legal" second-hand phone dealers through more flexible land use planning.

Generally, the proliferation of economic activities related to the digital revolution is not really recognized as a land use issue. Yet a lot of economic potential can be unleashed through proper urban design measures, and flexible land use controls, thereby also increasing access to ICT in the process.

The implicit dimensions of digital urban space

The interface between digital and physical is also *implicit* (and not just physical). This relates to the relations and negotiations that contribute to the production of space and where traders find themselves on the formal-informal continuum. It is more fluid, related to the exchange of information and negotiation, not the fixing of place. Yet it is intrinsically connected to place, since it represents the networks and interactions that underpin what happens in space. Interactions comprise:

- *Information sharing* on new markets, new spaces to trade, threat of evictions, police patrols for example.
- *Networking* between traders, traders and officials, traders and organisations

representing their interests, traders and their families, their creditors and suppliers.
- *Negotiation and bargaining* between traders, traders and officials, traders and potential gatekeepers.
- *Advocacy* and enlistment of support within the city, outside the city and internationally.

The type of technology used will no doubt enable different aspects of these functions. The use of web sites for advocacy work, for example, enables local issues to become global and then, local again, as a global support base often supports local action. The use of internet resources is limited by price, computer literary and access. Efforts to enable broader access through public intervention: making libraries digital hubs, investing in cheaper technologies (the eThekwini Council is currently investigating the development of cheap palmtops), enabling VOIP and public hot spots will yield a more equitable distribution of ICT opportunities. Given that ICT is less constrained by geography, and enables communication across country and continental boundaries, efforts to broaden access will enable exposure and support (in advocacy terms especially) globally.

Information sharing through e-mail and the web is documented and it is a particularly useful medium for deepening associational connections. Mobile phones, on the other hand, facilitate immediacy, mobility and real time communication that many street traders need in the face of possible harassment. They enable the personal and professional to be merged in time and space as kinship networks become more important the more marginal the economic activity is. They impact on space as traders are able to move the location of their activity as demand and conditions dictate. This is difficult to manage from a land use perspective, but requires a flexibility in approach without compromising the public good.

How the use of ICT translates into urban space, it is argued here, depends on how the transactions that enable access to space, the organization of that space and the degree of permanence is enhanced, extended and influenced by digital technology. Explicit public and physical representations of ICT lend themselves to a land use planning and urban design approach that recognizes the opportunities they bring economically. If done with insight, such measures can enhance the cultural and social dimensions of African "digital lives", creating spaces that are urbane, connected and functional. Recognition of the implicit dimensions of ICT-enabled networks, facilitating and enabling those interactions to flourish creates the less obvious "spaces" for empowerment and mobility. Informality is representative of the livelihood strategies, associational networks and creative energies of those at the economic margins of our cities. The street traders we see on our streets, in our public spaces are engaged in an ongoing endeavor to claim their part of the city. The use of urban space is an important input into the economic processes that underpin these activities. Access to, and use of, these spaces are constrained by many factors; some of them obvious (such as physical barriers and police harassment), others less apparent such as access to information, negotiation and bargaining. The point is to understand how ICT can enable the inclusion of these marginalized groups into the city by making these processes less threatening and onerous, or empowering them to become entitled urban citizens.

References

Amin, A and Thrift, N. (2002), *Reimagining the Urban* (Cambridge: Policy Press).

Benjamin, P. (2003), *The Universal Service Agency's Telecentre Programme: 1998 – 2000 Surveys, Analyses, Modeling and Mapping Research Programme, Occasional Paper 2* (Pretoria: Human Sciences Research Council).

Boraine A.; Crankshaw O.; Engelbrecht C.; Gotz G.; Mbanga S.; Narsoo M. and Parnell S. (2006), 'The State of South African Cities a Decade after Democracy', *Urban Studies* 43:2, 259–284.

Bridges.org. (2002), *Taking Stock and Looking Ahead: Digital Divide Assessment of the City of Cape Town,* 2002. Report for the City of Cape Town, available on www.bridges.org.

Brown, A. (Ed.) (2006), *Contested Space: Street Trading, Public Space, and Livelihoods in Developing Countries*. (Rugby: ITDG Publishing).

Carey, Z. (2004), 'Generation Txt: The Telephone Hits the Street', in Graham, S. (Ed.) *The Cybercities Reader*. (London: Routledge).

Castells, M. (1996), *The Information Age: Economy, Society and Culture: Vol. 1. The Rise of the Network Society*. (Oxford: Blackwell).

Day, P. (2005), 'Community Research in a Knowledge Democracy: Practice, Policy and Participation', paper presented at the *Community Informatics Research Network Conference*, Cape Town, 24 – 26 August, 2005.

Donner, J. (2005) Microentrepreneurs and Mobiles: An Exploration of the Uses of Mobile Phones by Small Business Owners in Rwanda, in Information Technologies and International Development, Volume 2, Number 1, Fall 2004, 1-21.

Gedye, L. (2005), 'Knysna's 'up-yours' to Telkom: Municipalities large and small race to provide Internet Access', in *Mail and Guardian*, November 18 to 24, 2005.

Gedye, L. (2005), 'Come on down, the Oysters and Wireless Internet, are fine', in *Mail and Guardian*, November 18 to 24, 2005.

Graham, S. (Ed.) *The Cybercities Reader*. London: Routledge.

Graham, S. and Marvin, S. (2001), *Splintering Urbanism: Networked Infrastructures, Technological Mobilities and the Urban Condition* (London: Routledge).

Hampton, K. (2002), 'Place-based and IT Mediated 'Community'', in *Planning Theory and Practice*, Vol. 3, No. 2.

Kopomaa, T. (2002), 'Mobile Phones, Place-centred Communication and Neo-community', in *Planning Theory and Practice*, Vol. 3, No. 2.

Liff, S. (2005), 'Local Communities: Relationships between 'real' and 'virtual' Social Capital' in Van Den Besselaar, P., De Michelis, G., Preece, J. and Simone, C. (Eds.) *Communities and Technologies 2005*. (Dordrecht: Springer).

Mohlageng Strategy Consultants, WITS LINK Centre and Sangonet. (2003), *Citizen Access to E-Government Services: Report for the Centre for Public Service Innovation* (Pretoria: CPSI).

Ralfe, K; Briginshaw, D. and Odendaal, N. (2007) *Report on the Location of Street Barbers in Durban (Isipingo, Clairwood and Durban CBD)*. Unpublished report to Siyagunda, Durban.

Republic of South Africa. (2000) *Municipal Systems Act*. Available on www.gov.za, accessed in May 2007.

Simone, A. (2001), 'Between Ghetto and Globe: remaking Urban Life in Africa', in Tostensten, A., Tvedten, I., and Vaa, M. (Eds.) *Associational Life in African Cities*. (Stockholm: Nordiska Afrikainstitutet).

Simone, A. (2004) *For the City Yet to Come: Changing African Life in Four Cities*. (London: Duke University Press).

Simone, A. (2005), 'Introduction: Urban Processes and Change' in Simone, A. & Abouhani, A. (Eds.) *Urban Africa: Changing Contours of Survival in the City*. (London: Zed Books).

Simone, A. and Gotz, G. (2003), 'On Belonging and Becoming in African Cities', in Tomlinson R., Beauregard, R.A., Bremner, L., and Mangcu, X. (Eds.) *Emerging Johannesburg: Perspectives on the Postapartheid City* (New York: Taylor and Francis).

South African Cities Network (2006), *State of the Cities Report* Available on www.sacities.net. Accessed in May 2006.

Swilling, M. Simone, A. and Khan, F. (2002), "My Soul I can See': The Limits of Governing African Cities in a Context of Globalisation and Complexity', in Parnell, S., Pieterse, E., Swilling, M., and Wooldridge, D. *Democratising Local Government: the South African Experiment*. (Cape Town: UCT Press).

Townsend, A. (2000), 'Life in the Real-time City: Mobile Telephones and Urban Metabolism', in *Journal of Urban Technology*, Vol. 7, No. 2.

www.durban.gov.za Accessed in November 2006.

Personal communication with:

Gaby Bikombo, Chair, Siyagunda Street Barbers Association

Jacque Subban, eThekwini Metropolitan Council

Nosipho Zuma, Ohlanga Digital Hub manager

Chapter 17

Woven Fabric: the Role of Online Professional Communities in Urban Renewal and Competitiveness

Eleonora Di Maria and Stefano Micelli

Introduction

In the past ten years regions and metropolitan areas have been selected as the core of policies for innovation and competitiveness. Specifically, metropolitan areas have been involved in renovation processes focused on the development of new economic value-added functions, new professional skills and on the attraction of new talents (OECD 2006). According to this perspective, policies oriented to foster growth at the local level should be improved by integrating the set of interventions with a new approach. The regeneration of cities or regions is not only oriented to reduce social disadvantages or unemployment; rather, those places – and specifically at the metropolitan scale – represents relevant *loci* where social dynamics merge with economic processes, by opening new opportunities for innovation rooted into creative contexts.

This chapter describes and analyzes how professional communities represent an important complement to traditional innovation policies at the metropolitan level. Professional communities do connect and create mutual support between professionals that could be at the time being, as well as in the future, an important resource for local enterprises and institutions. Our contribution aims at enriching the debate on how to promote the competitiveness of local areas in the scenario of ICT (information and communication technology), reshaping the policy thinking concerning the drivers of territorial economic development.

The relevance of these professional communities has dramatically risen with the diffusion of information and communication technologies that expanded the traditional boundaries of professionals' interconnections and increasingly allow the sharing of non-codified knowledge through innovative communication tools. Hence, in the policy framework those players become relevant as independent driving forces, able to push innovation and to support economic process improvement locally. At the same time, because of the emphasis on ICT, through the promotion of conditions for the creation and flourishing of professional communities in a specific area, policy makers may also sustain ICT adoption.

Although the diffusion of these professional communities has been, in Italy, mostly an emergent process, two important experiments developed in the North East region demonstrate how these communities can be the joint outcome of deliberate policy and of bottom-up grassroots effort. In the Treviso province, two professional communities, Treviso Designers and Club BIT, have been created to promote and consolidate crucial expertise in knowledge domains that firms and institutions do not necessarily invest upon. The two cases show how online professional communities at a territorial level can be implemented as a tool for local development and integrated in a more general economic regeneration policy.

Knowledge and Professional Communities Between Local and Global

Many scholars highlighted the strong connection between knowledge and local development. The Lisbon strategy stresses the need for a "knowledge-based economy" (Rodriguez 2002) and pushes national and local governments to identify policies focused on knowledge and innovation as drivers for sustainable growth. This perspective has put particular emphasis on the role of research (R&D) and training as the drivers for gaining competitiveness, stressing the relevance of science-based knowledge and codified processes of innovation (that is, patents). In such a scenario, large firms, universities and research centres have been considered as the main players in knowledge production and management (Antonelli 2005). Hence, national and regional authorities have promoted policies oriented to increase research activities within companies as well as collaborative relationships among firms and the research networks – also at the European and international level – to achieve economic growth and support local development.

However, this approach has often underestimated the social complexity of knowledge creation and diffusion, where knowledge creation, its use and diffusion are strongly interrelated with the social dimension of innovation processes (Latour 1987). Scholars referred to the notion of "stickiness" to stress the relevance of the connection between knowledge management and the social context in which such processes takes place (von Hippel 1994; Brown and Duguid 2001). Empirical research highlights how knowledge dynamics can develop in a twofold manner: on the one hand, they are the results of vertical procedures usually managed within codified frameworks (top-down) within R&D offices; on the other hand, they refer to horizontal linkages involving people informally, such as in the case of epistemic communities able to sustain knowledge creation in a very different manner compared to the former one (Knorr Cetina 1999). Despite having been neglected by research and scholars for many years, even the Italian experience of industrial districts can also be explained in terms of a successful example of accumulation and diffusion of knowledge rooted into local contexts (Porter 1989; Becattini and Rullani 1996), by coupling their specific economic and social fabrics.

Knowledge management, communities of practice and technologies

Within local places knowledge flows through face-to-face interaction and proximity: people and firms within the same context develop and share high absorptive capacity related to local knowledge. Opposite to large multinational companies – where the knowledge circuit is strongly codified – each place can offer an access to knowledge based on processes of capillarity, normally informal and often not voluntarily started. Such characteristic is crucial as it is difficult to transfer and to acquire from the outside. From this perspective, the territory can be considered as a platform that qualifies people to use social knowledge as well as to establish a local efficient division of labour in terms of learning and innovation.

Even if the role of tacit knowledge has indeed been considered as a relevant driver of competitive advantage (Nonaka 1994), literature on local manufacturing systems and industrial districts has given little attention to the ways cognitive processes are organized and take place within local contexts. Scholars have stressed the role of proximity as the key element in knowledge transfer among persons, but less emphasis has been put on the real going on at the local level, except for the "industrial atmosphere" mentioned by Marshall (1920).

Instead, many contributions can be obtained from scientific studies on learning, which offer interesting insights on how to increase the internal connectivity of territories to improve interaction among players at the local level. Some researchers (Brown and Duguid 1991, Lave and Wenger 1991, Wenger 1998) have identified communities of practice, considered as a strong and interdependent group of people sharing the same work and collaborating continuously, as one of the main social domain in with knowledge flows take place. Brown e Duguid (2001) stressed the relevance of the concept of "practice" in explaining how knowledge can overcome organizational boundaries ("leakiness") as well as in the difficulties of accessing knowledge within departments of the same firm ("stickiness").

The notion of "practice" supports the explanation of the mechanisms of knowledge creation and the domains in which such processes take place (or do not). Sharing the same practice as well as the engagement in reciprocity among community members represents the key factors to describe knowledge flows within the same organization (or place) or through different ones. Communities of practice - to be proper ones – require high interaction levels amongst their members, who share the same work context, where proximity and daily interaction as well as collaboration in work activities characterize relationships among experts.

Brown and Duguid (2001) also introduce the notion of "network of practice", to interpret the phenomenon of informal networks built amongst professionals who, though sharing the same practice, work for many different firms. We identify the networks of practice with the professional communities considered as spaces for aggregation of homogeneous professional profiles, who meet to find services – training, professional upgrade, problem solving – and to discuss about their professional competencies (Micelli 2000).

For example, empirical studies carried out in the mechanical industry district of Brescia (Lissoni 2001) highlighted the role of specific professional communities in supporting local competitive advantage. Mechanical or electro-mechanical engineers working within the district support on the one hand the use of explicit and codified knowledge offered by the research system and, on the other hand, give value to tacit knowledge developed during their working experience. Hence, it is exactly those professional communities that play a role in supporting the dynamics of knowledge creation and diffusion at the territorial (district) level as well as contributing to reinforcing the conditions that enable knowledge development for the manufacturing processes.

Information and Communication Technologies have a key role in supporting new aggregation in local contexts, by enhancing internal connectivity. ICT qualify through a technological framework the mechanisms of cognitive accumulation, by establishing virtual environments able to give value to distributed contributions (Jones 1995). These technologies define new digital spaces for meeting and discussion, in which professionals have the opportunity not only to share information and knowledge, but also to collaborate to solve common problems (von Hippel 2006)[1]. Specifically, those areas of online interaction are relevant as new tools for professional training as well as for innovation and experimentation. New technologies are also able to reconfigure communication flows and connections among heterogeneous players, by creating new experiences and new opportunities for interaction. Such processes are not competing with the traditional mechanisms. Instead, they must be considered as an integration of local interaction.

Some successful experiences show how the dynamics of giving value to new professional profiles at the local level can be supported through network technologies. As we describe later, a specific online project offered to four professional communities – information system managers, marketing experts, logistics and quality experts – in the North East part of Italy was a virtual service environment, oriented to sharing information and knowledge – by way of forum and chat technologies – as well as to providing professional training – both online and face-to-face. During the project, the intensity and frequency of interactions amongst professionals increased. Specifically, the ICT infrastructure does not simply help professionals to meet other colleagues, but it also consolidates the existing relationships within the community. Such relational strength leads to an intensive schedule of face-to-face meetings and seminars, focused on developing specific topics of interests for the community.

---

1 We define a professional virtual community as a group of people, sharing common practices, jobs or activities, who find through the net an environment of access to specific information, of meeting, of common interaction and sharing of experiences and knowledge. Members in professional virtual community could exploit opportunities given by information and communication technologies related to speed in identifying participants, to large scale of interactions (in a global context), to variety of experiences and to wide electronic links (i.e. D'Adderio 2001; Lacy 2005). Sharing established paradigms and models of thought, based on common languages and rules, becomes the milestone that allows people to create new knowledge and new practices rapidly (Orr 1995).

Each community of practice defines itself through what it is about, how it functions and what capability has produced; therefore internal activities and processes of a professional virtual community depend from its professional field together with expertise members can provide to the whole community (Wenger, 1998). Learning in professional virtual community emerges from a continual process of selection and re-elaboration of knowledge shared by members, combining informative and participation base. Successful knowledge management is due to a framework of formal and informal aspects: wide access to information and knowledge repositories, sharing tacit knowledge through social interaction, sharing of specific language and common practices, variety in motivations to participate, leader support (Davenport and Prusak 1998,Wenger et al. 2002).

Internal and external connectivity

For many years the success of local manufacturing systems has been explained in terms of strong internal connectivity. Connection at the territorial level has great importance in supporting knowledge flows among economic and social players: social, cultural and organizational proximity provide a common ground for the creation of shared practices embodied in strong network relationships. The exploitation of the innovation created within districts has been mainly managed through external market connections, where local knowledge is embodied into products and technologies. On the one hand, there are local networks, strongly interconnected, where knowledge diffusion is carried within a specific place and its surrounds. On the other hand, there are international networks, oriented to enlarge the scale of knowledge use.

However, especially recently, from a theoretical perspective research contributions put evidence on the role of discontinuity as a driver of local competitiveness, where knowledge and competence inputs from outside the locality can positively impact on local dynamics and evolution (Florida 2002). The entrance of new players, such as professionals, immigrants, and so on, can break the economic and social stability of the context through introducing new creativity inputs, specifically at the metropolitan level. In this perspective, innovation – as well as competitiveness – is not perceived as an incremental process. Rather, it can benefit from ideas, procedures and experiences developed in other places, which force local players to consider their activities and strategies from a different perspective (radical innovation based on discontinuity).

In this scenario, there is a twofold strong transformation emerging (see Table 17.1). On the one hand, districts are not able to sustain their competitiveness based on local innovation circuits, considered as closed local systems. Instead, the knowledge cycle has to extend to a wider scale that involves first of all the metropolitan area in which the district is located. As we describe later, this process is strongly linked with the evolution in the competencies and practices upon which the local competitive advantage is built: from industry-specific and idiosyncratic practices to horizontal ones (see Florida 2002). On the other hand, local social and economic communities should open to the international dimension, by approaching the global economy in terms of new opportunities for knowledge exploitation, but also for learning.

For a metropolitan area to build its competitive advantage a condition is the development and reinforcement of those two complementary networks:

- internal connectivity: the territory offers to local players a dense cognitive network able to facilitate local knowledge sharing;
- external connectivity: the territory is connected to global circuits of knowledge and innovation, where the value of local knowledge increases depending on the scale of its use. At the same time, such form of connection with the global economy offers access to global knowledge and gives the opportunity to build transnational framework of divisions of labour. The local area can benefit from this approach by increasing its specialization.

However, those processes cannot be easily obtained as endogenous and autonomous paths of local development. The large programs and policies focused on regions and urban areas, put in place at the national and international level, stress the relevance of a shared framework for economic development based on innovation (see Di Maria and Micelli 2005). The identification of drivers of competitiveness for regions and local areas requires long terms policies built within a participatory framework focusing on bottom-up processes, while often large companies have been the most active players, able at shaping the growth of specific areas around the World, according to codified processes of knowledge management (focus on top-down) (Sproull and Kiesler 1991).

Institutions are certainly one of the players involved in local development based on knowledge propagation, as this process strongly depends on the ability of people and social groups to activate knowledge creation and diffusion. From this point of view communities are important not just as the addressees of policies, but as key players in local innovation and renovation.

Communities can be considered in terms of citizens, but we can also look at professional communities. The former can contribute through their evaluation of local service offer and their knowledge and practice of local places and processes (i.e. Casapulla *et al* 1998). The latter can become the interface between the knowledge obtained from R&D, formal innovation circuits and the manufacturing practices. From this perspective, they support innovation adoption, while adapting it at the specific context of use (Boland and Tenkasi 1995; Brown and Duguid 2001).

Policies for local economic development should take into account those dynamics and contribute to reinforce internal connections, also through the use of information and communication technologies. Local authorities should not outline programs of interventions only on the basis of internal analysis, but also create opportunities for learning from local players – citizens, professionals – as a new way of designing policy consistently with the demand and with its participation in the process.

Table 17.1 A New Scenario for Competitiveness

	Past	**Present**
Territorial level	Closed local system (district, community)	Metropolitan area
Connectivity	Internal	Internal / external
Knowledge domain	Industry idiosyncratic	Horizontal

Online professional communities as the link between local and global

Local economic systems, considered in terms of communication infrastructure that sustain the knowledge circuit, have had and still have an enormous importance for collective learning processes and the reduction of uncertainty. Informal and tacit knowledge related to work activities and complex problem-solving processes is shared on the basis of dynamics rooted in the local environment, where people share the same language and use their own semantic framework to manage relationships at the local level (Micelli 2000). Innovation within local economic systems is, thus, the result of specific characteristics of the local systems: high specialization and cooperation. Moreover, innovation becomes a distributed process built collectively, where the pivot is the same social environment of embeddedness.

Network technologies open new opportunities for online cooperation and interaction in completely new environments, partially in competition with that confined system of social and cultural references identified as the territory. Competition refers to the cost and the quality of knowledge one has access to. As regards the first element, the social systems of relationships within which the entrepreneur – or the worker – acts is costly to maintain. In the past, the territorial level was a competitive infrastructure for coordination when compared to large organizations. Now it has to face the technological solutions available, which are able to link specialized but complementary contexts as well as supporting knowledge circulation among contexts.

From such a perspective, the real potential lies in the quality of the knowledge one can access to through the local area and through the network. The more the knowledge required is linked to the specialized competencies developed in a context, the higher the quality of the knowledge available (Davenport and Prusak 1998; Rullani 2004). The transmission of such knowledge is in fact controlled, evaluated and selected at the local level. Hence, it gives the opportunity to access local knowledge at the basis of innovation processes, where the distinctive competencies and skills have been developed locally through learning-by-doing inter-organizational processes. The more this knowledge is far from specific contexts, the easier the opportunity to transfer such knowledge outside the local context, through networking high technologies. From this point of view, the net can become a threat only for those territories not able to renovate the drivers of their competitive advantages. Rather, a strategic use of information and communication technologies oriented to community building and management can represent an innovative approach for economic development policies at the local level. It becomes thus important and beneficial to reorganize the traditional training processes as well as the mechanisms that sustain innovation, by leveraging the potentialities of network technologies as the drivers of the changes at the local level – for both professional communities and local institutions. These new communities of practice can thus reinforce their roles of drivers for competitiveness, as knowledge management systems at the local as well as international level.

Once the professional community develops the original and distinctive knowledge related to the valorisation of specific competencies of the local network, it could promote this knowledge worldwide, outside the local context. Communities can have a proactive role in translating global knowledge towards the local environment as well as in sharing knowledge obtained locally – focussed on key knowledge and local excellence – with other international players. From such a perspective, communities of practice cannot function autonomously. Instead, coordination among the many players located in the same context is important to achieve this goal. Local authorities should consider the relevance of those professional communities from two perspectives. On the one hand, these players support innovation processes at the local level through the specific learning mechanisms described. Hence, they are key drivers for competitiveness of local economic systems. On the other hand, professional communities represent a *trait d'union* between the territory and the global economy. Sharing the same practices becomes the glue to facilitate knowledge processes and to support a rapid inclusion of foreign players within the local boundaries.

Renovating Competitiveness Through Communities: the Italian Case

The Italian economy has demonstrated to be able to achieve great success at the international level through the competitiveness of its local manufacturing systems, where about 200,000 firms and more than 65% of the total Italian employees refer to "made in Italy" industries. During the '90s such productions – fashion, furniture and home products, mechanical products – have become the main drivers of Italian export, by stressing the value of the Italian form of economic organization – the "industrial district" model (Corò and Micelli 2006). This spatial agglomeration of economic activities at the territorial level characterizes especially Central and North East Italy (the so called "Third Italy"). Those regions have been able to gain significant growth in the last fifteen years through a positive mix of specialized and distinctive local competencies, quality of product and flexibility linked to incremental innovation (Becattini 1991; Piore and Sabel 1984; Porter 1989).

The local form of organization exploits the advantages of the network, where economic players are strongly embedded into local contexts and social and economic processes reinforce one another. Moreover, the overlapping between the urban and the entrepreneurial fabric facilitates internal cohesion and supports economic development. Manufacturing knowledge as well as competencies related to economic processes nurture innovation at the local level, even in the case of low or medium-tech industries, which involves learning-by-doing instead of R&D (Becattini and Rullani 1996).

However, the recent scenario of a globalizing economy increases competitive pressure. The declining share of Italian trade at the international level, with negative results in 2004 after almost a decade of positive outcomes as well as the difficulties encountered by "made in Italy" products in foreign markets, open new issues for competitiveness and policy interventions. In the new competitive market scenario new competencies are required at the local level, compared to traditional manufacturing skills. Firms need competencies related to new areas such as marketing, communication and design (Bettiol and Micelli 2005). New professional skills are related to logistics and information management systems, and these are relevant to support extended commercial and manufacturing networks scattered internationally. Investments in both those areas – logistics and ICT – can leverage local economic activities and upgrade firms operating in traditional contexts. At the same time, they require a more focused vision about the division of labour taking place at the local and international level, where the coordination among many specialists is crucial also for knowledge management aims.

Because of the small size of the firms characterizing local economic systems, those changes cannot occur easily and autonomously, especially in a breakthrough scenario of competencies and technologies. Hence, the development of professional communities – supported by networking technologies – appears to be a fundamental priority for many reasons. First, small and medium enterprises (SMEs) are unable to offer targeted and customized training programs to their professionals. The dual need of SMEs for cost reduction and work stability reduces firms' interests in training activities. On the one hand, many professional profiles are not considered as crucial for companies' competitive advantage and, hence, they do not benefit from the offer of training programs. On the other hand, those professionals who are important for competitiveness cannot easily be taken off work processes during their training. The opportunity for such profiles to access knowledge and interaction electronically can be fundamental in terms of professional upgrading with less impact on companies' organization. The same issues can be raised also when it comes to considering public administrations.

Second, as stressed earlier, professional communities can have a key role in local economic systems. The difficulties in facing training needs for firms' human resources shift the focus onto the single worker, who becomes aware of the relevance of customized lifelong learning activities as a gateway for better job chances. The opportunity to share information and resources online allows such profiles to become more visible and to increase their actual competencies and skills. By decreasing isolation, professional communities increase local mobility and, hence, reinforce the local labour market.

Third, training offers are, at the local level, strongly limited. The variety of problems and needs those professionals are facing, their different degrees of competency as well as the quality of their work environment are factors that impact on the heterogeneity of their training activities. At the same time, such issues impact on the real ability to identify and offer user-oriented training programs. Such demand can be faced only under the condition of high prices, in line with consultant activities. However, SMEs are not able to pay high prices for training. Hence, only bottom-up processes rooted in professional practice contexts may sustain professional upgrading, especially for young workers. Those processes should be rooted in a mix of established training programs and on line environments.

In this scenario metropolitan areas are even more at the core of such transformations (Turri 2004), as catalysts for creative talents and places for knowledge sharing depending on their quality of life and dynamism (Landry 2000). Here cities are expected to promote opportunities for innovation and knowledge management linked to those professional profiles, consistently with the manufacturing fabric that characterizes their territories. From this perspective, the professional community model can help governments and local authorities to develop new strategies for local competitiveness. By focusing on discontinuity, those programs may not just invest in traditional competencies already available at the local level, but mainly consider (online) communities as new tools to build competencies and enhance knowledge management in the new selected domains of competitiveness (see Table 17.1). The real chance is to enlarge the innovation circuit beyond the local context, by connecting professionals, markets, firms, research and manufacturing places. Moreover, professional communities can take part in the innovation process, but also in the identification of priorities at the local level – also following inputs coming from abroad (network of practice).

On the one hand, professional communities may contribute to increase the internal connectivity within the territory, as a communication and relational infrastructure that links professionals not identified at the core of the traditional specialization of the local context – for instance designers with respect to jobs related to manufacturing processes such as goldsmith or wool weaver activities. On the other hand, professional communities provide visibility of local competencies worldwide, supporting internationalization processes and offering cooperative networks (external connectivity). They can push local institutions towards redesigning their services and building a new vision for the competitiveness of territories, based on their contribution and input on the strategic planning.

Two innovative professional communities

In the Veneto region, the province of Treviso has been remarkable for its economic and social growth. As one of the most active areas in manufacturing productions related to furniture, sport systems – shoes, textile and apparel – machinery and wine, Treviso had obtained widespread success at the international level thanks to the quality of its products, the flexibility of small firms and the excellence of their workers. The "industrial district" model is important to explain the local dynamics of innovation. However, as other territories, even in the province of Treviso internationalization processes and increased competition have been pushing firms to reinvent their drivers of competitive advantages, and local administrations to support such transformation towards the intangible economy.

We present and discuss two relevant cases of professional communities developed in the province of Treviso and aiming at supporting knowledge improvement of the members. These have obtained an interesting role as participants of local economic development, with local institutions.

Club B.I.T.: network technologies and SMEs.

Club B.I.T. (Backup Informatico Trevigiano)[2] is a community of professional people operating in the field of information systems, who belong to firms located in the province of Treviso. Its main goal is to increase members' knowledge related to operating and information systems, through training and virtual meetings. Based on interaction, discussions among members and training, this community improves professional knowledge sharing within firms localized in Treviso.

Built through a bottom-up approach in 1994, Club BIT wants also to become the point of reference for all the professionals working as CIO (Chief Information Officer) or in the field of network technologies, sharing the same practice.

Even if it is linked to the local industrial association (*Unindustria Treviso*), Club BIT works autonomously, by offering services to its members and to the territory. It is engaged in:

2 <http://www.Club-bit.org>

- promoting the membership among professionals working in the ICT domain;
- developing a better recognition and reputation of the function of the information system;
- creating a center for information and training for the members, even in an international framework;
- developing actions of cooperation and promotion to increase members' professional skills and competencies, as well as the creation and diffusion of knowledge;
- promoting collaboration with Italian and foreign institutions and associations with the same goals;
- organizing and promoting workshops, debates and conferences on specific topics and in any domain related with the practice of information system manager.

Club BIT has now more than 100 members working locally in small and medium enterprises. The community organizes formal and informal events ("Meet-BIT") where discussions about the needs and interests of the members take place. Moreover, annual company visits are organized to merge the professional dimension with the organizational sphere of the work context.

After ten years of activity, in 2004 Club BIT started, in collaboration with Venice International University, a study on the transformation of the role of the CIO within the organizations to evaluate the changes in the professional profile. This is one example of the many activities carried out by the community, which has invested in the participation in projects and collaboration with research centres as important opportunities for knowledge improvement. Local institutions such as the Treviso Chamber of Commerce focused on Club BIT as one of the key players at the local level to involve in the development and implementation of projects where innovation and competitiveness are central.

Networking technologies have a relevant role not only as the core topic of the professional community, but also as the infrastructure that sustains members' interaction and discussion. The web site of the community is used for communication purposes and for knowledge sharing among members – in a reserved area – while the mailing list ("BITlist") is the main tool for interaction and discussion. Club BIT took part in an important national project on the role of online professional communities as a tool for personal training, funded by the Ministry of Labour and coordinated by the *Fondazione Cuoa* with the collaboration of Venice International University, *Fondazione G. Brodolini*, AGFOL, and S.I.A.V. Based on innovative training solutions, as the other communities – quality controllers, logistics, managers – Club BIT becomes an active partner in identifying the knowledge needs of its members and developing a sophisticated demand for online and face-to-face training sessions. At the same time, as members are also key workers in their firms, through projects like the one mentioned, the community is able to impact on the competitiveness of local firms as well as to potentially increase their knowledge assets.

Trevisodesigners: New Competencies for Old Economic Systems.

Different from the Club BIT experience, the professional community of Trevisodesigners has been developed thanks to the support of an institutional project promoted by the Treviso Chamber of Commerce and its special agency *Treviso Tecnologia* (in cooperation with Venice International University).

In 2004 *Treviso Tecnologia* promoted a research on design in the sport system district of Montebelluna (eKM-Dicamo) aiming at understanding the link between competencies in design and competitive advantages of the district – and province – area. In previous years many companies located within this old district developed delocalization strategies, oriented to take advantage of cheaper labour cost markets in Eastern Europe. Such processes and the transformation in the competitive scenario raised new issues about the renovation of the local system in terms of new competencies and new knowledge – from manufacturing to marketing and design. In this context, the research work was important in identifying the design-related potential of the Treviso area, something in which also the IUAV (Architecture University Institute of Venice) invested by offering design curricula in the city of Treviso.

The project helps the designers working in the local area to become aware of their role for the competitiveness of Treviso's system. More specifically, the metropolitan area of Treviso despite being far from Milan - the worldwide recognized capital of design – could rethink its specialization as becoming based on design and not only on manufacturing processes and skills. Coherently with the inputs coming from the Florida's idea of creative class (Florida 2002), local institutions supported the rise of a formal association among designers, which became in 2006 the *Trevisodesigners* community. With more than 226 members and over 12 events organized, the community started to play a role of aggregation and debate at the local level, aiming at increasing the visibility of designers in the Treviso area as well as nationally. From eighty-three designers involved at the time of the project, the numbers of the community have more than trebled but with an increasing focus on young designers (under 30s have grown from 20% to 55% in three years).

By exploiting networking technology, the community has created a new online environment[3] different from the one of the initial project – which is anyway still in place as the new Treviso portal on design[4]. The community web site is dedicated to discussion and sharing of knowledge and experience, and has now more than forty daily accesses. The site also hosts a blog, a technology solution which, focusing on multimedia content, is coherent with the professional approach of designers to their work. Moreover, the community is able to support young designers through its online database. It offers a faster access to the network of competencies and, at the same time, it allows young professionals to spread information about their projects and activities. In a professional domain where reputation and seniority are key drivers for success, the community can sustain both the goals at the local and international level. Membership includes not just designers working in local small

3 <http://www.trevisodesigners.it>, recently renamed as <http://www.designpeople.it>

4 <http://www.design.tv.it>

Figure 17.1 Designpeople.it website
Source: www.designpeop;e.it

and medium enterprises, but also professionals operating in design companies, as well as a small group of designers linked to university.

Through the community, local institutions are able to identify a new important player with whom to develop projects and interact to focus on new emerging needs. More importantly, the community also becomes aware of the potentialities in collaborating with university and research centres. In addition, the community creates a place for discussion, aggregation and knowledge sharing not available before at the local level.

Conclusions

The two cases of professional communities in the Treviso area highlight how such initiatives may become active participants for local competitiveness. Those communities can offer to local governments – especially in metropolitan areas – important inputs in terms of new services to be promoted at the local level. Based on their members' knowledge, communities of practice are not only the beneficiaries of local policies, but may have a key role in identify strategic trajectories for local competitiveness.

There are areas of dialogue of particular urgency on which experimentation is to be welcomed in the near future. Those areas concern local services: transport, infrastructures for knowledge access and sharing (customized training) are all areas of intervention in which the empowerment of demand can support significant improvements in the quality of the service, with relatively minor costs being transferred to the supply side. Because of their specialization as well as their ability to become interface between the local and the global contexts, professional communities may require to public administration, local authorities as well as other local players new services not available at the local level or supporting programs to augment their space of action – in terms of physical as well as relational connections internally or with outsiders.

In order to promote the rise of professional communities local authorities or public bodies should first create favourable conditions for talent attraction from outside – local attractiveness oriented to specific professional practices – in terms of local services such as specialized training services, mobility facilities, innovation centres, and more; local labour market support, as well as a stimulating cultural environment – relevant for instance in case of design communities. For local bodies approaching professional communities means not only considering the firm or the single worker as unit of analysis and addressee of policy, but also a group of professionals that act into an "augmented space" across firms, across cities in metropolitan areas and across regions.

Second, the promotion of projects targeted to selected professional practices carried out by policy makers and local institutions may increase the aggregation process of single professionals locally as well as enhance the diffusion of ICT tools for coordination purposes on a broader level. Above all, also through those projects, professional communities may be recognized, over time, as formal partners of local bodies and contribute to the definition of policy goals within participatory dynamics.

Moreover, those institutions should also leverage on the presence of leading firms – or attracting them – at the territorial level for identity-building purposes and local reputation: the case of STMicroelectronics in the Catania metropolitan area of Southern Italy, or Tiscali in the Sardinian city of Cagliari, are two interesting examples of firms that transform the outside perception of local specialization – from Catania to the Etna valley – by legitimating the presence of ICT professionals in those areas.

Information and communication technologies become the infrastructure for community aggregation and interaction, even in local areas which have been traditionally characterized by face-to-face relationships. As the two case studies described in the chapter point out, the role of the net is to offer visibility, support and identity even at the local level – and at low costs – mainly for professional aims, but with potential connections also to local government activities and services. Online communities can constitute an instrument to promote the capacity for self-organization of demand and the lever to reconstruct a communicative context wherein individuals and professionals in difficult circumstances can find the necessary resources to support problem solving and propose solutions, also with positive impacts on local competitiveness.

References

Antonelli, C. (2005), 'Models of Knowledge and systems of governance'. *Journal of Institutional Economics* 1:1, 51–73.

Becattini, G. (1991), 'Italian industrial districts: problems and perspectives', *International Studies of Management & Organisation*, 21:1, 83–90.

Becattini, G., Rullani, E. (1996), 'Local systems and global connections: the role of knowledge', in Cossentino F. et al. (eds.), *Local and regional response to global pressure: The case of Italy and its industrial districts.* (Geneva: International Institute for Labour Studies).

Bettiol, M., Micelli, S. (eds.) (2005), *Design e creatività nel Made in Italy. Proposte per i distretti industriali.* (Milano: BrunoMondadori).

Boland, R.J., Tenkasi, R.V. (1995), 'Perspective Making and Perspective Taking in Communities of Knowing', *Organization Science* 6:4, July-August, 350–372.

Brown, J.S., Duguid, P. (1991), 'Organizational Learning and Communities-of-Practice: Toward a Unified View of Working, Learning, and Innovation', *Organization Science* 2:1, February, 40–57.

Brown, J.S., Duguid, P. (2000), 'Mysteries of the Region: Knowledge Dynamics in Silicon Valley', in William Miller et al. (eds.), *The Silicon Valley Edge*. (Stanford: Stanford University Press).

Brown, J.S., Duguid, P. (2001), 'Knowledge and Organization: a Social-Practice Perspective', *Organization Science* 12:2, March-April, 198–213.

Casapulla, G. et al., 'A citizen-driven civic network as stimulating context for designing on line public services', *Proceedings of the Fifth Biennial Participatory*

Design Conference 'Broadening Participation', Seattle, WA USA, November 1998 <http://www.retecivica.milano.it/paper/pdc98.htm> accessed 7 December 2006.

Castells, M. (2000), *The information age: economy, society, and culture*. (Cambridge: Blackwell).

Corò, G., Micelli, S. (2006), *I nuovi distretti produttivi: innovazione, internazionalizzazione e competitività dei territori*. (Venezia: Marsilio).

D'Adderio, L. (2001), 'Crafting the virtual prototype: how firms integrate knowledge and capabilities across organizational boundaries'. *Research Policy*, 30, 1409–1424.

Davenport, T.H., Prusak, L. (1998), *Working Knowledge. How Organizations Manage What They Know*. (Cambridge: Harvard Business School Press).

Di Maria, E., Micelli, S. (2005), *On Line Citizenship. Emerging Technologies for European Cities*. (New York: Springer).

Florida, R. (2002), *The Rise of the creative class*. (New York: Basic Books).

Jones, S.G. (ed.) (1995), *Cybersociety. Computer-Mediated Communication And Community*. (Thousand Oaks: Sage).

Knorr Cetina, K. (1999), *Epistemic Cultures. How the Sciences Make Knowledge*. (Cambridge: Harvard University Press).

Lacy, S. (2005), 'LinkedIn Expands Its Connection', *BusinessWeek*, (updated 16 March 2005)<http://www.businessweek.com/technology/content/mar2005/tc20050316_1715_tc119.htm>

Landry, C. (2000), *The Creative City: a Toolkit for Urban Innovators*. (London: Earthscan).

Latour, B. (1987), *Science in Action*. (Boston: Harvard University Press).

Lave, J., Wenger, E. (1991), *Situated Learning*. (Cambridge: Harvard Business School Press).

Lissoni, F. (2001), 'Knowledge codification and the geography of innovation: the case of Brescia mechanical cluster', *Research Policy*, 30:9, December, 1479–1500.

Marshall, A. (1920), *Principles of Economics* (London: Macmillan, Eighth edition).

Micelli, S., (2000), *Imprese, Reti, Comunità Virtuali*. (Milano: Etas).

Nonaka, I. (1994), 'A Dynamic Theory of Organizational Knowledge Creation', *Organization Science*, 5:1, 14–37.

OECD (2006), *OECD Territorial Reviews: Competitive Cities in the Global Economy*. (OECD Publishing).

Orr, J.E. (1995), 'Ethnography and Organizational Learning', in Zucchermaglio C. et al. (eds.), *Organizational Learning and Technological Change* (Nato ASI Series Series III, Computer and Systems Sciences).

Piore, M.J., Sabel, C.F. (1984), *The Second Industrial Divide. Possibilities for Prosperity*. (New York: Basic Books).

Porter, M. (1998), 'Cluster and the new economics of competition', *Harvard Business Review*, 76:6, Novembre-December, 77–90.

Rodriguez, M. J. (ed.) (2002), *The New Knowledge Economy in Europe. A Strategy for International Competitiveness and Social Cohesion*. (Cheltenham: Elgar).

Rullani, E. (2004), *Economia della conoscenza*. (Roma: Carocci).

Sproull, L., Kiesler, S. (1991), *Connections. New Ways Of Working In The Networked Organization.* (Cambridge: MIT Press).

Turri, E. (2004), *La megalopoli padana.* (Venezia: Marsilio).

Von Hippel, E. (1994), '«Sticky information» and the Locus of Problem Solving: Implications for Innovation', *Management Science*, 40: 4, 429–439.

Von Hippel, E. (2006), *Democratizing innovation.* (Boston: MIT Press).

Wenger, E. (1998), *Communities of Practice: Learning, Meaning, and Identity.* (Cambridge, Cambridge University Press).

Wenger, E. et al. (2002), *Cultivating Communities of Practice.* (Cambridge: Harvard Business School Press).

Chapter 18

The Digital Urban Plan: A New Avenue for Town and Country Planning and ICT

Romano Fistola

City and Technology

The focus on the relationship between city and technology must necessarily refer to that between human settlement and tools of transformation of the natural environment or maybe, earlier than that, to that between man and "techniques".

Firstly it should be clarified what we mean when using the term "technology".

It seems useful to adopt the meaning shared by a huge part of the modern philosophical critique and recently proposed by Umberto Galimberti in the definition of the term "techniques".

For Galimberti (2002, 41) "techniques" represent: 'the universe of the means which as a whole compose a technical apparatus'. Technologies represent therefore the instrumental elements of a process aimed at pursuing a purpose. In the specific interest of this study the purpose is represented by the adaptation of the natural space in order to frame an anthropical habitat in which human activities can be located. In this case it is possible to think about technology at the base of the development process of urban settlements. This assumption leads towards a new consideration of the relationship between man and technique.

Many contributions coming from different scientific fields deal with this dichotomy.

The book by Galimberti underlines how the relationship between *psyche and tecné* has progressively modified itself through the transition from the pre-technological and humanistic era, in which Karl Marx theorised the human alienation caused by the technical production of the capitalistic system, into the post-industrial and digital one for which, considered the annulment of one of the two subjects of the man/machine dichotomy, alienation becomes "irreversible" determining the survival of only one subject of the dichotomy: the machine.

Close to the position of Marx was Max Weber who, at the beginning of the 20th century, described modernity as the catalyst of a process founded upon the calculability and impersonality of the human behaviour. Among the technology-averse positions the one by William Morris should be highlighted as he considered the advent of the industrial society as a major human calamity. He used to say: 'the principal passion of my life was, and it is still, the hate for the modern civilization'[1] and by this he synthesised his own aversion for the transformation processes determined by new technologies. Among such transformations Morris observed the massive urbanisation and the consequent urban degrade determined by the inadequacy and structural brittleness of the British urban system at the end of the 19th Century. Furthermore he attributed to the technological progress (and to the consequent development of the industrial production) the beginning of the first occurrences of environmental pollution.

Therefore technology, which had represented an essential condition to the human progress in the pre-technological era, was considered the cause of urban system degradation.

In the following years, technology started belonging to a process of inclusion in the human ethics becoming an essential component of it.

The use of a great variety of means has allowed, as mentioned, the modification and the adaptation of the natural environment in order to locate human activities. In the concentration of such activities must be recognised the genesis factor of the city.

Nowadays technology represents a basic component of modern society. We will not analyse here whether the actual social dimension is better or worse compared to the pre-technological one. We can confirm that now there is no society without technology and technology is a dimension of society (Castells 2002). Without technique (and without machines which are the products of technique) human beings would not have been able to activate some cultural progress. But technology itself modifies the behaviours and the life habitat of human beings. Such a claim, emphasised in the visions of William Gibson at the beginning of the 1980s, in movies like Terry Gillian's *Brazil* , or in the recent screen-plays by the Wachowski brothers, brings the focus onto the relationship between human settlement and technology. The relationship between technological innovation and territorial transformations is subject to sudden accelerations which will produce effects and deep modifications on the way human beings act and interact inside the city.

The diffusion of the Internet and the construction of "virtualized activity systems"[2], is modifying the cultural heritage, the social ethics and the ways citizens use the city.

This process is hardly visible but evolves with strong accelerations, due to the introduction of user-friendly technologies at low cost. Such modifications are involving all the territorial structures and infrastructures as well, and produce deep transformations within the urban social system.

---

1 Pronounced at the Second International Conference of Socialists at Paris (1988).

2 This refers to the system of those urban functions (dwelling, commerce, production and so on) which undergo a process of "virtualization" as they get partly performed by the telematics network (see Fistola 2001).

The Technological Betrayal

New technologies are becoming essential for urban functions. The digital flows are transferred along the nets and can be elaborated, managed and turned into services, thanks to the electric power.

In such a sense the urban system is inflexible. Today there is no alternative to the use of electric power to support urban activities and to make the city work. If no electric energy is available the urban functional system collapses to the point that it becomes difficult managing to reach one's own home if the traffic light system is out of order. Moreover, if black-out occurs without warning, catastrophic effects can be generated.

This is what happened in Italy at the end of June 2003. The supply of electric power was suddenly interrupted in a number of Italian cities. Six million Italians found themselves "unplugged". A great number of firms (particularly food industries) and companies suffered enormous damages because of this unplanned interruption of electric energy. All urban activities, banks, administration offices, postal offices, universities, hospitals and also business/stock exchange were paralyzed.

Furthermore residential activities and life suffered as all domestic devices – from the fridge to the air conditioner – nowadays essential for our everyday life, were hit by the black-out.

Without energy the city doesn't work. The Italian black-out produced an economic damage of hundreds of millions of euros. This problem was not determined by any complicated phenomenon involving electric field, but simply by a tree which had fallen down onto a major electric line.

So, in order to talk about digital networks, we have to think about the availability of electric power and, maybe, think about a new, more sustainable way to get energy.

It is now possible to update our thinking on earlier considerations on perceivable effects of the new technologies on human settlement. And this is anyway necessary in order to check the trends and the scenarios that many researchers have formulated some years ago. In other words it could be interesting to understand if the new information and communication technologies (NICT) have really caused a functional decentralization inside the city, a weakening of the meaning of spatial proximity, a drastic reduction of physical movements, etc. as predicted by many scientists (Mitchell, 1995). When the NICT started to show their enormous potentialities in terms of connecting people and groups located in remote sites, one of the most remarked possible impacts of the electronic medium was its ability to allow a new form of digital communication, hence replacing the traditional support for the transmission of information: paper. The digital support would allow a far richer and more informative transmission than that allowed by the layers of cellulose. Furthermore the digital revolution would have manifested its own environmental vocation contributing to save enormous quantity of trees which are annually sacrificed to the human necessity of graphically memorizing and transferring information.

But this has not been the case. The consumption of paper is in constant increase all over the World. The demand for paper in 2010 should grow to 31% and Italy is in seventh place for paper consumption in the world. The case of paper is symbolic for the wider technological anomalies occurred in the relationship between NTIC and city.

As already recalled, we can identify at least three visions about the urban effects of NTICs which have not been confirmed by facts so far.

In extreme synthesis the NTICs would have:

- Caused a sprawl of the urban activities on the territory thanks to the possibility to develop these in remote different places;
- Allowed (consequently) the abatement of the spatial proximity inside the city giving rise to new urban morphologies;
- Reduced drastically the necessity of physical movement through urban space thanks to virtual mobility.

But this has not been the case either. Whilst the NTICs seem to be able to affect the distribution of urban activities inside non-metropolitan areas, a threshold in terms of urban complexity seems to exist (Fistola 1993) beyond which the effects of the NTICs are hardly predictable. Some urban activities show a consistent inertia to being subject to new (dis)placements. And those activities which can be easily re-located by digital technologies, such as financial ones, firmly remain situated to Midtown (New York), Shinjuku (Tokyo) or inside the City of London. In such sense the book *The Global City* by Saskia Sassen is enlightening when it explains how the central business districts of the main World economic centres – New York, Los Angeles, London, Tokyo, Frankfurt, Sao Paulo, Hong Kong and Sidney – have reached an unprecedented penetration of advanced telecommunications. Sassen notes how the expansion of the number of companies locating themselves inside the core of these cities during the 1980s, contradicts most predictions of territorial sprawl (Sassen 2001).

NICT act on the functional system of the city activating processes of "virtualization" (Fistola 2001). New functional systems are then generated incorporating the "resistant functions" (lowly sensitive about the impact of the NICT), producing something new in the physical space (the "generated functions") and transferring other functions completely into cyberspace (the "mutant functions"); yet the structure of the city, or its spatial order, does not seem to change. NICT act on the sub-systems of the wider urban system producing effects on the arrangement of the located activities (Moss and Townshend 2000). In such a sense the digital network topology can facilitate useful synergies for urban development, if it is deployed in harmony with the town masterplan.

The Digital Osmosis: from the City to the Net - from the Net to the City

The shifting of urban activities from the urban functional system to cyberspace and vice versa (from the digital dimension to the physical city), is characterizing our cities more and more. This phenomenon could be defined as "digital osmosis".

A good example of this digital osmosis process from cyber to physical space has been developed within the city of Naples: the "telematics plaza" (in Italian: *piazza telematica*).

The telematics plaza (generated function) represents a case in which digital activities have found their place in a real site of the city, built for that purpose.

The Municipality of Naples has launched a plan for e-government that foresees a number of interesting initiatives among which the realization of a network for the provision of demographic services through the *Lottomatica* (the digital infrastructure dedicated to the game of Lotto, something very popular with Neapolitans) in order to issue certificates and to allow the payment of fines and taxes.

One of the first documents[3] about the telematics plaza states:

> The use of the word "piazza" is meant to underline the semantic significance of the urban area *par excellence* traditionally dedicated to communication, meetings, and the exchange of goods (material and immaterial) and services. At the same time, the project has to be seen as a web of urban sites (physical and mult-icentric telematics infrastructure) connected by means of telematics networks and sufficiently distributed throughout the territory (city, rural villages, etc.) to ensure new urban functions.

The concentration of high-level, technologically innovative services offered by the telematics plaza together with its potential for social contact, could thus exercise a strong force of attraction capable of triggering the development of a new poly-centric urban structure.

The 'piazza telematica' could be thought both as:

- A physical place for the concentration of services based on powerful hardware, software and telecommunications tools;
- A public place of access to the knowledge and utilization of the "raw material" of information for the purposes of work, study, creativity and enterprise, as well as an interface with the public administration and with other agents for purposes regarding the following sectors:
  - entrepreneurial/professional/commercial/financial
  - cultural/cognitive/informational
  - administrative/fiscal/social
  - security/authorization
  - entertainment/leisure/social activities
- New urban polarity enjoying sufficient critical mass, economies of scale, and environmental liveability to activate an aggregative process based on a mix of interests capable of triggering a virtuous spiral of attraction towards the new polarity;
- A pole of attraction for daily commuter flows providing new capacity for:
  - in the short term, re-orienting the occasional and permanent flows of mobility;
  - in the medium/long term, modifying the type and quality of the structures, infrastructures and services gravitating around the new polarity.

---

3 Web site of the Associazione Piazze Telematiche <http://www.piazzetelematiche.it>

In light of the above, the *Piazze Telematiche* idea constitutes a fresh opportunity to implement a project aimed at the rehabilitation of abandoned buildings and areas and at helping to combat urban fragmentation and the loss of identity typical of metropolitan outskirts, thus triggering processes of re-rooting and integration, based upon the following:

- A self-centred development making the best possible use of local resources (human, entrepreneurial, environmental, cultural, etc.);
- A new stimulus for local authorities to undertake an entrepreneurial role by launching projects of urban regeneration and upgrading based on conditions of project financing negotiated between public agencies and private concerns".[4]

Figure 18.1 The 'Sails' of Scampia, Credit: Author

4 From the web site of the Associazione Piazze Telematiche <http://www.piazzetelematiche.it>

Figure 18.2 The Urban Context of Scampia, Credit: Author

The telematics plaza (TP) is a pole of a network that unfortunately, in the Neapolitan experience, has not yet succeeded in shaping up properly and fully, the experiment being limited at the moment as a safe "pole" in the district of Scampia.

Scampia is one of the new districts of the Neapolitan outskirts, counting around 90,000 inhabitants, though only 45,000 are officially registered in the Census, and in which 70% of the residences are represented by public housing. Most of Scampia's population is composed by evacuees of the 1980 earthquake which hit the Campania region. In the Scampia-Secondigliano area a big "camorra" (the Neapolitan organised crime) war saw 76 people killed from just January 2004 to April 2005. The district was shaped from the division with the district of Secondigliano and it is well known for the story of the public housing buildings nicknamed "the sails" (see Figure 18.1).

Such buildings, known because their original building typology recalls the principles of Le Corbusier's *Unitè D'Abitation*, quickly became the symbol of the social degrade of the Neapolitan outskirts. Two building complexes have recently been dejected. This is the urban context in which the intervention of the TP has been carried out (see Figure 18.2).

The project has been partially supported by the European Commission to the second phase of the Urban Pilot Projects (PPU), financed by the Art.10 of the European Fund for Regional Development (FESR). The objective of such fund aims at reducing the differences in development among the regions and sustaining, in such a way, the economic and social cohesion of Europe.

The project for the plaza has been one of 26 selected among the general 506 proposals introduced from European Union countries. In the Telematics Plazas documentation the benefits and aid that the project would deliver to the local communities are listed, and particularly:

> For the productive community: services of tele-working are individualized, with beneficial effects on the general mobility, services of tele-commerce, as well as services of tele-conference.
>
> For the community of young people: services, convivial activity and other various activities (games, music and shows), as well as educational services and e-learning.
>
> For the tourist community: information services on itineraries, places and suggestions, as well as services of tele-booking.
>
> For the community of the citizens: privileged and reasoned access to municipal online services and those of the Public Local Administration; access to services of electronic shop-window to facilitate matching demand and offer, with particular reference to the job market.
>
> The TP will be also the place for events and demonstrations that somehow materialize the virtual meetings and events taking place on the net: the plaza becomes the threshold place between material and immaterial.
>
> Among the services dedicated to the productive community, great importance assumes the Laboratory of Production (LabNet) for which a considerable amount of the plaza space has been reserved.
>
> The laboratories, Open Space with 4 to 6 workstations, will be dedicated to:
>
> Web design, with PCs equipped for the design of web sites and for graphic elaboration;

Figure 18.3 The Structures in the District of Scampia, Abandoned for Over Ten Years before the Intervention for the Installation of the Telematica Plaza, Credit: Author

Figure 18.4 The New Spatial Arrangement of the Telematica Plaza in Scampia, Credit: Author

Multimedia and CAD, with some workstations suitable to the use of CAD;

3D animation and multimedia stations for the production of hypertexts and DVDs;

Workstatons dedicated to tele-working and an equipped room devoted to the enterprise incubator, in order to make the plaza as a centre where it becomes possible to start a new business.

To serve this purpose the LabNet has to constitute a unit of incubation with two main commitments:

to provide legal and fiscal consultation services for the start-up of new professional and entrepreneurial activities;

to perform a systematic search on the chances of funding both for the formative activity of the centre and for entrepreneurial initiatives.

Other function of the incubator is the realization of a node of exchange among job seekers and employers. This must be supported by a special web application already connected with other existing similar services".[5]

In other terms the TP, with its organization and structure, develops a functional role covering the following aspects:

- social (improvement of quality of life and new juvenile aggregation through the use of the NTIs, support in the new place);
- cultural (encouraging literacy);
- economic (as support to the local entrepreneurial initiatives);
- urbanistic (retraining of the urban space and interaction with the present functions);
- institutional (as a direct channel between citizens and institutions);

A group of young people will manage the internet-café. An interesting initiative that will immediately be activated is in fact related to the creation, by the students of the Hotel Management School, of a cooperative enterprise that will manage the internet-café of the TP.

From a spatial point of view the TP is a structure of around 3,000 square metres. It is articulated on two levels: the ground floor represents the place of permeability and meeting of the flows where there is equipment used by municipal public relations offices. The first floor has lecture rooms, laboratories, modular spaces which can be re-configured to meet the demands of specific consumers as well as small companies.

In the process of realization, the greatest difficulties have concerned the material aspect of the TP, in consideration of the need to regenerate a strongly degraded space, which needed disposal of asbestos, etc. (see Figure 3). Nevertheless the Local Authority and numerous operators that have worked for the realization of the TP

5 From the Telematics Plaza plan of the Municipality of Naples.

have succeeded in overcoming every difficulty and now Naples sports the first Italian telematic plaza (see Figure 18.4).

However, for every innovative piece of work a risk exists for the initiative not to take off because of some resistances, identifiable in:

- low support from the municipality "limiting" managerial ability;
- scarce demand from the local communities (particularly of young people);
- limited ability in sponsoring the initiative from the managers;
- lack of preparation of the operators to the enrolment and fidelization of the users:
- scarce demand from the local entrepreneurial groups;

Where such resistances create knapsacks of inactivity the TP becomes just another space of accumulation of functions and, probably, another town office where quite different activities will be developed respect to those which were expected to happen.

The Digital Urban Plan

Considering what has been explained so far, the need to prefigure an urban "tool" able to mediate the different needs shown by actors, managers and citizens in general seems to be necessary.

At present there is no codified procedure (or tool) which allows to properly co-ordinate the intervention of deploying ICT infrastructures – wired or WI-FI – within the city, able to take into account all the issues, questions and expectations coming from the different groups of actors/managers active in the place. Within this perspective, an urban government tool has to be prefigured in order to foresee a "digital zoning" of the land and establish urban "wiring" action rules. This plan can be called a "Digital Urban Plan" (DUP).

Summing up the details of the DUP (Fistola 2001), it can be assumed that this should define a set of indications/regulations which should be followed by the new actions of territorial transformations governance. The Digital Plan is not a further urban planning tool, but intends to be an overlapping field of concerted development between ICT and infrastructure, on the one hand, and the urban system considered in all its sub-system components, on the other.

The basic elements for the digital plan are:

- Present and in-progress definitions of urban planning with indications about destination and distribution of activities over municipal territory;
- The wiring plans submitted by the telecommunications companies with indications about the net layout and the schedule for the deployment of its different sections;
- The map of the potential virtualization of urban ambits embedding functions being strongly sensitive to digital transformation;
- The actions to be implemented in order to effectively contrast digital divides

and allow a non-selective access to the net for all urban communities, with particular reference to the disadvantaged ones;
- Charting of the urban macro-functions existing and in-progress with indications about supply units representing the "sensible receptors" (schools, hospitals, social centres, local bodies offices, etc.

The map of "urban function virtualization" allows to identify the urban ambits containing the most sensitive activities toward digitalization processes (mutant functions) and which can be considered as zones in which, being broadband infrastructure present and accessible, there will be a progressive reduction of use intensity (Fistola 2001).

The map of the "digital divide" is aimed at identifying the areas (Census sections) showing more sensitivity to this problem. The relative information layer will be built by considering several indicators such as: income, literacy degree, population density, crowding index (inhabitants/rooms), unemployment index, and so on. In those areas the presence of homogeneously distributed network segments, and ways of granting easy access to video-services should be planned. At the same time, the plan will have to define "digital standards" that refer to the town planning standards concept, and describe the net supplies to be expected in each area according to the functional characteristics, population typology, presence of sensitive functions and so on.

The digital standards will have to be implemented in a compulsory way and to the advantage of the public manager who will include them into the Integrated Cabling Operating Plan (ICOP). In this plan, apart from the standards on wiring (materially represented by those network segments that the companies assigned to wire a determined municipal zone will have to realize under favourable conditions and free of charge) there will be indications about network topology, the segments and the digging techniques (cuttings, micro-cutting, no-dig, etc) and the skylight passages to be realized, the bandwidth to be reserved for public use, etc.

The interventions program fixes the temporal segments for the DP realization, defining an opportune sequence of the works to be implemented.

The rules for DUP implementations will have to be linked, within possibly a Geographical Information System (GIS), to the different identified digital zones, and represent an articulation containing all the indications for territorial wiring and digitalization.

The map of network topology will represent the reference document describing the tracing of the optic fibres network and the technical indications for the deployment.

The financing program will show all the funding sources available to the municipality to realize the public network as well as the possible mechanisms of financial participation by private actors (TLC companies). These, by participating in the funding, will be able to obtain the right of pre-emption for the concession of the network segments of their interest.

Finally, the regulations will comprehend a detailed articulation of specific rules on the network realization techniques, the concession of the network segments to the TLC companies, installation of public hot spots, technical aspects of access certification, and so on.

An Italian standard urban masterplan (the *Piano Regolatore Generale – PRG*) divides the city's territory into homogeneous zones according to the type of activity foreseen for each area. The PRG prescribes the articulation/zoning of the city under six homogeneous categories. Particularly zones A, B and C are those forecast as residential; zone D is forecast as a production/manufacturing area (either handicraft or industrial); zone E is forecast as agricultural, and zone F as urban facilities (high schools, universities, sport centres, etc.).

In order to make a first test of the procedure and considering that Naples has recently approved the new masterplan (see Figure 18.5), this city can be used as a benchmark.

As said before the DUP should represent an urban tool able to harmonize territorial development with the action of wiring it (Bobbio 2002). What represents the objective of the procedure is the individualization of the areas in which it would be opportune to foresee the presence of wired networks in order to support their undergoing activities.

To such purpose it is necessary to analyse the followings territorial characteristics:

- Density of population;
- Foreseen activity;
- Presence or forecast of location of strategic functions;
- Presence or forecast of location of sensitive functions;
- Presence or forecast of location of executive functions;

For the first characteristic the number of inhabitants for each zone as acquired in the Census areas has been noted. Referring to the needs of wiring, the classification is: high for the areas A., B, C; medium for D and low for E and F.

Qualitative judgments have been formalized quantitatively using the normalization of the numerical staircases through the fuzzy-sets, already used in other studies (Fistola and Urciuoli 1994).

The whole procedure is built inside a Geographical Information System (GIS) whose database includes all the data previously recalled. The graphic component of the GIS is mainly represented by two subdivisions of the urban territory: the census zones (see Figure 18.6) and the zoning from the masterplan.

The location of the sensitive/strategic functions has been noted and geo-referenced on the territory, including hospitals, schools, universities and other similar entities. (see Figure 18.7).

Inside the GIS a based procedure has been defined on an algorithm that connects the five indicators described above and calculates and produces a set of values indicating, area by area, the necessity of telematics infrastructure.

Besides, some poles are identified, generally barycentre to a certain number of strategic functions (see Figure 18.8), that are essential to reach by optic fibre connections.

All these prescriptions end up defining a sort of zoning of the DUP (see Figure 18.9), and the results will support the management of the telematics infrastructure

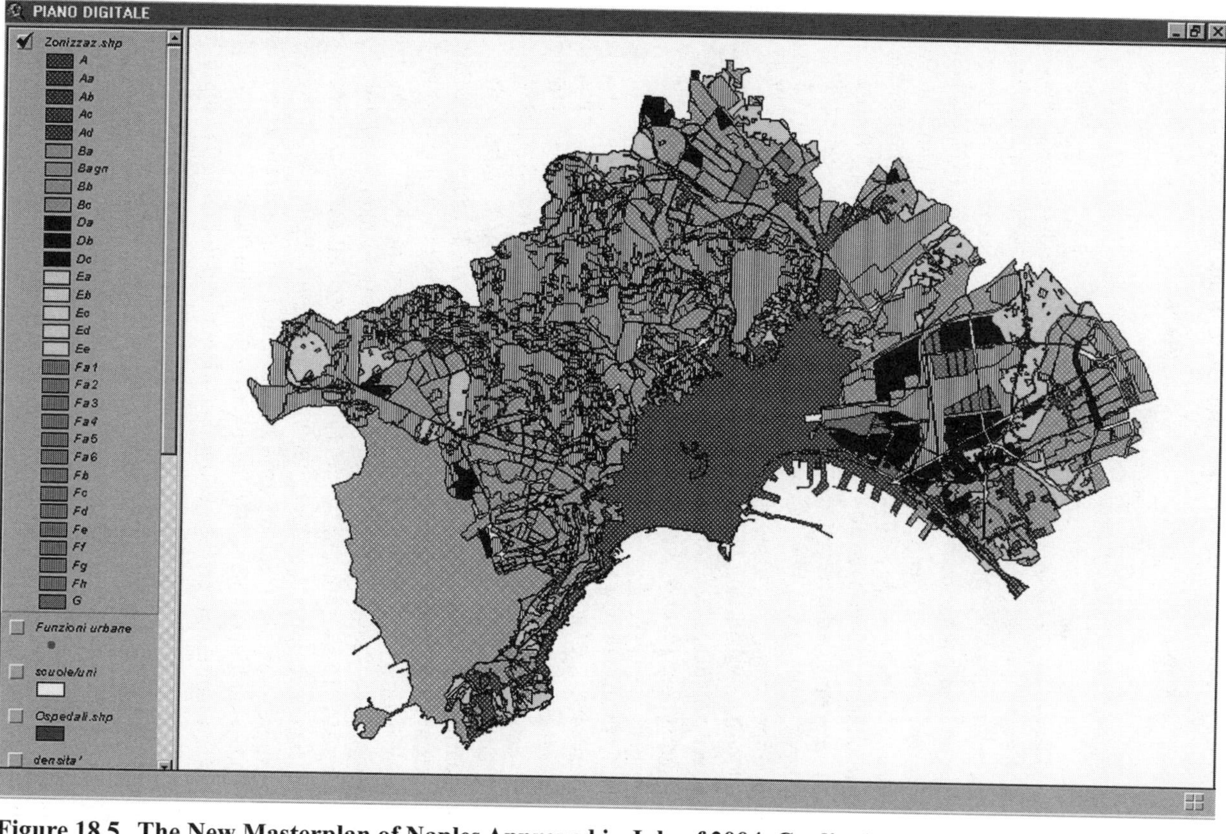

Figure 18.5 The New Masterplan of Naples Approved in July of 2004, Credit: Author

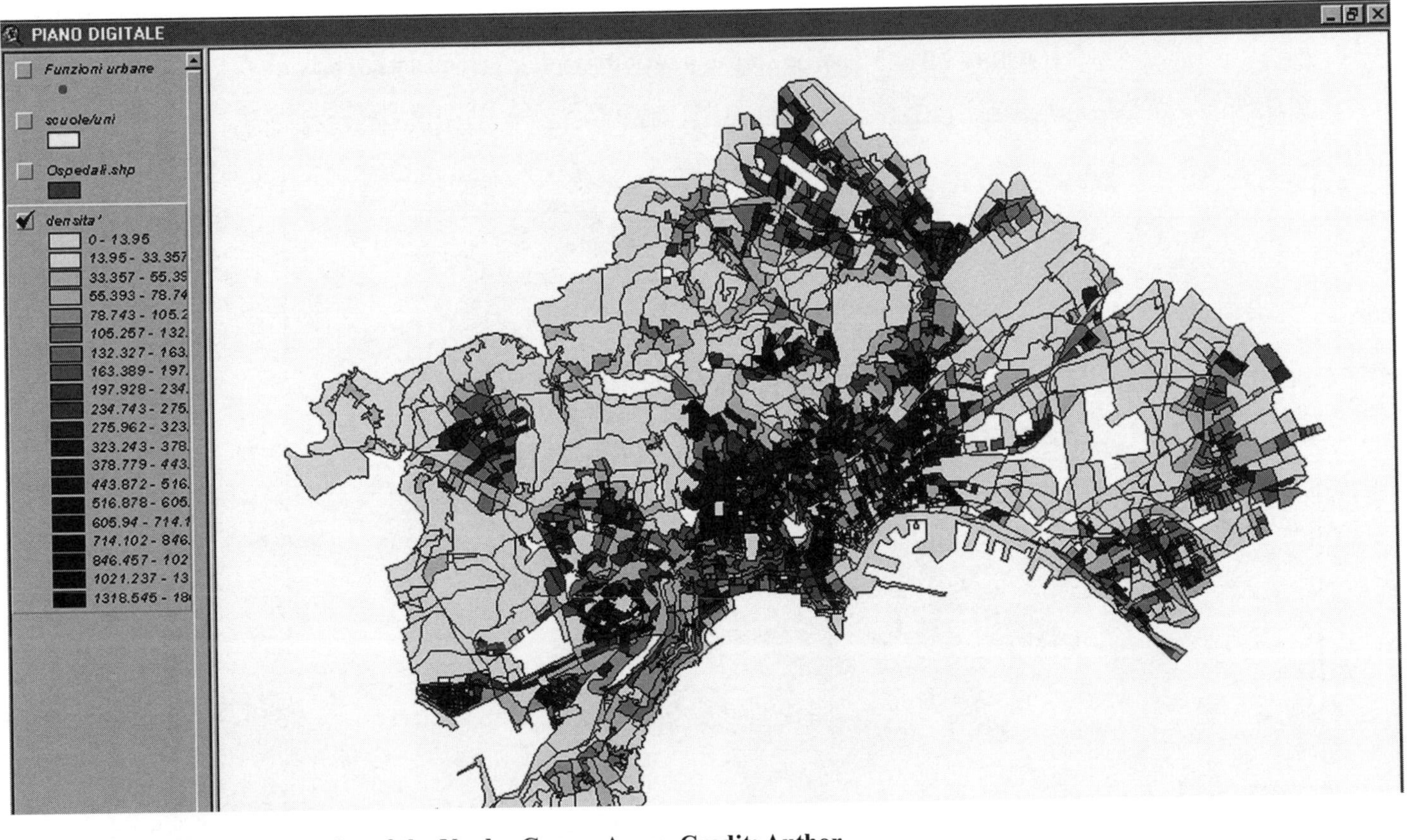

Figure 18.6 Population Density of the Naples Census Areas, Credit: Author

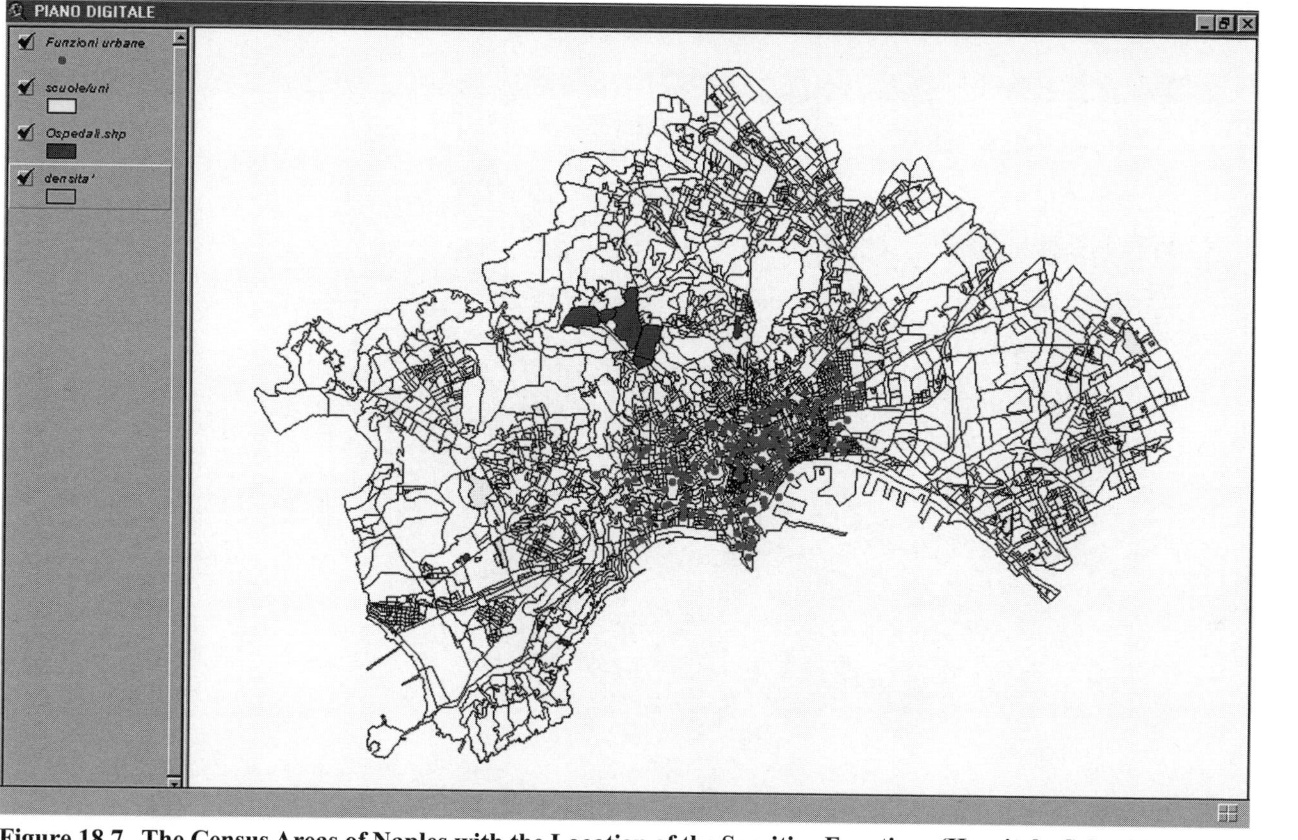

Figure 18.7 The Census Areas of Naples with the Location of the Sensitive Functions (Hospitals, Schools), and the Main Strategic Urban Functions, Credit: Author

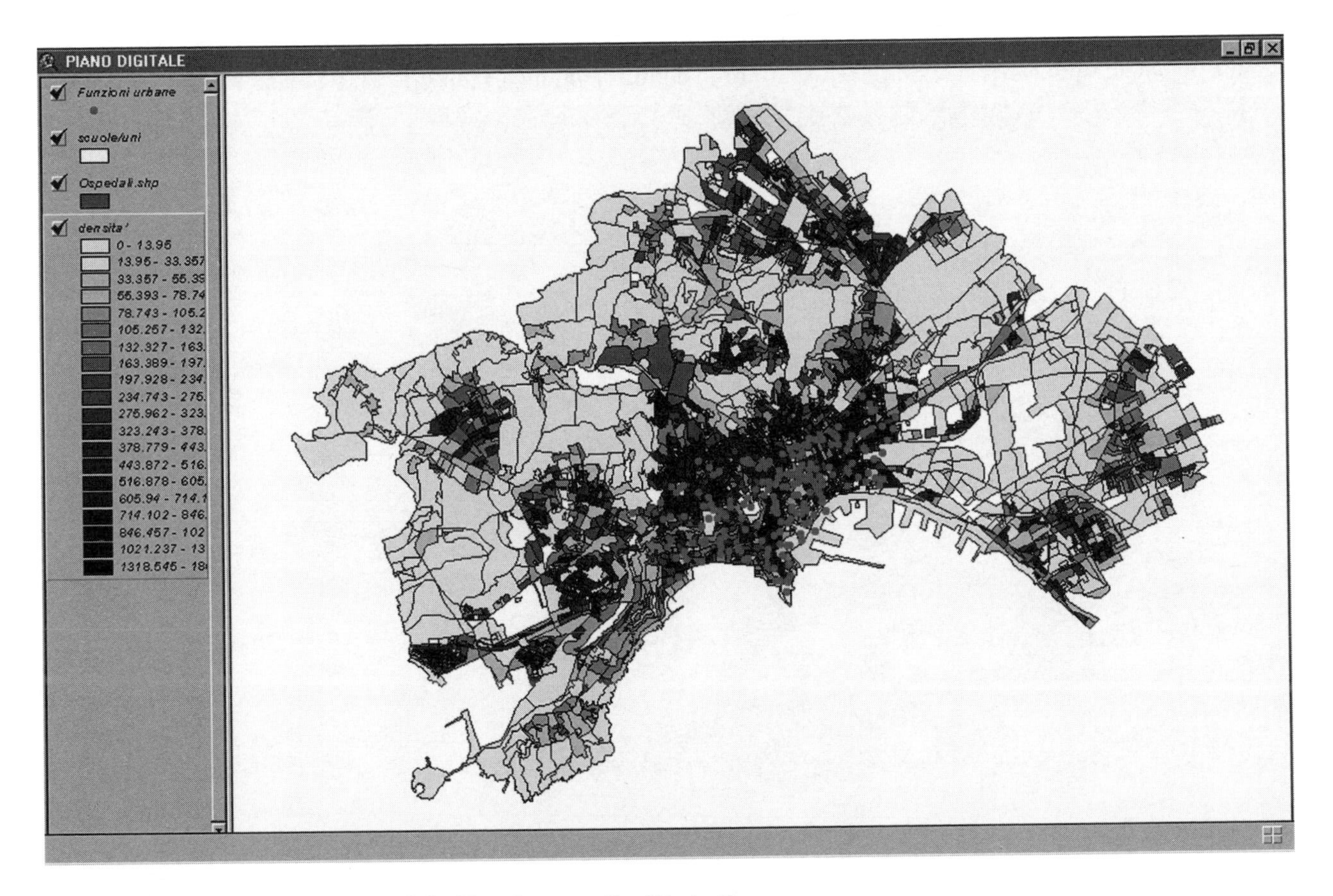

Figure 18.8 The Overlapping of the Two Layers, Credit: Author

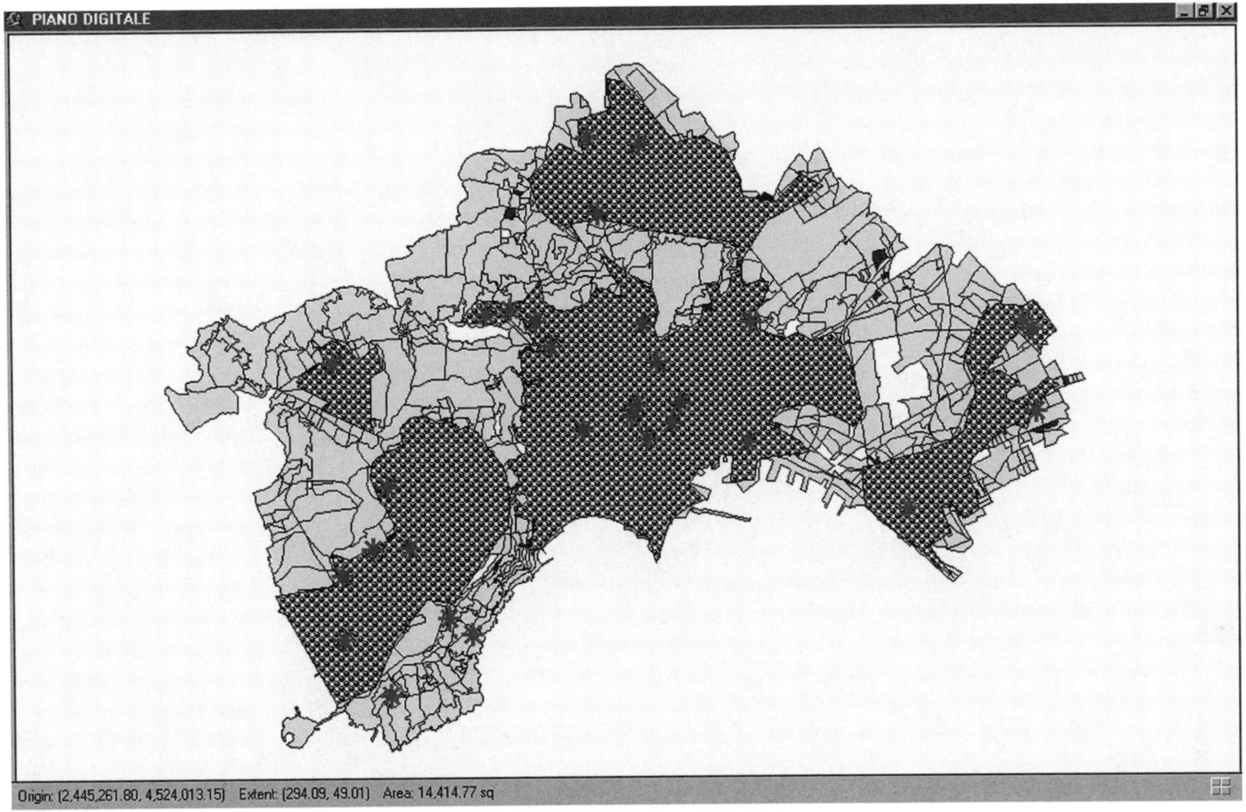

Figure 18.9 The Zoning of the Digital Urban Plan, Credit: Author

interventions on the territory, and can be a useful tool in order to successfully deal with TLC companies.

The local government can use such a tool in order to bring harmony between the development of the new urban technological infrastructures and the foreseen (social, functional and spatial) development of the territory.

Conclusions

The relationship among new technologies and human settlement is now getting to a mature phase in which the city has to incorporate NICT inside its own structure. The city economy is increasingly based on the elaboration and transfer of data. The majority of the processes of urban regeneration, development and competition cannot ignore the issue of the availability of telematics facilities.

The relationship between cities and NICT is constantly evolving and produces transformations in the functional, but also physical, urban system (Graham and Marvin 1996), perceivable by the urban community. The ways by which citizens behave and interact within urban space are rapidly changing. Rampant innovation constantly introduces new communication technologies that spread straight away throughout urban society. The systems of relationships change as well as the activities (which get partly virtualized), the communication codes and the meeting places.

Setting up tools that could harmonize the actions for the governance of territorial transformations and address the modifications produced by NICT seems to be a duty for town planners, though these are often not completely up-to-date with the changes under way. Nevertheless, it should be stressed that within urban planning a new awareness of this phenomenon is developing, and this is proven by that fact that the 2003 congress of the European Planning Schools Association (AESOP) was titled "The network society: a new context for planning". It is necessary to keep investigating the phenomenon, deepening our knowledge of its characteristics and testing new ideas, which can lead towards the development of a new perspective on the governance of urban and territorial transformations.

The NICT finally have to be accessible by everybody both from an economic, cultural, as well as spatial point of view (Carcani 2003).

Finally, and beyond the scope of this chapter, what can be pointed out from the trends of development of the NICT is that a major new challenge for the development of urban high technology is introduced by the role played by the contents transmitted on the networks; such "information" can produce new ethics and systems of relationship inside the urban system and, consequently, a new dimension of the city.

References

Bobbio R. (2002), 'Nuove reti e nuovi impianti di urbanizzazione', in *Urbanistica Informazione* 182, March-Apr 2002.

Castells M. (2002), 'Communication technology as material culture: Internet and autonomy building in the network society', Annemberg colloquia, USC – School for communication, 14 November 2002.

Fistola, R. (1993), 'Servizi innovativi e funzioni urbane' in Beguinot, C. and Cardarelli, U. (eds), *Per il XXI secolo una enciclopedia. Città cablata e nuova architettura*, Università degli Studi di Napoli "Federico II" (Di.Pi.S.T.), Consiglio Nazionale delle Ricerche (I.Pi.Ge.T.), Napoli, vol. II, cap. 1, pp. 346-368.

Fistola R. (ed.) (2001), *M.E-tropolis funzioni innovazioni trasformazioni della città*, I.Pi.Ge.T.-CNR, Napoli: Giannini.

Fistola R. and La Rocca R. (1997), 'Cybercities: a new way of thinking about the town planning of the future',Geospace & Cyberspace, Contiguous territories, network territories, International Geographical Union Annual Meeting, , Universitat de Les Illes Balearis, Palma de Mallorca, 26-29 May 1997.

Fistola, R. and Urciuoli, P. (1998), 'Mesuring functional urban polarisation', in Beguinot, C. and Cardarelli, U., *La città cablata, un'enciclopedia*, Napoli: Giannini.

Galimberti U. (2002), *Psiche e techne l'uomo nell'età della tecnica*, Milano: Saggi Feltrinelli.

Graham, S. and Marvin S. (1996), *Telecommunications and the City: Electronic Spaces, Urban Places*.London and NewYork: Routledge.

Mitchell, W. (1995), *City of Bits: Space, Place and the Infobahn*, Cambridge MA: MIT Press.

Moss, M. L. e Townsend, A. M. (2000), 'How telecommunications systems are transforming urban spaces', in Wheeler, J. O., Aoyama, Y. e Warf, B. (eds.) *Cities in the Telecommunications Age: the fracturing of geographies,* London: Routledge.

Sassen, S. (2001), *Le città nell'economia globale*, Bologna: Il Mulino.

Internet-based references

Carcani G., (2003), 'Il digital divide', <http://www.peacelink.it/dossier/divide/dossier.pdf> last accessed Jan 2008.

Chapter 19

Planning and Managing the Augmented City: ICT Planning in Medium-sized Cities in São Paulo, Brazil

Rodrigo J. Firmino

Introduction

It seems to be common sense, after two or three decades of speculation and studies about the intensification of communication through the use of increasingly sophisticated Information and Communication Technologies (ICTs), that we have exponentially increased our ability to interact and expand our personal and collective boundaries. And so has urban space. The intangible relations between physical space and what Manovich (2002) calls "dataspace" – data, information, and all kinds of flows that invisibly populate spaces and places – are contributing to the existence of what can be called augmented reality, and analogously, augmented city.

This expansion of our ability to communicate has been compared to unlimited extensions of our own bodies and boundaries (Mitchell 2003). Particularly within urban studies, concepts, ideas, predictions, models and metaphors have been mushrooming, as researchers have tried to re-conceptualize the city as the ultimate physical support to material and immaterial exchanges which characterize nowadays way of living.

Augmented reality and augmented spaces are not novelties in human history. Religion, magic, metaphysics and art have always provided means for augmenting the immediate material worlds of our existence. What are the sacred moments of prayers and meditation if not an immaterial augmentation of quotidian reality? What are the uses of symbols through human history and the pictorial representations of places and historical facts since Renaissance or even the abstractions of modern art if not an attempt to create extensions of time and space? 'Alberti's window, once merely an artistic device, has since become a "style of thought, a cultural perception, a way of imagining the world"' (Romanyshyn 1992, quoted in Wyeld and Allan 2006, 615).

What seems to be novel, are the amount and speed of information exchanged today, along with new ubiquitous ways of interaction. Thus, I believe we are witnessing, in fact, the intensification of a phenomenon that already existed, or put in other words, the further augmentation of augmented spaces.

However, when it comes to understanding and managing today's augmented space, many studies (Graham and Dominy 1991; Spectre 2002a 2002b; Aurigi 2005;

Firmino 2005) show that proactive planning initiatives related to ICTs, tend to appeal to the ill-grounded utopianism of technological deterministic approaches. This, in turn, tends to create more distrust and scepticism from other municipal departments and civil servants about the involvement of planning in urban-technological strategies.

Recently in Brazil, there has been a wave of public initiatives regarded as best-practices of policies for a so-called digital inclusion strategy which, local authorities argue, is to be followed by a broader process of social inclusion. Public participation is one of the main elements in such policies, and is one of the first assets to be cited in local authorities' discourses. As ICTs have become an economic competitive advantage for attracting inward investments, many of these initiatives tend to be supported by what Graham and Marvin (1996) call "cosmetic reason" (the use of ICTs chiefly as an investment attractor). And so, possible strategic roles of ICTs within the urban agenda could be compromised by a more immediate and superficial application.

At least two major questions arise from these conditions: how do we organize and build the private and public spaces of our cities, given the escalating use of increasingly pervasive technologies which contribute to the constitution of augmented spaces? And what are the challenges posed to urban planning and governance while attempting to plan and manage the augmented city?

This paper tries to answer these questions in three main sections. First, it sheds some light on the theoretical assumptions that underpin the existence of an augmented city, by briefly juxtaposing major research fields and trends of the interplay between ICTs and cities. It also looks into the challenges posed to architecture, planning and governance in the way these activities are historically missing the pace of technological developments in the city. In the second part, it questions our ability to manage the augmented city through the observation of some concrete cases in which planners and local authorities of medium-sized and well-developed cities in Brazil have been faced with these challenges. The major interest here is to see how (if so) they have managed to address the challenges of dealing with a city augmented by mobile, networking and wireless technologies, while designing and planning public initiatives. Finally, a third part draws some conclusions and recommendations that might help us better understand the augmented city.

Augmented Augmentation:Space and Time in the Contemporary City

Are augmented realities, and consequently, augmented cities parts of a new unprecedented phenomenon? One of the key aspects in the complex process of the production of space (Lefebvre 1991) today, is the understanding that, in history, we have repeatedly been trying to augment our immediate reality. Since the appearance of the most ancient technologies, tools have been emulating or expanding our capacity to execute tasks, communicate and interact.

In more complex ways, religious, metaphysics and art have been used, not only to expand our physical ability to perform tasks, but to create new worlds of existence by augmenting reality through mind-powered situations and environments. Attempts to depict this augmented reality challenge our imagination at least since the revolution of perspective-based representation of space in the fifteenth century Italy. According to Wyeld and Allan, though, the principles behind the representation of cities remain almost the same, ever since:

> Like della Francesca's *Ideal City*, the nineteenth century panorama, Le Corbusier's *Voisin*, and the filmic city vision, we find today's architects, planners, and developers using contemporary perspectival technologies – 3D computer modelling – to forecast growth and development in their utopian visions of the city. (Wyeld and Allan 2006, 618)

The notion of augmented spaces comes from the overlaying of physical space with information through the use of images, decoration, art, and any kind of media or architectural technique that adds information, and I would argue intention, to the built environment. In this way, it can be said that we have always used architecture, and the city itself, as a medium to communicate ideas (Venturi 1996; Venturi and Scott Brown 2004). Even modern architecture, with the abolition of ornaments has, ironically, used architecture to pass on its very ideology (Manovich 2002).

The uniqueness of today's relationship between physical space and information, is that we are augmenting an already-augmented space via the addition of extra layers of information, or the invisible flows of data and signals that run through all kinds of environments, known as cellspace or dataspace (Manovich 2002). Besides, the experiencing of augmented spaces has become an increasingly world wide interconnected phenomenon of communication and interaction happening at an unprecedented speed.

What can then be learnt from this process of augmentation of the augmented city? Specific relations between time and space might help us understand some local and global patterns of this phenomenon.

Time, space, territory, control and the boundaries of urban life

The ubiquity of ICTs and the parallel development of transportation links constitute technological changes that have served to minimize the frictional effects of distance and physical dislocation. ICTs and improved transport also enabled a virtual or a parallel urbanization where physical location and contiguity were no longer the only important factors for interaction between places (Firmino 2004).

In general terms, this possibility of a virtual urbanization, reinforced by physical links, serves to emphasize the paradoxical relationship between decentralizing and centralizing forces in the contemporary city. Splintered across different locations, manufacturing activities require sophisticated logistics (Graham and Marvin 2001), and this in turn can lead to spatial processes such as that described by Geyer (2002):

- Deindustrialization of important regional centres of the modernist industrial era, sometimes followed by economic decline
- Specialization of urban functions in regional and global scales, like cities which are economically specialized in tourism, production of specific goods,

distribution, consumption, knowledge, technologies etc
- Increased growth of urbanized areas, forming complex and fragmented metropolises with different and scattered 'centres' and different horizontal and vertical relations with other areas and cities.

Cities and today's world have been flooded with a wave of new, miniaturized technologies that have been changing our lives in the past two or three decades. These technologies are intermingling with old and traditional elements of urban space and of our daily lives to form robotic and cybernetic beings, cities, objects, spaces, and so on. ICTs have been served as means through which we have been able to extend our bodies and boundaries, augmenting the limits of our personal and collective influence.

As traditional elements still dominate both our perspectives and the construction of spaces and places, most of us take new technologies for granted. In addition, the technical and physical structures that constitute these technologies are invisible, too small or too hidden to be perceived as new elements in the space.

What is important about this conception of a hybrid symbiotic and augmented space is the integrative and pervasive way in which technologies are considered as part of the space as a whole. Rather than the deterministic arguments which refer to urban space as an empty container and technologies as aseptic instruments, this integrated vision of the city starts considering technologies as part of the process of a social construction of the space. Within this framework, cities are thus understood to result from a dialectic coexistence of and relationship between people, objects, territory, institutions, as well as material and immaterial flows from distinct eras of urban history. Space is in an unstoppable and complex process of building and re-building itself (Santos 1997). Or as Mitchell (1995, 163) puts it: 'at each stage [in history], new combinations of buildings, transportation systems, and communication networks have served the needs of the inhabitants.'

What makes cities even more augmented than ever is the profound pervasiveness of ICTs. Dana Cuff distinguishes past technological developments from the current one according to three basic characteristics:

> They [pervasive technologies] can be distinguished from past developments by the fact that this new technology can be both everywhere and nowhere (unlike the automobile that is mobile but locatable); that it acts intelligently yet fallibly, and its failure is complex (versus the thermostat, which is responsive but singular and unintelligent); and that intelligent systems operate spatially, yet they are invisible (unlike robots). (Cuff 2003, 43)

As already pointed out, the element of information and communication technologies was added to the already dense 'soup' of other elements which forms the contemporary city. This does not turn the city or urban life into new unknown entities, but rather redefines some of our relations with space, time, the city and consequently urban life itself.

This redefinition tends to become blurred due to the invisibility, complexity and simultaneous vagueness of new telematics technologies. A range of factors define the relations between ICTs and cities: ubiquity, pervasiveness, invisibility, remoteness, complexity, intrusiveness and integration, to name just a few. In other words, as Page and Phillips put it:

> The city is a result of many players and forces that operate through global, regional, and local spheres of influence, utilizing tools and techniques that are permanent, ephemeral, invisible, and strategic. They are knitted together through both real and virtual networks of physical connection, telecommunications, social relationships, and political positioning. (Page and Phillips 2003, 9)

Control – or the possibility to control – is one of the aspects of space and cities considerably improved with the advance of ICTs, and an area of significant interest for many scholars. Cuff (2003) and Page and Phillips (2003) are in no doubt that we live in increasingly controllable environments.

Pervasive and invisible technologies have made it possible to turn Bentham's and Foucault's panopticon (seeing without being seen), and Orwell's big brother (controlled society) into potential reality. The limits between what is public, what is private and where the semi-public boundary is, have become blurred. Because of the development of these unnoticeable technologies (ICTs), control 'once held through straightforward spatial boundaries, has become more subtle, uncertain, and fragmented' (Page and Phillips 2003, 2).

Cuff (2003), however, links control to choice and information as the principles controlling the construction of today's public spaces. She believes, on the other hand, that more information made available to the public by architects, designers and planners could give more choice and bargaining power to the people. This, in turn, could then enable the giving back of certain levels of control over the public space.

Additionally, Page and Phillips (2003) are convinced that mobility and context are the other characteristics to be related to control. We have never been so mobile, they argue; commuting has become the great routine of our time. New technologies allowed us to travel more in time and distances, again refusing the hypotheses of physical substitution and replacement constantly raised by utopians. Consequently, our context – meaning the social and spatial environment we are personally involved and affected by –has been expanded and diversified by the increased ability to move more and faster.

> There is little evidence pointing to the total consumption of the physical by the virtual. Thus far, indications point to a hybridized reality where the physical and virtual compete, complement, and splinter each other [...] The rapid expansion of time-space as facilitated by our hyper-mobility is clearly warping and displacing our sense of context. Today, our contexts are highly customized and greatly expanded. (Page and Phillips 2003, 3)

These characteristics or potential principles of augmented spaces are influencing or should influence our ability and capacity as architects, designers and planners to understand, assess and intervene in the space as we seek to even up the balance between commercial and civic interests (and also the boundaries between private and public spaces) of the augmented city.

Designing and planning in the augmented city

In *Digital Places: Building our City of Bits*, Thomas Horan (2000) draws an analytical framework and proposals for grounding William Mitchell's ideas of a recombinant architecture. He justifies his intentions of bringing recombinant ideas to the place-making process by saying:

> Building in this notion on how technology facilitates the fragmentation and recombination of places, my aim is to analyze digital placemaking in homes, workplaces, libraries, schools, communities, and cities. (Horan 2000, 6)

Grounding the recombinant concept – which is a very similar idea to that of augmentation, applied to the practices of architecture, design and planning – is a multidisciplinary task which involves improving social actors' awareness of the symbiosis between dataspace and physical spaces as well as the direct and indirect consequences for every aspect of their normal daily lives (citizens, politicians, developers, governments, officers, civil servants, architects, planners, technicians, the private sector, depending on the specifics of each local strategic urban-technological development process). The concepts of cellspace, dataspace, or even digital places seem crucial here as spatial elements of augmented reality.

To try and create the right environment for a more aware and conscious development, Horan (2000) analyses recombinant architecture/design/planning at local, community and regional levels according to four different strands: fluid locations, meaningful places, threshold connections and democratic designs. Im sum, these terms try to, first, relate the new paradigms of space and time to the traditional elements of our daily life – strongly supported by material exchanges – to, then, establish important aspects which designers must be aware of.

> There is no doubt that digital technologies will have an impact on our social and community relations, but how well they integrate with these relations will depend on how well we build our city of bits. (Horan 2000, 132–3)

Interestingly, Page and Phillips (2003, 10–12) list a number of examples which they call 'acts of urbanism' that capitalize on the hybridization of dataspace and physical space, and the constitution of the augmented city.

As only one of many examples, a public art project called 'welcome to America's finest tourist plantation' in San Diego (USA), took the theme of illegal immigration and the local authorities' policies on this issue.[1] The aim was to spread critical interventions across many parts of the city through the use of photo-montages in highly visual places.

According to Page and Phillips, these actions 'incited intense debate about the problem and attracted national attention' (Page and Philips 2003, 10–12). This simple example shows how a combination of community actions and digital technologies can affect policy-making. Other social actors could of course also have been mobilized and called in to participate.

1 This project can be followed at http://crca.ucsd.edu/~esisco/bus.

Improved and more coordinated projects of this nature, using even more artefacts, instruments and technologies, spread to other important local issues, and involving a wider variety of social actors could be linked to the points raised by Horan (2000).

In this way, considering different aspects of the augmented city and recombinant development would involve integration in a more comprehensive urban-technological strategy for cities and a closer participation of architects, designers and planners in these matters.

Although it appears that planners and city-makers have lost control of urban space, it is down to these professionals to understand this moment of spatial redefinition, and to embody information, mobility, integration and other characteristics of the symbiotic and recombinant space in strategies for the augmented city.

As the survey below will support, and according to other studies (see Graham and Dominy 1991; Firmino 2004), the very activities of planning and designing have been fragmented. Particularly in Brazil, planning as an attribution of the public sector has changed its name and, many times, its ties with urban design and territorial control. Among other things, this shows how important is to architects, designers, planners and city-makers, an awareness of the complex, heterogeneous and inter-related issues that define or influence the production and design of urban spaces.

Only the process of increasing awareness would allow city-makers to consider what Horan (2000, 20) refers to as recombinant landscape, or 'a collage of settings which, properly designed, will advance the symbiotic relations between people and technology'.

The Appropriation of ICTs for Urban Development in Middle-sized Cities in São Paulo

Planners and city-makers do not look specifically close to these developments in terms of space, time and technology. There seems to be a certain incompatibility between the real ways in which space is evolving and being socially constructed and the ways in which it is being understood and assessed by planners and local authorities.

The power of "real city-makers" once attributed to planners under the modernist aspirations of the industrial city, is now shared by other professionals and forces. One could say that planners have lost both power and interest as regards the factors that are shaping the physical form of today's cities. Other specialists, such as IT consultants, demonstrate greater control of certain elements (for instance telecommunications) which are contributing to the reshaping of spaces and places.

Cities and technologies have already been addressed as a socio-technical phenomenon by Aibar and Bijker (1997) in a very specific and different account. They address a unique framework of social construction of technologies to explain the rival plans for the extension of the urban area in Barcelona in the nineteenth century. This study, *Constructing a City: The Cerdà Plan for the extension of Barcelona,* is particularly interesting in analyzing urban planning through the lens of social constructivism.

According to Aibar and Bijker, the issues of cities and technologies started to attract the attention of urban researchers in 1979 with a special issue of the *Journal of Urban History*. The first striking vision of this phenomenon could not be other rather than one dominated by technological determinism or, as the authors put it:

> Researchers studied the role of technologies like street lighting, sewage, of the telegraph in the processes of geographical expansion of cities and of suburbanization. Technology was analyzed as a force that shaped society and the cities, but its own character and development were regarded as rather unproblematic and even autonomous. (Aibar and Bijker 1997, 5)

By considering the approach of planners and policy-makers to technologies historically, the authors found evidence of a grotesque dissociation between the social, economic, spatial and technological spheres.

Space was commonly seen – and still is, in many cases – as an aseptic container for social activities, while technology was taken to have little influence over the shape of local societies inhabiting these spaces. Space was rarely considered by planners and city-makers to be a social event itself, complexly interrelated with everything else (including history); nor was technology itself seen as socially shaped. This created a methodological and conceptual gap between the development of technologies in cities on one side and planning interests and activities on the other.

Yet, according to Graham and Marvin (1996), the development of information and communication technologies has worsened this complicated relationship between planning and technologies even further because of their invisibility, fast evolution and novelty in terms of application and consequences for social and territorial configurations.

Telematics technologies have opened the way to further confusion about their relationship with the urban milieu. As Graham and Marvin state it, 'urban studies and policy remain remarkably blind to telecommunications issues' (Graham and Marvin 1996, 7).

A survey carried out between January and August 2005, was planned and designed to show, at a first glance, how integrated were issues related to information and ICTs with traditional urban and policy matters in the municipal administration of 54 medium-sized cities in the state of São Paulo, Brazil.

This is an original unprecedented study in urban-technology in Brazil, which is to be extended soon with two case studies, as part of the same research project. One of the aims of this study is to verify the conceptual gap between the practice of planners and their understanding of the augmented city, as recent studies have well documented, especially regarding European cities (Graham and Dominy 1991; Spectre 2002a 2002b; Aurigi 2005; Firmino 2004).

A recent piece of research titled *Atlas do Mercado Brasileiro* [The Brazilian Market Atlas] shows a remarkable presence of the so-called medium-sized cities among the first 300 most dynamic cities in Brazil (Gazeta Mercantil 2005). The following variables were considered as parameters for ranking cities according to their dynamism: Consumption Potential Index (CPI); Human Development Index (HDI); evolution of bank deposits and investments per head; number of newcomer companies; number of new residences; and number of bathrooms per residence. On the other hand, the criteria for selecting medium-sized cities respected the size of their population, ranging from 100 to 500 thousand inhabitants. According to *Gazeta Mercantil* (2005), out of the 300 most dynamic cities in Brazil, 178 or sixty percent corresponded to 'medium-sized' category.

This shows the relative importance of medium-sized cities for urban dynamism and development in the country, which was decisive for sampling the cities for this survey. This was even more the case in the state of São Paulo, where medium-sized centres form an articulated network of well-developed cities, spatially disperse but homogeneous in terms of urban and geographic structure.

The process of data collection privileged some crucial information, in order to offer a broad view of planners and local authorities about the appropriation of information and ICTs while planning and executing public strategies and policy-making. It was giving priority to information such as:

- Each and every public initiative making use of ICTs
- Methods and practices (policy and technical-wise) through which ICT- and knowledge-based projects were conceived, developed, implemented and maintained
- Cities' political and administrative environment
- The commitment of planners and urban planning departments with information, knowledge and ICT issues
- Public policies under which ICT- and Knowledge-based projects were implemented
- Motivations for the development (or not) of such initiatives
- The different social groups interested in these issues
- The existence or non-existence of a clear public strategy related to urban matters affected by the use of information, knowledge and ICTs.

Planning, public administration and ICTs

With sampling defined, a mixed questionnaire with both close and open-ended questions was produced. This was aimed at generating a general view of local authorities' perceptions of ICTs, the use of information and knowledge in making decisions, and to find out what these municipalities were doing in terms of real projects and initiatives on the ground. All concepts addressed by the questionnaire were described in endnotes to avoid confusing the respondents with jargon.

The questionnaire was made of 26 questions and an additional section for open commentaries. In order to facilitate responses, it was divided in four independent sections that could be filed up separately, being: (A) about the city; (B) about electronic government policies; (C) about initiatives on the Internet; and (D) about infrastructure and management of information and ICTs.

The responses were first grouped in thematic categories that could better express the reasons and ideas for each of the four sections of questions. Then, they were finally grouped into 11 units of analysis which translated the whole preoccupation of the survey. Below follows a brief description of each one of these 11 units, grouped into three major thematic subsections:

Mapping ICT initiatives

- Types of ICT initiatives
- Types of network infrastructure
- Initiatives involving mobile and wireless technologies

Managing technologies and technological development

- Development of ICT initiatives
- Influential factors to ICT initiatives
- Specific legislation for information and ICT issues.

Planning and designing for the city of bits

- Institutionalization of the relations between ICTs and the city (constitution of an official body)
- Hierarchy involving the ICTs official body
- Involvement/commitment of the urban planning department with ICT initiatives
- Concerns regarding the influence of information and ICTs in urban development
- Importance of Internet for municipal administration

Mapping ICT initiatives

With regards to *types of ICT initiatives*, assumed by respondents as part of broader ICT municipal policies, there was a strong tendency on pointing Internet portals and websites as the most important action of a public strategy of usage and diffusion of information and ICTs. Every participant city confirmed that municipal portals on the Internet were being developed as a major part of urban-technological projects. It is also remarkable the highlighting of other two types of initiatives. Electronic government was the choice of eighty-six percent of respondents, meaning the use of tools and strategies for optimizing administrative activities, while seventy-seven percent of the cities were offering public physical Internet access, whether through cybercafés or via public libraries and schools.

Other types of projects cited by respondents included: development of municipal networks (by forty-five percent of respondents); extensive use of Geographic Information Systems, GIS (forty-one percent); and surveillance in public places (nine percent).

A very homogeneous scenario appears in the *types of network infrastructure* used by participant cities for the internal administrative structure. Most cases confirmed having experienced more than one type of infrastructure, directly related to the municipal administration, being: twisted pair cables, optic fibres, wireless technologies, and intranets.

Initiatives involving mobile and wireless technologies maintained a general tendency showed throughout the survey, of local authorities demonstrating little or no interest for a strategic view of urban-technological developments. Only nine percent of participant cities admitted the use of mobile or wireless technologies in public initiatives. A staggering ninety-one percent of cities either were considering plans to use this sort of technology or were completely ignoring this issue. This shows an intriguing contrast to the previous unit of analysis, that is, while most cities admitted experiences with mobile or wireless networks for their internal affairs, an absolute minority is actually using these technologies for broader initiatives.

Managing technologies and technological developments

For the unit of analysis *development of ICT initiatives*, two kinds of answers were considered, being projects developed by demand and projects developed in anticipation of a certain scenario. These were created to clearly distinguish the level of preparation of local authorities while dealing with situations involving the use of information and ICTs for local urban development.

A majority of fifty-seven percent of respondents assumed that projects are developed in anticipation to specific scenarios, such as the provision of public cybercafés in order to avoid digital exclusion. This number contrasts with previous research which showed that European cities tend to respond to demands of problems in course rather than anticipate them (Firmino 2004). In fact the Brazilian case could be interpreted in two different ways, considered the broad nature of this survey. On the one hand, it could mean that previous experiences in developed countries have prepared Brazilian local authorities to look ahead and consider information and ICTs as strategic issues on planning the future of their cities. On the other hand, it could imply a level of imprecision from respondents in trying to make their initiatives look more proactive.

This seems to be an expected problem when using survey. Robson (2002, 230) suggests that one of the main disadvantages of this type of study is what he calls a 'social desirability response bias' where 'people won't necessarily report their beliefs, attitudes, etc. accurately … responding in a way that shows them in a good light'. For this matter, a group of other units of analysis were considered out of different questions before general conclusions were made.

In the next unit of analysis, *influential factors for ICT initiatives*, the general tendency showed by other studies in Europe and USA is maintained in the medium-sized cities of the state of São Paulo. It is remarkable the role of two main factors of influence. Shortage of resources was pointed out by sixty-eight percent of respondent cities as the main limitation against the use of information and ICTs in public initiatives. Lack of interest and awareness by politicians and civil servants were considered by forty-one percent of respondents as strong influential factors. This is connected to the former since a lack of interest in using information and ICTs for planning and governing, would possibly implicate in reduced budget for initiatives of this kind. Other factors cited included: lack of a strategic view (twenty-three percent of participant cities); lack of human resources (thirty-two percent), and technical limitations (nine percent).

With a recurrent lack of interest from local authorities on issues of urban-technology, the unit of analysis *specific legislation for information and ICTs* is totally concentrated on the issue of mobile phones transmission towers and their possible location within the urban fabric. All forty-eight percent of participant cities that confirmed the existence of municipal legislation about information and ICTs had it only for the regulation of transmission towers location. The other fifty-two percent had no laws for that matter. No other kind of technological development has appeared to influence legislations within the participant cities.

Planning and designing for the city of bits

Two other important units of analysis – the *institutionalization of the relations between ICTs and the city* and the *hierarchy involving the ICTs official body* – have to do with the creation of special institutional bodies dedicated to dealing with these issues. Mostly, though, these bodies become too specific and specialized, which turn them into no more than technical support units for ICTs.

Regarding the creation of institutional bodies for dealing with ICTs and the management of cities' information, responses were dominated by cases in which, whether, there is no dedicated department or when there is such unit, it is limited to technical support – in the fashion of old data processing centres (DPC). The sum of these two cases was up to sixty-four percent of responses received. Meanwhile, the number of cases in which there is a dedicated department with strategic importance in the municipal administrative structure achieved thirty-six percent of participant cities.

There are a few cases worthwhile mentioning here, all of them with dedicated departments, in the cities of: Atibaia, Catanduva, Guarujá, Mauá, Praia Grande, Presidente Prudente, São Caetano do Sul and São Carlos. These cities have within their administrative structure either secretaries, directories or departments nominally designated to deal with the management of information and ICTs.

One of the most interesting cases was presented by the city of Catanduva where there seems to be a direct connection between ICTs, planning and urban development, through an institutional body called Secretary for Planning and Informatics. This is an isolated but symptomatic testimony of the strategic links between information, technology and urban development being considered by local authorities and expressed within the city's administrative structure.

Planning is very rarely involved in more strategic urban actions with regards to the implementation of ICTs and more interrelated uses of information. The next two units of analysis give evidence of this separation between planning practice and information and technological policy-making.

As for the *involvement/commitment of the planning department with ICT initiatives*, it is important that we make a distinction before showing results. The notion of urban planning activities in Brazilian cities is not homogeneous, although it is considered a solid and consistent field of studies.

The history of urban planning in Brazil is marked by a trajectory of ascendancy (in the 1960s and 1970s) – which coincided with the dictatorship period in Brazil and many others Latin-American countries – and decline (from the 1980s) regarding its relative importance within municipal governance as a whole. During the survey, many different activities and bodies were found as being reported as urban planning, from secretaries responsible for the whole cities' territorial management, to divisions dealing only with public infrastructural work, to others responding by financial and administrative planning.

As far as this unit of analysis is concerned, municipal bodies labelled as *urban planning* secretaries, directories, divisions or departments, independently of their specific attributions were considered.

In accordance to what has been found in European studies, in only nine percent of responses, urban planning was reported as occupying a leadership position on the strategies related to urban-technological development. A significant number of cities, thirty percent, declared that urban planning divisions have little or no participation in these issues. Meanwhile, the majority of cases, or sixty-one percent, stands in an intermediate position, when planning contributes to urban-technological strategies in the same proportion as any other municipal body in the administration.

This seems to be indicative of, at least, two main factors regarding the parties involved, and the overall urban-technological approach. First, it suggests that planning, in general, is not regarded by the municipalities as more important than anything else. I am not arguing that planning should be regarded as the most important division (above all others), but it should certainly be given more importance. Planning has, after all, a central role in directing public interests to public initiatives through policies and plans.

The second indication is that, within the planning department, planners themselves have little or no interest in understanding the physical effects of ICTs in the city. These technologies in the planning division are normally restricted to the instrumental use and implementation of Geographical Information Systems (GIS) or the division's own web-site.

To a certain extent, these indications seem to be causally related to one another. In general, if planners do not show an interest in participating and using ICTs at all, it is natural that other divisions will not see them as key players.

The unit of analysis *concerns regarding the influence of information and ICTs in urban development* corroborates this situation. It was made of questions related to the ways in which the use and development of information and ICTs in urban issues are discussed within the municipal administrative structure and turned into actions and policies. Answers to these questions were grouped according to themes representatives of the main concerns regarding information, ICTs and cities. Five main themes described the sort of importance urban technology has on respondents' minds: importance for administration, digital inclusion, information security, public services, and urban development. Absence was also considered, along with cases in which there is no discussion of these issues at all.

Urban development was one of the least concerns of respondents together with digital inclusion, each one mentioned by seventeen percent of participant cities, superior only to information security (four percent). Not surprisingly, the most popular themes were administration (forty-six percent) and public services (twenty-one percent), along with cities that admitted not having discussions and concerns about urban technology (twenty-one percent).

These results are symptomatic of the way in which urban technology is considered by local authorities and civil servants. Besides the high percentage of cities that admitted not having concerns about urban technology involving information

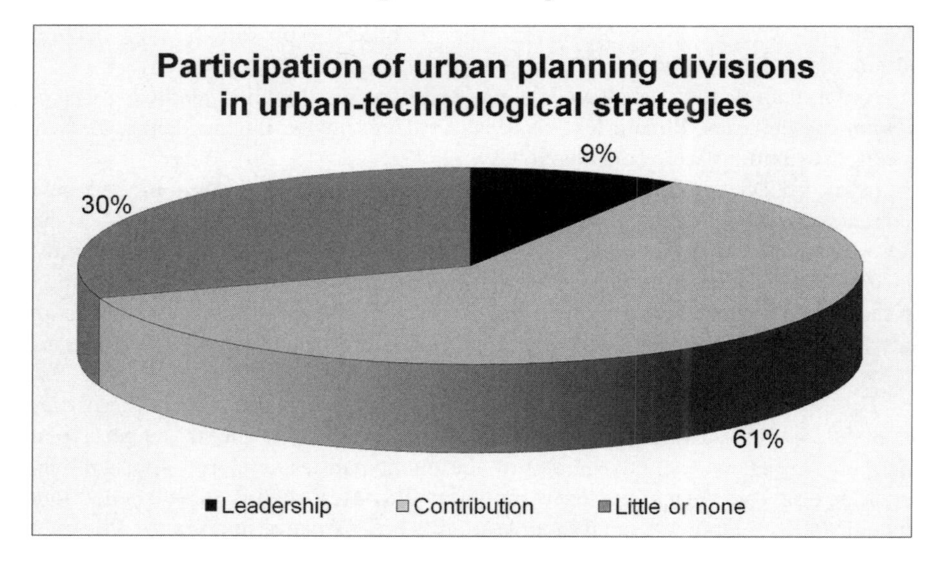

Figure 19.1 Participation of Urban Planning Divisions in Urban-technological Strategies, Credit: Author

and ICTs, it is important to highlight the close relationship between the two most popular themes – administration and public services – with a worldwide tendency of preoccupation with electronic government as a way to improve back-office activities and customer relationship. This is closely connected to the first unit of analysis showed above (types of ICT initiatives), where electronic government figured as the most important ICT initiative for eighty-six percent of respondents.

With regards to the *importance of Internet for municipal administration*, electronic government, again, dictates the symmetry between responses, being: communication (cited by sixty-one percent of participant cities); public services (fifty percent), back-office (forty-three percent); and information (forty-three percent). In fact, this homogeneity also appears in a complementary survey carried out with all the participant cities, to analyse their virtual or digital counterpart on the Internet. Very few differences were found among the majority of municipal portals studied, with the dominance of provision of information over other characteristics such as: provision of channels of communication, navigability, and provision of public services (Coelho 2007).

For the survey as a whole, it is evident the existence of at least five structural limitations involving the appropriation of information and ICTs for urban development in the medium-sized cities in the state of São Paulo, not to say limitations involving the planning and management of the augmented city in the state:

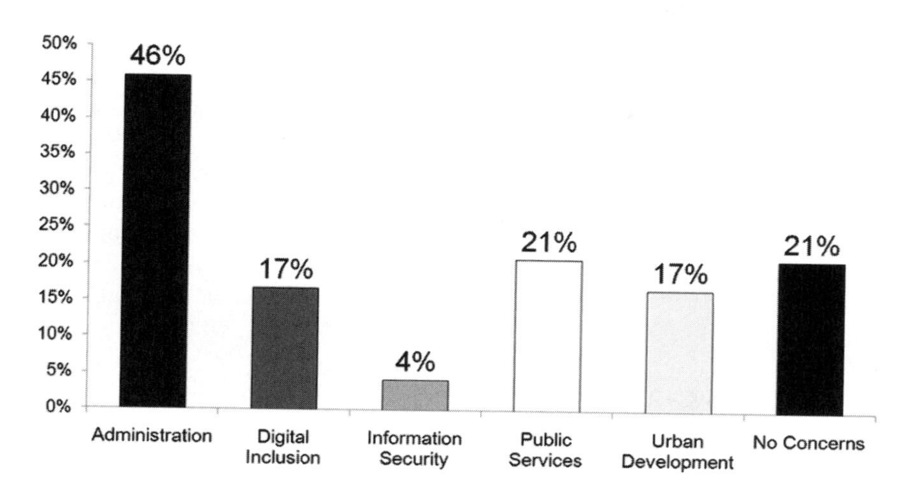

Figure 19.2 Concerns Regarding the Influence of ICTs, Credit: Author

a. Lack of knowledge and awareness of the possible uses of information management, ICTs, and their potential impacts on urban development
b. Lack of interest for these issues
c. Lack of discussions and debates inside the municipal administrative sphere
d. Lack of ability and structure to discuss and reflect upon these matters, while considering the impacts of contemporary forms of social organization on cities
e. Timid proximity between urbanism and the participation of urban planning in issues involving ICTs and information management, essential aspects of contemporary urban development strategies.

Just a few cities had showed being on track for overcoming, at least partially, the above-mentioned limitations in public administration, planning and governance, which highlights the importance of increasing awareness among architects, planners and other city-makers about the relevant dimensions involved with the augmentation of spaces in the contemporary city.

Conclusions

We saw how the possibilities for communicating and interacting have been dramatically increased in the last two or thee decades, with the parallel advance of ICTs. This is part of a phenomenon called by many as augmented reality, and accordingly, augmented space. We also saw that augmentation *per se* is not a new event, but that communication, interaction, information flows and the experience in imaginary worlds have reached proportions never met before.

This new sort of augmentation has confronted architects, planners, local authorities and urban scholars with the challenge of practicing in and understanding what is known now as the augmented city. As we saw, this late augmentation of reality, spaces and the city has been made possible by the use of ICTs that allow us to break some classic paradigms of time and space.

Historically, for competitive or purely economic reasons, governments around the world have found themselves in an unsustainable situation where old Public Switched Telephone Networks (PSTN) could not resist the rapid evolution of the market, and pressures for modernization and privatization. As a result, many western countries have already handed over their standard public national systems to the private sector, sometimes to more than one company (Graham and Marvin 1996).

From a sociological point of view, these economic and political changes have meant a crucial shift in the way that public interests are dealt with. The combination of new regulatory systems and technological advances has splintered the telecommunication provision system that national states had taken years to establish. A minimal level of control in terms of patterns and maintenance has been replaced by very fragmented and biased systems, with, for instance, high-tech control systems allowing the private sector to 'cherry-pick' the best customers and locations for their services (Graham, 2001).

These changes represent a significant historical moment on the fragmentation and "fragilization" of planning, managing and governing what today is being called the augmented city. Public administration has lost monopoly upon telecommunication systems and services since the 1980s, and has been losing power increasingly ever since. The initial aim to offer good standards for the whole population through the homogenization of services controlled by single national monopolies was turned into a fierce competition between telecom companies for more customers and higher profit margins.

As for today – after almost three decades of privatized systems – several studies show the prominent influence of information, knowledge and ICTs to the way we plan, design, use and govern space and cities. The challenge here is to understand that the subject of urban planning is being transformed, to find the best way to comprehend and analyze it, and to re-think the methods and instruments for intervening in it. To Aurigi (2005, 8), this is hard not 'just because contemporary cities have reached levels of extreme complexity, but because the notion of "city" itself is facing a crisis never encountered before.'

In the same way, ICTs are seen as no more complex than any other technical element, with no account taken of the implications they might have upon the complex chains of political and social relations in the city. In general, planners seem to have been reluctant to recognize that new technologies have profound relations with the spatial organization of our cities.

This has been found by other studies on European cities (see Firmino 2004) and has been accordingly verified to be the case for some of the most dynamic and proactive cities in Brazil (represented by the medium-sized cities in the state of São Paulo). In general, the fragmented approaches in these cities still treat ICT initiatives as an alien phenomenon in terms of public sector strategy, where these initiatives are related to the economic development, urban competition and administrative modernization in the cities. Many cases showed little or no concern at all about the augmentation of urban experience and to the possibilities of an integrative strategy for planning and managing urban development.

This also reveals a fragile set of circumstances in which the shape of urban-technological strategies depend on the way local authorities assume responsibility for adopting either a defensive or a proactive attitude. In the former, little room for anticipation is allowed, due to a diminished power of the public sector in overseeing the overall strategy. In the latter, the public sector has the power to oversee projects and initiatives and foresee possible long-term impacts of ICTs development in order to take its decisions.

The distance between urban planning, urban development and ICTs could be critically analysed within an approach that considers the different visions from planning officers and from other ICT entrepreneurs in the city. This could be also related to a kind of planning for the augmented city in which conventional methods intermingle with new interpretations of the contemporary space. In turn, this could open the door to studies focused on the development of augmented public spaces.

Acknowledgements

I would like to express my gratitude to *Fundação de Amparo à Pesquisa do Estado de São Paulo – FAPESP* and *Professor Azael Rangel Camargo*, for providing the support that made this research possible.

References

Aibar, E. and Bijker, W. (1997), 'Constructing a City: The Cerdà Plan for the Extension of Barcelona', *Science, Technology, & Human values* 22:1, 3-30.

Aurigi, A. (2005), *Making the digital city: the early shaping of urban Internet space* (Aldershot: Ashgate).

Castells, M. (1989), *The Informational City: Information Technology, Economic Restructuring, and the Urban-Regional Process* (Oxford: Blackwell).

Coelho, R.A. (2007), 'Portais Municipais na Internet: estrutura e tipologia dos portais das cidades médias do Estado de São Paulo', *Study Report* (São Carlos: E-urb, Escola de Engenharia de São Carlos, Universidade de São Paulo).

Cuff, D. (2003), 'Immanent Domain: Pervasive Computing and the Public Realm', *Journal of Architectural Education* 57:1, 43-9.

Firmino, R. (2005), 'Planning the unplannable: How Local Authorities Integrate Urban and ICTs Policy-Making', *Journal of Urban Technology*, 12:2, 49-69.

Firmino, R. (2004), *Building the Virtual City: the Dilemmas of Integrative Strategies for Urban and Electronic Spaces*. Unpublished doctoral thesis, University of Newcastle.

Gazeta Mercantil. (2005), *Atlas do Mercado Brasileiro*. 6:6, February 2005.

Geyer, H., Ed. (2002), *International Handbook of Urban Systems* (Cheltenham: Edward Elgar).

Graham, S. (2001), 'The City as Sociotechnical Process: Networked Mobilities and Urban Social Inequalities', *City* 5:3, 339-349.

Graham, S. and Dominy, G. (1991), 'Planning for the Information City: The UK Case', *Progress in Planning* 35, 169-248.

Graham, S. and Marvin, S. (2001), *Splintering Urbanism: Networked Infrastructures, Technological Mobilities and the Urban Condition* (London: Routledge).

Graham, S. and Marvin, S. (1996), *Telecommunications and the City: Electronic Space, Urban Places* (London: Routledge).

Horan, T. (2000), *Digital Places: Building Our City of Bits* (Washington, DC: ULI – the Urban Land Institute).

Lefebvre, H. (1991), *The Production of space* (Oxford: Blackwell).

Manovich, L. (2002), 'The Poetics of Augmented Space', In: Anna Everett and John Caldwell (Eds.). *Digitextuality* (London: Routledge, 2002).

Mitchell, W. (2003), *Me++, the Cyborg Self and the Networked City* (Cambridge, MA: MIT Press).

Mitchell, W. (1995), *City of Bits: Space, Place and the Infobahn* (Cambridge, MA: MIT Press).

Page, S. and Phillips, B. (2003), 'Urban Design as Editing', *Mimeo*. 18.

Robson, C. (2002), *Real World Research: A Resource for Social Scientists and Practitioner-Researchers* (Malden, MA: Blackwell).

Santos, M. (1997), *A Natureza do Espaço: Técnica e Tempo, Razão e Emoção* (São Paulo: Hucitec).

Spectre (2002a), *Strategic Planning Guide - Dealing with ICT in Spatial Planning: A Guide for Pratictioners*, Haarlem, Provincie Noord-Holland.

Spectre (2002b), *Vision on ICT and Space - Vision on the Relationship Between Information and Communication Technologies and Space*, Haarlem, Provincie Noord-Holland.

Venturi, R. (1996), *Iconography and Electronics upon a Generic Architecture: a View from the Drafting Room* (Cambridge, MA: MIT Press).

Venturi, R. Scott Brown, D. (2004), *Architecture as Signs and Systems: for a Mannerist Time* (Cambridge, MA: Belknap).

Wyeld, T.G. and Allan, A. (2006), 'The Virtual City: Perspectives on the Dystopic Cybercity', *Journal of Architecture* 11:5, 613-20.

Chapter 20

Epilogue: Towards Designing Augmented Places

Alessandro Aurigi

Introduction

On the 8th of September 2007, Beppe Grillo, an Italian comedian beloved by many and well known for his satirical portrayal of the establishment, managed to bring thousands of people into many Italian squares, to support three proposals for changing the laws regulating the election and tenure of Italian MPs. The event, whose main stage was in the city of Bologna's main central square – the Piazza Maggiore – had been named by Grillo "V-Day", where the "V" – far from meaning "victory" – was an abbreviation of something very similar to the "f-word" insult in English.

All of this had been meticulously prepared for weeks through Grillo's blog on the World Wide Web, one of the most viewed websites in Italy. A few days later, The Independent newspaper in the UK described and celebrated, as much of the press and media, the event, by highlighting how powerful Grillo's impact had been, and how it had managed to get beyond the virtualised debates of cyberspace:

> F-Off Day seemed ambitious for a movement confined to the blogosphere. The comedian called for one big rally in Bologna, home of Mr Prodi, and 250 in other cities around the country (and 30 abroad). But he knew what he was doing. Fifty thousand assembled in Bologna. Hundreds of thousands more gathered around the country (The Independent 2007).

Beyond the obvious thoughts on the power and impact of virtual communities and public fora on the Internet, this story makes an interesting point for the perusal of the readers of this book. Grillo's blog[1] had been in existence, and updated daily, for nearly three years, but its possible impacts had not been taken very seriously by press, media and the political establishment. Articles about it had been written, and comments made, given the public profile and popularity of its owner, but as the quote from The Independent somehow reveals, these had tended to interpret the blog – and the apparently sheer number of its users – as something "virtual" which would not have a solid grounding in reality and therefore any concreteness.

What changed everything was physical space. Years of blogging materialised in one of the most symbolically-laden squares in Italy. To paraphrase Allucquère Rosanne Stone (1991), the real bodies actually stood up – and ended up queuing

1 <www.beppegrillo.it> last accessed January 2008.

Figure 20.1 "V-Day" in Bologna: the Stage, Credit: Piero Tasso

up at marquees to sign a petition too. And this combination of digital community-making and physical space and place had an instantly powerful impact, sending ripples through the political party-dominated press and media, and forcing them to address the issue. In the days that followed the event, journalists stopped talking about Grillo as a naughty but fundamentally innocuous commentator, and started describing him in ways that ranged from the founder of a new and innovative political movement, to a potential destroyer of party-based democracy creating a platform for dictatorship. Political leaders appeared much less defiant of Grillo's criticism than before, and some started reacting by attacking the comedian and describing him as a dangerous populist. A government official proposed in October 2007 a controversial law, basically against the phenomenon, to heavily regulate and control blogging, which was criticised amongst others by Bernhard Warner on The Times Online (2007), highlighting how forcing bloggers to officially register with the state would be nothing short of censorship as well as something impossible to implement after all. One way or another, digital and physical space and community had clearly "augmented" each other. Digital technologies had established and fostered links, relationship and ideas, but these would have never been as effective and "real" without the grounding of physical space and the sense of place and empowerment Piazza Maggiore could provide. Needless saying, the same rally would not have been as meaningful had it taken place in an out-of-town industrial estate.

So, space and place matter, and they do even when it comes to considering digital technologies and somehow lifestyles. Seeing physical and digital as combined, as the two sides of the same "spatial" coin, is not just a very realistic stance, but one on which we can invent and design for. Space can become proactively "augmented".

Augmented Space

However, it can be argued that the city is already being augmented, regardless of any focused efforts to design it so. In a way, it has been incrementally augmented throughout history, and we are still adding features – some of which are currently "digital" – to the city and urban space, without threatening the uniqueness of space itself. Several authors – for instance Mitchell in his 2003 book *Me++: The Cyborg Self and the Networked City*, Haythorntwaite and Wellman (2002) and Cuff (2003) have already pointed out at how ubiquitous computing on the one hand, and the increasingly everyday character of the Internet and high technologies on the other, should suggest how previously popular conceptual divides (space Vs cyberspace, real Vs virtual and so on) should be regarded as an outdated way to look at things.

Augmented urban space is here and now, it is immanent. As well as for more conventional, historical urban configurations and designs, it has a "spontaneous" side. It "happens" as a consequence of our urban space-bound usage of mobile or internet technologies, and its articulations go well beyond the simple direct impacts of some specialist – and not necessarily very widespread – devices. For instance, amongst digital technologies, it is not just GPS and "sat-navs" that can affect our movements. More simply, choices to move somewhere – to a certain shop or meeting venue instead of another for instance – are likely to be guided by information retrieved digitally and relationship enhanced or sustained through the use of the internet both from home as well as through mobile devices.

High technologies have become more mobile, but also increasingly located or anyway dependant on location. Bluetooth communication – which is eminently local, spanning roughly just 10 metres in its basic configurations – is now embedded into most mobile phones, and can be used to foster place-bound information retrieval or interactions. For example it can support blind-dating or match-making software for people sharing the same spot, providing a platform which operates only through spatial proximity. It can therefore become an instant though unplanned and transient add-on feature to that specific area. Augmenting technology needs a context, despite it often seeming context-neutral, and by being personal it will create certain "meanings" for certain people, but not for others, so that the same "space" can even more than before, become different "places" to different people. The "meaning" in these cases will result from the intersection of physical and digital or, looking at space as a whole, we might be allowed to talk of an augmented *genius loci*.

Mobile telecommunication companies understand to an extent the power of this combination of digital and physical features of space, and are flirting with – and commercially exploiting – the ability of digital and mobile networks to extend concepts like local community and meeting place, especially when they refer to the young end of their customer base and market. In Italy for instance TIM (*Telecom Italia Mobile*) – but Vodafone seems to have similar plans in other countries – has been exploiting the metaphor as well as the specific benefits and sense of belonging of an urban "tribe" and has been pushing these as a package and brand to young mobile users through the "Tim Tribe" concept. Your mobile is not to be seen just as a standalone device to ring people up, but the key to associating and keeping the ties with an increasingly large number of friends. This is pushed to the point of extending the remit of the mobile phone, which is a device allowing fluidity and freedom of movement, yet ensuring belonging – and possibly customer loyalty – to a network/framework. This extends into the virtual social spaces of *My Space*[2] as well as into discrete, tribe-dedicated events happening at physical venues, such as parties, concerts and contests, and links these all together. The "tribe" is then fluid, transient as far as its use of physical spaces is concerned, but well grounded within them. It could not fully exist without them as well as without the "real" people which can be encountered there. As an entirely virtual community it would be inevitably much less appealing. Layers are added to the city, and the physical and the digital interact more seamlessly. They become just two aspects of the same ritual.

Augmenting the City

Cities themselves can proactively implement high technologies aimed at wirelessly layering their spaces to offer more and better facilities. This can go beyond, and extend, the simple use of web sites as information and service platforms – something strongly pushed by the bureaucratic innovations and benefits driving e-government

---

2 <www.myspace.com> last accessed January 2008.

Figure 20.2 Synergy Between Mobile Networks, Blogging and Physical Events through Tim "Tribe"

initiatives in many Western as well as developing countries. To keep with the same exemplar technology used in the previous section, if Bluetooth can be spontaneously used as a way to link up within the same area, it can also become the means to add planned and public digital layers to the town's fabric. One of the most active European cities for ICT-based innovation, Bologna in Italy, is implementing a coordinated sequence of Bluetooth transmitters along one of the busiest and most typical tourist and shoppers' routes in the city's historical centre, creating a digital/informational pathway which augments specific spots, adding layers of extremely local/grounded information to the physical environment. The municipality – and it is just one of many similar examples – is also gradually casting a Wi-Fi 'shroud' all over the city centre, to deliver its services right in its physical public spaces.

But "proactive" does not necessarily mean "institutional". Creative, unexpected intervention in augmented space-making is indeed possible. Colonising the space, especially public space, with facilities as well as digital layers of information can become an empowering and controversial practice, as it can mean controlling and influencing part of that experience and some of the meanings associated to that location. Space can be planned and augmented in a top-down way, but very local action can still be imagined: more layers can be added, multiplying space's potentiality. The "Wi-Fi Bedouin" backpack system conceived by Julian Bleecker[3] is an example of how ad-hoc information and software facilities can be deployed instantly to support specific, transient activities which would not be able to rely on public or semi-public networking infrastructures in public space or to contrast other, more institutional or commercial forms of space augmentation.

So, urban spaces can get augmented spontaneously as well as in a planned, designed way. They can be augmented by institutions and with more general, public purposes – whether these could be agreeable or not – as well as by individuals and more or less closed groups acting with their own, particular agenda. They can create more transparent or more hidden spatial enhancements.

All of these manifestations of augmented space are relevant and should be looked at by the planners and designers of the city. The "spontaneous" aspects of urban augmentation need to be understood as well as possible. Any proactive intervention will have to take into account of, and designed for, the context and the existing relationships – spatial as well as social – of the "site" being looked at, a principle that designers of urban spaces know well. But designers also need to look at the so-called 'precedents' for inspiration, reference, and learning – from success as well as mistakes – and this means considering very carefully what has already been proactively attempted in urban augmentation, as well as the principles that can be drawn from existing design, urban and social theory.

Adding "Value" to Spaces

Digital technologies are clearly able to augment space in a quantitative way. Informational layers and facilities can be virtually infinite – or at least numerous

3 <http://www.techkwondo.com/projects/bedouin/> Last accessed January 2008.

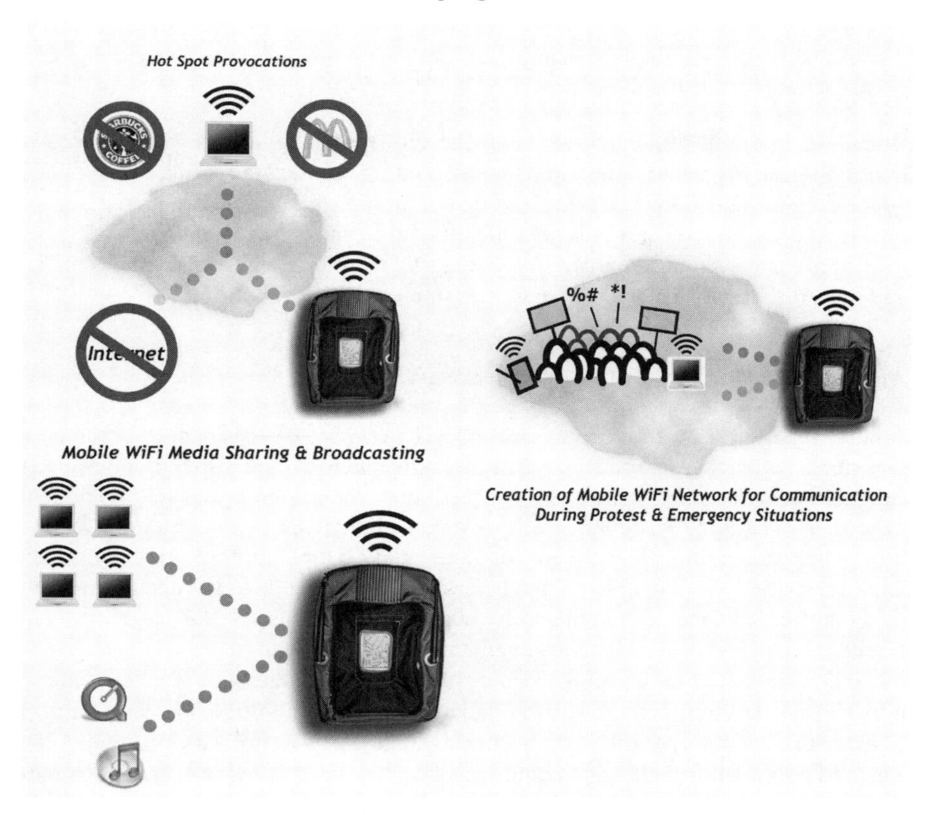

Figure 20.3 Julian Bleecker's Wi-Fi Bedouin, Credit: Creative Commons 2003, Julian Bleecker, Near Future Laboratory, www.nearfuturelaboratory.com

– within the same physical location, hence in a way supporting many different variations of the same urban space.

It can be interesting, however, to reflect for a moment on the definition itself of the expression "to augment". A dictionary, for instance, defines it as 'to increase the size or value of something by adding something to it' (Cambridge Advanced Learner's Dictionary). This suggests another important aspect of the phenomenon: the – possible but certainly not guaranteed – increase in the "value" of what gets augmented. This – it is argued here – should not be missed in our thoughts and debates, at least because it is an important factor related to design and proactive interventions. We do not design just to add – and sometimes not to add at all – in a quantitative way, but to, somehow, add "value", and we do it in ways that transcend the strict economics and monetary aspects of the word:

> For example, the social benefits of a high quality public realm and the productivity gains arising from well-designed urban spaces and workplaces occur in the form of externalities. This is a distinctive characteristic of 'public goods' that have no immediately identifiable monetary exchange value and are therefore not usually considered important by the

market. Such goods can easily be undervalued in public and private investment decisions. Yet their true value can be much greater than the supply price or the cost incurred in making them available" (CABE and DETR 2001, 15).

When we promote initiatives to make the city more digital, are we also really improving its augmented spaces? Is space getting "better" by being augmented? Answering the question is far from trivial, as defining what "value" or "better" could or should mean in themselves is certainly not either an easy or a univocal task.

But a basic – though not certainly exhaustive – set of reflections can be made, and to start these off a Newcastle-based example could turn out useful.

An Example: The Value of Public Terminals

A fairly typical application/deployment of ICT in physical urban spaces, something that stems strongly from the e-government "movement", is the installation of public information terminals in semi-public spaces such as libraries and local government offices, as well as fully public spaces – streets and squares – where digital kiosks have been appearing. These can be 'threshold' (Horan 2000) points between physical space and some of its digital layers, and a way to allow universal fruition of those layers. It is one of the potentially many manifestations of the usage of urban "screens", and one of the elements augmenting a city.

In the summer of 2005 a pilot study of the public kiosk facilities – called "I-plus" – in Newcastle upon Tyne, UK, was undertaken (Kalouti 2005). The study aimed to carry out and analyse a limited – in scope and time – set of observations of the usage of public information and communication kiosks installed in several locations within the Newcastle and North Tyneside areas. To what extent did this facility fulfil its aims of providing and encouraging access to public information as well as basic digital communication? How often were the terminals used, by whom and to do what?

The results were, in a way, not very encouraging. Few people were using the terminals. The disadvantaged social groups who were supposed to benefit the most from the presence of these devices were not really doing so – for instance only 4% of users were elderly and only 5.3% belonged to ethnic minorities – and the "socially useful" contents were amongst the most ignored. The communication facilities which were expected to facilitate public participation – another major aim related to the deployment of terminals – did not do well either: 'The communication applications, such as the free email, the council feedback and the I-Poll, that are supposed to enhance e-democracy and the decision-making process are not used in reality' (Kalouti 2005, 56).

Several reflections were prompted by the consideration of these results and the related observations. Quite obviously, as it often happens in these cases, this appeared to be an entirely "top down" solution, in which the local communities which could have benefited had been "interpreted" as mere end-users of the technology. The terminals were nicely designed objects – though only to an extent, as usage by disabled or elderly people could have been helped with dedicated seating for

Figure 20.4 I-Plus Terminal in Central Newcastle, Credit: Author

instance – but had been both placed and configured in ways which were un-related and indifferent to their context, both spatial and social.

We started asking ourselves: what if these terminals had been located where people, and mums and kids really go and stay? What if they had been designed with the meeting of people in mind, providing seating, making them part of an already existing gathering place, maybe putting them in play areas? What about exploring all of those fairly basic issues someone would do if they were dealing with a place-based piece of urban design, rather than the positioning of strange alien objects? What if their functions had been tailored to what those communities do or need in their own places, which can be different from what other communities do? What – to cut the story short – if they had been part of a place-making strategy, which certainly they were not? And of course the Newcastle case is just one of many similar examples that could be considered.

So, one thing to reflect upon is that the majority of initiatives – often municipal – trying to deploy ICT in the city, tend unfortunately to create much less "value" (that is being used and useful, making a difference, changing the city for the better, adding meaning to locations) than expected. Many – initially funded by the European Community for instance – end up having made basically very little difference. Amongst possibly other factors, this is also because ICT are still looked at on their own, seen as an add-on, sometimes even a fairly inconvenient one, to physical space. Much can be – and has been – said about public space and the public realm and the role of high technologies in this, but very rarely has a holistic strategy been seen to inform this, based on an approach focused to create/reinforce a sense of place, and facilitate the production of some "value" in the augmented space. So, when it comes to envisaging as well as deploying space-augmenting initiatives, some awareness that what we are dealing with is – in fact – an extended version of more traditional practices of place-making and urban design can help greatly, so we do not forget about the city and the need to see high technology as simply part of it. An augmented city needs an approach which is sensitive to place.

First Reflection: the Importance of Place-making

One of the key factors for people interested in finding ways to produce value – and sustainable one, something that may last – out of the augmentation of cities, is to shift our perspective – and action – certainly from the dualist vision of competing physical and virtual spaces which might replace each other – and which they very often do not – to one of an urban space which is whole, one, enhanced by ICT. But the next step towards value-making is to consider the not easy challenge of place – as opposed to space – making.

Why so much importance to this distinction? Because place – not space – is what people actually experience, what can prompt them to have an active role and participate, and what can make them own – and therefore make more sustainable – initiatives and ideas of urban augmentation.

But "place" is complex, challenging, hard to grasp. What will an urban planner sitting at their desk with traditional tools and pieces of information like maps, diagrams and so on in front of them, really understand about the land-use of a place? And what will not be that easy to see or capture?

It will be possible to have a fairly clear idea of the urban morphology, maps and drawings will illustrate these aspects. The ways buildings are being used might be known, at least to an extent. Data on demographics and possibly results of other studies on the area might be available. But, as Nancy Odendaal (2008) has noted within her own chapter of this book, much will be missed and/or will be extremely hard to grasp and plan for. All sorts of informal relationships, aspects of the economy, local knowledge, interpretation and use of place will hardly be part of the picture.

This becomes even truer when it comes to introducing high technology in urban spaces. Firstly, this is because of the ability of telecommunications to amplify fluidity, mobility and the transitory aspects of what already is a set of informal economic and social relations. Activities can – and maybe will – be moved and rearranged quickly to respond to changes in the context, or to take advantage of more suitable or welcoming environments. Land use can become even more dynamic; the same plot can accommodate different fragments, and can be used differently at different times. Distant plots can become literally one in terms of their usage. A multi-dimensional set of relationships is further increased.

Secondly, whilst all of this would call for a sophisticated and fine-grained approach towards the conception and deployment of "proactive" ICT-based initiatives, in order to understand and respond as best as possible to this complexity, such an attitude is often absent within these projects. In fact, whilst planning and urban design theory and practice have been responding to a growing awareness on the complexity of arenas and "place" with innovative ways to look at participative design, governance and institutional capacity, participant observation research approaches and in general tactics to fill this gap, urban high technology has mainly been pushed by governments as well as private entrepreneurs as IT "solutions" into urban space with a general indifference – or indeed ignorance – towards any principles of place-making.

The literature on the differences and relations between the concepts of space and place is far too vast to properly analyse it here. In his very successful book *Digital Ground*, McCullough (2004) has rightly made the connection between interaction design – which is of course the building bricks of urban augmentation – and the need to look at the importance of place-making. Tensions between the notion of space, or spaces, as an analytical and fragmented way to look at aspects of our environment, and place as the holistic, meaning and identity-rich – though diverse – humanisation of the environment itself, have been highlighted in different ways by many commentators.

A converging definition of the difference between space and place does not exist, and systematically analysing all literature about the topic would be a task well beyond the scope of this paper. However, this chapter would like to put forward the very basic reflection that while space is divisible, place is not. "Spatial" are all of those different ways to analyse discrete and partial aspects of environments. Place is whole, complex, inextricably multi-dimensional, lived, experienced, meaningful – and characterised of course by multiple meanings. And it is also the "real thing", what makes people "go there", "remember that", "like the location". Place is – in one way or another – owned.

The problem is that dealing with a "whole", designing holistically, is far too difficult and – when it comes to shaping something with "place" in mind – we end up needing to manage the process by looking at its different aspects, de-constructing place into spaces. The urban design text *Urban Spaces, Public Places* by Carmona et al (2003) draws a very useful distinction amongst different "dimensions" of urban space. These are defined as the morphological, perceptual, social, visual, functional and temporal dimensions of our cities. In a practical way, dealing with place can mean trying to negotiate designing all of the dimensions listed above, while at the same time acknowledging that the "experiential", interpretative, meaning-making activities are certainly not all "plannable". We cannot ignore places' holistic character, and yet we know that we cannot control it entirely. So, making "place" is an approximate though highly strategic job. We tend to care about as many aspects of "space" as possible – and as deeply as possible – trying to coordinate these to work successfully as a whole, and we hope to facilitate the shaping of meaningful places. As mentioned before, such an approach is particularly rare when it comes to urban ICT projects, even those which have much to do with physical space and place.

Figure 20.5 is a coarse and somehow naïve vision for taking ICT into public spaces that the author produced back in 1993. Not many people were talking about

Figure 20.5 A "Terminal" Multi-modal Square

the Internet and the World Wide Web was in its infancy, but the nexus between telematics and the city was already very fascinating. However roughly, the terminals in the prototype square presented were trying – if nothing else – to engage much more with place than the ones analysed in the aforementioned 2005 study. They were trying to be a designed part of the environment by shaping an artificial, technological "wood" in the public space. They were aiming to combine themselves with different functions and possibilities through a wider strategy – this hypothetical *piazza* was also an amphitheatre with a stage for performances and a multi-modal transportation hub – into a multi-dimensional place with an identity, where it hopefully would be nice and meaningful to stop, hang around and meet friends (Aurigi 1993).

Similar principles and ideas can be found in, and have been informing the shaping of, the hyper-modern Seoul's Digital Media Street,[4] where ideas of augmented space and place have been extended to a fully-blown, larger-scale urban design project. There, social activities and ties are meant to be supported by augmented places, where physical space and high technology complement each other.

Several other projects have been appearing in the past few years which proactively try and design enhanced spaces. In the much-mentioned Zaragoza's Digital Mile project[5] digital technologies become main urban design features and street furniture

4 See <http://dmc.seoul.go.kr/english/jsp/about/dms.jsp> Last accessed January 2008.

5 See <http://www.milladigital.es/ingles/home.php> Last accessed January 2008.

elements. It is an evocative and interesting project, combining physical space and ICT. However this design might end up combining with "society" and "community" less than it should, and could be seen – as most projects in this area – as an evocative but limited pilot experience. It deals with "place" but probably not very exhaustively, possibly serving mainly a city-marketing and image-reinforcing approach, much like certain landmark buildings and spaces – the Guggenheim in Bilbao for instance – than the specific needs of local communities.

And people, communities, local culture and knowledge are indeed very important. Combining purposefully physical and digital can be a first step towards innovative place making, but will not necessarily be enough. Ultimately, as mentioned previously, people create the meaning, the "place", and if they have not been properly included and considered, they can deny the designers' intentions, and make augmentation of urban spaces weak and unsustainable.

Determinism and Place Do Not Go Together Well

Although not an "urban" space in itself, the story of the brave, fascinating but eventually rejected innovative design of the TBWA Chiat/Day Los Angeles and New York-based offices is exemplary of the importance to distinguish between space and place, and focus on the latter. Company's boss Jay Chiat – as well as architect Gaetano Pesce for the New York premises – had envisaged at the end of the 1990s a fluid and augmented environment that would unleash creativity and support the mixed modes of work – partly located, partly virtual/nomadic – which many people nowadays have adopted. But the concept, together with the idea of a paperless way of working, had been taken so uncompromisingly far to deny employees any form of personal or private space, any chance – in a way – to create, even physically, their own meanings. After a while employees reacted by changing the rules of the cybercafé-looking workplace themselves:

> After six months, a counterrevolution was in full swing in both offices. In LA, people took to using the trunks of their cars as file cabinets, going in and out to the parking lot, in and out. There had been discouragement against this, 'but people just ignored it' says one staffer. Rabosky took over an entire meeting room, declaring it 'my office until somebody fires me'. Eric McClellan, the New York creative director, did the same back east. The LA office eventually started using sign-up sheets for assigned spaces. People stopped returning their portable phones and PowerBooks, stashing them in their lockers at night. Gradually, makeshift desks were put in place. Desktop computers began arriving in the LA office. The media kept gushing about Chiat's virtual adventure, but by the end of year one, the whole 'grand experiment' was already wobblier than a Gaetano Pesce chair. (Berger 1999)

This could be labelled as a "deterministic" attempt to agument space. Although this was a bold and very interesting experiment of spatial and functional IT-driven recombination, it failed to win over the staff having ignored their needs, aspirations and habits. It was very possibly a brilliant space, highly celebrated in architecture and interior design books and magazines, but failed to be the "right place" or to create a positive sense of place. In "place" – and we could say in the "real world" – people play an active role and reject or modify technologies and innovation, hence the failure of many otherwise apparently good ideas within the recent history of urban ICT developments. An attention to place-making is therefore crucially important not just for "traditional", mainstream urban design and planning, but for those initiatives that aim at augmenting the city. Indeed, the gap between "traditional" and "augmented" is very simply nonsense. We design within a complex, augmented, scenario anyway. The danger lies in believing that, because of high technology, new rules are established that allow us to disregard or override basic design and planning principles.

Second Reflection: Towards the Design of Augmented Urbanity

Several authors have already carried out some phenomenal early work on the definition of guidelines and issues for the effective, meaningful design of digitally-enhanced spaces. Just to name a couple of them, McCullough (2004) has formulated principles and identified issues for the design of interactive spaces and the fostering of digitally-endowed ecologies. Horan previously had also reflected on the challenges that should be faced when engaging with what he defined as the digitally-endowed 'transformative design' (2000) of space.

Drawing from the conclusions of the previous section on place-making, this paper suggests that a degree of critical engagement with issues that relate to mainstream urban design is important for two concurrent reasons: keeping projects and initiatives deploying ICT in cities grounded in place-making debates and practice, and by doing so drawing ideas for designing and shaping these projects in a more effective, place-wise, way. The next few sections reflect on urban design issues versus augmentation of urban space, and do so by trying to ground these reflections with the help of the general theories explained in the *Public Places, Urban Spaces* textbook by Carmona et al (2003).

Is ICT a "De-urbaniser" or is it a Contributor to "Place"?

The emergence of the information society has been seen by many theorists as a force that would increase de-localisation and fragmentation of cities and society further. A very good account of dystopian positions on the urban impacts of the information age remains part of chapter 3 of *Telecommunications and the City* (Graham and Marvin 1995), but many other authors, among whom some design theorists, have expressed their reservations about the potential of ICT – and indeed anything greatly increasing mobility and fluidity – to keep people apart from each other and indeed from physical, meaningful places. Christian Norberg-Schulz, for instance, at the beginning of the 1970s, would frame his views and theories on the relationships between architectural 'structures' and 'existential space'. He would end up 'refusing to accept that these principles lose their significance because of television and rapid means of communication' and add that 'as a matter of fact, a stable system of places offers more freedom than a mobile world. Only in relation to such a system can a "milieu of possibilities" develop' (Norberg-Schulz 1971, 114).

Most of these perplexities hold well as long as digital technologies are seen for their global, large-scale character and possible effects, and as long as we deem them a "virtualising" and displacing agent. However, once we start considering augmentation of specific localities by space-embedded technologies, the perspective changes. Arguing for their potential to contribute to – better – places has a rationale, and can reconcile theories – and care – for meaning-rich place with ideas to use high technology to facilitate the deveopment of a "milieu of possibilities".

Carmona et al (2003), though not addressing directly any issues on the spatial articulation of physical and digital, explain how 'the significance of the physicality of places is often overstated: activities and meanings may be as, or more, important in creating a sense of place', and that 'seen from a temporal perspective, the physical dimensions of places are most salient in the short-term, being displaced in the longer-term by sociocultural dimensions' (p.98). They also refer to Sircus (2001, 31) when he claims that:

> Place is not good or bad simply because it is real versus surrogate, authentic versus pastiche. People enjoy both, whether it is a place created over centuries, or created instantly. A successful place, like a novel or a movie, engages us actively in an emotional experience orchestrated and organised to communicate purpose and story (quoted in Carmona et al 2003, 105).

Again, although in the text which is being referred to, this argument is not used to justify the potential of augmented spaces, it clearly can support a positive reflection on these. Enhancing space with high technology is not going to make it less "authentic" – and indeed the notion of "authenticity" should be reviewed itself – useful or memorable. There can be – many – ways in which place is reinforced, rather then denied, by careful, sensitive and grounded digital augmentation.

Augmentation, Legibility, Permeability and Psychological Access: What Public Space has to do with ICT

Beyond the potential – certainly not deterministically guaranteed – of high technologies to reinforce the public realm and public discourse through offering virtualised arenas for debate, can augmentation also improve – and increase the value of – actual public and semi-public space? Again, ICT's "displacement" potential has been denounced many times as something weakening physical public space and people's ability to embrace its usage: 'the use of public space has been challenged by various developments and changes, such as increased personal mobility – initially through the car and subsequently through the internet' (Carmona et al 2003, 110).

But it can be argued that spatial digital augmentation – rather than replacement – can be designed to contribute positively towards making urban space more permeable. Permeability is defined in various ways, and it is a concept that can be closely related to the one of "legibility", as the ability to 'grasp' both a place's form and its uses (Bentley et al 1985, 42). Carmona et al (2003, 65) extend this and define visual permeability as 'improving people's awareness of the choices available'. This awareness can definitely be affected by the augmentation of space. First, the 'choices available' can be multiplied, as we have mentioned when talking about the possibility of multiple – virtually infinite – digital layers characterising a space and adding functions, facilities and possibly meanings to it. Secondly, a sensible use of those layers, as well as of a place-bound internet presence, can increase awareness of real places and hence enhance their visibility even when physical "form" alone – given by previously weak design and planning efforts – could fail to do so. The four key attributes of successful public spaces defined by the Project for Public Space (1999, quoted in Carmona et al 2003, 100): comfort and image; access and linkage; uses and activity; and sociability, all can be digitally augmented, both in quantity and quality. ICT can add functions and services to public spaces, increasing comfort and usability and can multiply the ways spaces are used and make certain functions more accessible. In particular, spaces characterised by digitally-enhanced information and communication could increase their users' knowledge of the place itself and of other people using it, and encourage them to use the spaces with confidence, gradually making it their own. Place-related information and navigation systems already perform – though partially – this function, increasing the visibility of those features – shops, leisure facilities, events and indeed anything that is addressed by those systems and accessible through them. Even remote locations can become "permeable" to each other through communication technology, and this could be harnessed to bring otherwise physically separate communities closer. In this sense, permeability does not just remain "visual" – however wide our notion of "visual" here can be – but can become "social", making people know other people more and/or better. The technology from Vienna-based company *Tholos Systems*, for instance, aims to implement full-size videoconferencing kiosks – very much like proper "windows" open onto other locations – in European cities' public spaces. The company claims that this form of augmentation – and linkage – of urban spaces will have the potential to 'become a tool of global as well as local meeting' to the extent of helping overcoming social and cultural barriers (Wallner and Fieder nd). Although the social benefits of the international dimension of the linkage could be debated, such a technology could be used in interesting ways within the same urban area. It could link neighbourhoods which otherwise have suffered from physical and social divides, and it could be exploited to increase the "public-ness" of a place, by making it observed – and surveilled, it could be argued – by more people. It could obviously be used to create "hyper-environments" for sharing and collaboration, linking classrooms, assembly and community rooms, and other similar semi-public spaces by transforming one or more of their walls into a threshold to another – remote – similar place.

The concept of the "third place" overall, defined among other by Oldenburg (1999) as an open, playful and discourse-supporting neutral ground, can be reinforced by digital augmentation, and indeed some third places can find in digital technologies an ally for their basic purposes. For instance, Anthony Townsend's chapter in this book (Townsend, 2008) describes how the traditional Korean 'bang' has been re-shaping itself with and for the presence of high technology, ensuring in a way its own survival and acceptability for younger generations, and still offering the same essential benefits for after-school or off-work activities and socialisation.

Augmentation and Neighbourhoods

People meet up and socialise in third places as these provide – by being somehow thematic or characterised by specific functions – an interest to share, whether it is in books, videogames, food or whatever else. As far as neighbourhood design is concerned, it has also been noted how resident interaction

> can be influenced by provision of opportunities for increased visual and eye contacts. Visual contacts, however, stimulate a relatively superficial form of social interaction. For more in-depth, enduring interaction, those involved must have 'something in common'. As Gans (1961a; 1961b) noted in his studies of residential environments, while propinquity may initiate social relationships and maintain less intensive ones, friendships required social homogeneity. (Carmona et al 2003, 115)

It seems therefore that the process of making meaningful neighbourhoods could involve facilitating – by design – visual contacts and meeting opportunities, whilst letting any natural affinities do the rest whenever possible, in creating more cohesive and interactive urban areas. But it is also important to highlight the need to avoid the 'easy fix' of pursuing socially homogeneous neighbourhoods and places, which of course can quickly become ghettoes – however poor or wealthy. Carmona et al, quoting Jupp's research (1999), note that a very fine grain of tenure mixing – at street more than block or larger scale – can yield benefits (Carmona et al 2003, 119).

Digital augmentation can play an important role in all of this, and become one of the "tools" – though certainly not the only one – which can be used in shaping better neighbourhoods. Much has been written on the potential of ICT to boost community-based communication, including several of the chapters in this book. It is interesting to consider how powerful it can be to "ground" within specific places, and at the small scale of the street, the idea of an augmented – rather than simply "digital" – community. This could support urban design practice and theory; it could actually become – again – part of it. Some relevant past research can help in appreciating this potential.

Hampton and Wellman for instance have described their study on a wired, newly built neighbourhood in the suburbs of Toronto in the late 1990s. There, a population of prevalently lower-middle class families with no higher-than-average ICT skills moved into homes which had been equipped with broadband connection and communication facilities, some of which exclusively geared at facilitating local exchanges. Some residents decided to take advantage of these facilities, whilst others were not interested, and could represent a "control" group for comparison. After an initial period of acquaintance with the place, the researchers noted how 'Wired residents recognise almost three times as many neighbours, talk with nearly twice as many, and have been invited, and have invited, one and a half times as many neighbours into their home in comparison to their non-wired counterparts' (Hampton and Wellman 2000, 205). As mentioned earlier, the permeability – at the neighbourhood scale – was augmented by ICT both in visual/contact terms – there were more chances to "see" or "notice" other people and to be seen – as well as social.

Another earlier but very interesting example of ICT augmenting neighbourhood at the very small scale is the "MUSIC" (Multi-User Sessions in Communities) described by Alan and Michelle Shaw (1999). Low-specification computers and an ad-hoc network had been installed within just some of the households of a low-income neighbourhood in Newark, NJ. Two features have to be highlighted here. Firstly, families had to commit – for the privilege of being assigned a computer – to allow their neighbours access to the facility and neighbours were encouraged to access the network. Secondly, the network was entirely local, grounded at the small scale. This somehow materialised a form of augmentation which matched – or nearly so – the street scale advocated by Jupp, and although the neighbourhood itself was allegedly not very mixed – socially speaking – the experience shows how reciprocal trust and social capital were greatly improved, and the neighbourhood was – as a result – a better place.

As cities and municipalities around the World have engaged heavily – and with alternate results – in the practice of civic networking and in delivering e-government initiatives, it can be argued that the idea of zooming in to deal with streets and neighbourhood when deploying ICT and augmenting place, can be something to consider seriously. Beyond masterplanning at the wider urban scale, digital augmentation can have an innovative role if seen as part of urban design efforts, and when – of course – urban design principles are taken on board.

Conclusion

Augmented place-making stems from both understanding and practicing the articulation of physical and digital together as one space, as well as from the augmentation of community – and hopefully by community.

Understanding the nature and different aspects of this articulation is necessary, but it is often neglected as it would slow down developments and make proponents less competitive in a fast-moving "market" and arena. But all development as well as design work needs research and a deep understanding of the context to which the designed scheme will have to relate. As a thorough analysis of "site" and context is vital in designing successful spaces – and hopefully making places – digital augmentation should not avoid such a process. Confining urban ICT initiatives to mere – however successful in themselves – technical R&D efforts and cutting the corners of a deeper understanding of space and place is a recipe for weak, unsustainable augmentation. The nexus between social, design and technological research is key here, and something that academia should be proactively pursuing, and technological entrepreneurs should invest into.

What this paper has tried to do has been highlighting how – when we get to consider ways to proactively augment urban space – many of the principles of good spatial urban design and planning come – or should come – into play and help directing our efforts. Existing guidelines for place-making and urban design theory and practice are needed as soon as we stop looking at ICT as an 'exogenous' element being "dropped" into space, but fundamentally alien from it and abiding different rules. They can provide a much-needed reality check for hi-tech interventions, though this paper argues that eventually we should revert to thinking about the fact that we are indeed designing and modifying space, simply with more tools available: we hold a sort of an augmented palette, and we need to find the best ways to use it. Urban designers could start considering augmenting space permeability; the changes in the tension between spaces of movement and social spaces of quality usage, given by the increased fluidity and overlapping of – virtual – mobility and socially-conducive spatial settings; hi-tech innovation in third place and public space design; and how the self as well as proactively facilitated augmentation of community can contribute to a more up-to-date definition and fruition of successful places.

This chapter and the book it belongs to have only highlighted a few of these possibilities. Their list, and related research, ideas and deployments can be much longer and compelling as the efforts of space planners and designers and social informatics developers combine and converge towards the shared aim of improving place.

References

Beppe Grillo: The New Clown Prince of Italy, *The Independent*, 18 September 2007.

'To Augment', *Cambridge Advanced Learner's Dictionary* <http://dictionary.cambridge.org/> Cambridge University Press.

Aurigi, A. (1993), 'L'intelligenza scende in piazza', *Paesaggio Urbano* n.6/93, Maggioli Editore, pp. 124-126.

Bentley, I., Alcock, A., Murrain, P., McGlynn, S. and Smith, G. (1985), *Responsive Environments: A Manual for Designers*, (London: The Architectural Press).

Berger, W. (1999), 'Lost in Space', *Wired*, 7.02, Feb 1999. Online at <http://www.wired.com/wired/archive/7.02/chiat.html>, last accessed Jan 2008.

CABE(Commission for Architecture and the Built Environment) and DETR (Department of Environment, Transport and the Regions) (2001), *The Value of Urban Design: A research project commissioned by CABE and DETR*, (Kent: Thomas Telford).

Carmona, M., Heath, T., Oc, T. and Tiesdell, S. (2003), *Public Places – Urban Spaces: The Dimensions of Urban Design*, (Oxford: Architectural Press).

Cuff, D. (2003) 'Immanent Domain: Pervasive Computing and the Public Realm', *Journal of Architectural Education*, 57, pp. 43-49.

Hampton, K.N. and Wellman, B. (2000), 'Examining Community in the Digital Neighborhood: Early Results from Canada's Wired Suburb', in Ishida, T. and Isbister, C. (eds.) *Digital Cities: Technologies, Experiences and Future Perspectives*, LNCS, (Berlin: Springer-Verlag), pp. 194-208.

Haythornthwaite, C. and Wellman, B. (2002), 'The Internet in Everyday Life: An Introduction', in Wellman, B. and Haythornthwaite, C. (eds.) *The Internet in Everyday Life*, (Malden, MA: Blackwell), pp. 3-41.

Horan, T.A. (2000), *Digital Places: Building Our City of Bits*, (Washington D.C.: The Urban Land Institute).

Kalouti, A. (2005), 'Terminals in the Street: A Study on the iPlus Technology and its Usage', Unpublished Master thesis, School of Architecture, Planning and Landscape, Newcastle University UK.

McCullough (2004), *Digital Ground: Architecture, Pervasive Computing, and Environmental Knowing*, (Cambridge MA: MIT Press).

Mitchell, W.J. (2003), *Me++: The Cyborg Self and the Networked City*, (Cambridge, MA: The MIT Press).

Norberg-Schulz, C. (1971), *Existence, Space and Architecture*, (London: Studio Vista).

Odendaal, N. (2008), '(D)urban space as the site of collective actions: Towards a conceptual framework for understanding the digital city in Africa', in Aurigi, A. and De Cindio, F. (eds.) *Augmented Urban Spaces*, (Aldershot: Ashgate).

Oldenburg, R. (1999), *The Great Good Place: Cafes, coffee shops, bookstores, bars, hair salons and the other hang-outs at the heart of a community*, (New York: Marlowe & Company).

Shaw, A. and Shaw, M. (1999), 'Social Empowerment through Community Networks', in Schon, D.A., Bish, S. and Mitchell, W.J. (eds.) *High Technology and Low-Income Communities*, (Cambridge MA: MIT Press), pp. 315-335.

Sircus, J. (2001), 'Invented Places', *Prospect*, 81, Sept/Oct, pp.30-5.

Stone, A.R. (2001), 'Will the Real Bodies Please Stand Up?: Boundary Stories About Virtual Culture', in Benedikt M. (ed.) *Cyberspace: First Steps*, (Cambridge MA: MIT Press), pp. 81-118.

Townsend, A. (2008), 'Public Space in the Broadband Metropolis: Lessons from Seoul', in Aurigi, A. and De Cindio, F. (eds) *Augmented Urban Spaces*, Aldesrhot: Ashgate.

Wallner, B. and Fieder, M. 'Tholos: A Roundabout of Conversation', Tholos Systems, AT INVENTURE Forschung & Entwicklung GmbH, Online at <http://www.tholos.at/files/4_Science.pdf>, last accessed January 2008.

Warner, B. (2007), A Geriatric Assault on Italy's Bloggers, *The Times Online*, 24 October 2007.

Index

Note: bold page numbers indicate figures. Numbers in brackets preceded by *n* are footnote numbers.